Coastal Planning and Management

Second edition

Robert Kay and
Jacqueline Alder

Taylor & Francis
Taylor & Francis Group

LONDON AND NEW YORK

First published 1999 by E & FN Spon
This edition published 2005
by Taylor & Francis
2 Park Square, Milton Park, Abingdon, Oxon OX14 4RN

Simultaneously published in the USA and Canada
by Taylor & Francis
270 Madison Ave, New York, NY 10016

Taylor & Francis is an imprint of the Taylor & Francis Group

© 1999, 2005 Robert Kay and Jacqueline Alder

Typeset in Sabon by
HWA Text and Data Management, Tunbridge Wells
Printed and bound in Great Britain by
TJ International Ltd, Padstow, Cornwall

Every effort has been made to ensure that the advice and
information in this book is true and accurate at the time of going to
press. However, neither the publisher nor the authors can accept any
legal responsibility or liability for any errors or omissions that may
be made. In the case of drug administration, any medical procedure
or the use of technical equipment mentioned within this book, you
are strongly advised to consult the manufacturer's guidelines.

British Library Cataloguing in Publication Data
A catalogue record for this book is available from the British Library

Library of Congress Cataloging in Publication Data
Kay, Robert (Robert C.)
   Coastal planning and management / Robert Kay and
      Jacqueline Alder.–2nd ed.
      p. cm.
   Includes bibliographical references and index.
   1. Coastal zone management. I. Alder, Jackie, 1954– II. Title.
HT91.K36 2005
333.9 ′7–dc22 2004014309

ISBN 0–415–31772–X (hbk)
ISBN 0–415–31773–8 (pbk)

It is to our daughters Elizabeth and Hilary, who will inherit the coast from us, that we dedicate this second edition

Contents

Contributors

Case study/technical contributors and general editorial assistance

Tom Bigford
Chief, Habitat Protection Division
National Ocean and Atmospheric Administration
1315 East West Highway
Silver Spring MD 20910
USA

John E. Hay
Professor, and Director of Professional Training
International Global Change Institute
University of Waikato
Private Bag 3105, Hamilton
New Zealand

Mick Kelly
Climatic Research Unit and Centre for Social and Economic Research on the Global Environment
School of Environmental Sciences
University of East Anglia
Norwich NR4 7TJ
United Kingdom

Richard Kenchington
Visiting Professor
Center for Maritime Policy
University of Wollongong
NSW 2522
Australia

Vincent May
Independent Consultant and Emeritus Professor of Coastal Geomorphology and
Conservation
School of Conservation Science
Bournemouth University
Talbot Campus, Fern Barrow
Poole BH12 5BB
United Kingdom

Alan White
Chief of Party
Coastal Resource Management Project, Tetra Tech EM Inc
5th Floor, CIFC Towers
North Area, Cebu City, 6000
Philippines

Case study/technical contributors and general editorial assistance to the first edition

Graham King
Director, Coastal Zone Management Associates
2 Newton Villas, Newtown
Swansea SA3 4SS
United Kingdom

Case study/technical contributors

John Cleary (contributor of 4.3.3 Landscape and Visual Assessment and Planning)
Landscape Management Consultant
PO Box 496
Claremont WA 6910
Australia

Sapta Putra Ginting
Director General (Bangda)
Ministry of Home Affairs, Directorate General for Regional Development
Jl. TMP. Kalibata No. 20
Jakarta 12740
Indonesia

Bruce Glavovic
Associate Professor, Resource and Environmental Planning
Massey University
Palmerston North
Private Bag 11222
New Zealand

Ursula Kaly and John Morrison
School of Earth and Environmental Sciences
University of Wollongong
NSW 2522, Australia
Also: Ursula Kaly, Tautai International Consultants
PO Box 913, Port Moresby, NCD
Papua New Guinea

Emmanuil Koutrakis
Fisheries Research Institute
National Agricultural Research Foundation
640 07 Nea Peramos, Kavala
Greece

Arif Satria
Department of Fisheries Socio-Economics
Faculty of Fisheries and Marine Science
Bogor Agricultural University
J. Lingkar Kampus
IPB Darmaga
Bogor 16680
Indonesia

Wayne Schmit
Program Coordinator, Parks, Recreation, Planning and Tourism
Department of Conservation and Land Management
PO Box 104
Como WA 6012
Australia

Jenny Stratford
Communications and Education Officer
Thames Estuary Partnership
Remax House
31/32 Alfred Place
London WC1E 7DP
United Kingdom

Mark Zacharias
California State University Channel Islands
One University Drive
Camarillo, CA 93012
USA

Case study/technical contributors to the first edition

Ian Dutton
The Nature Conservancy
4245 North Fairfax Drive, Suite 100
Arlington, VA 22203-1606
USA

Greg Fisk
Queensland Department of the Environment
PO Box 155
Albert Street
Brisbane QLD 4002
Australia

Simon Gerrard
Project Manager, EMS Club
School of Environmental Sciences
University of East Anglia
Norwich NR4 7TJ
United Kingdom

Kathy Kennedy
Independent Consultant
Carpenters Cottage
Burdock Lane
Great Melton
Norfolk NR9 3BN
United Kingdom

Preface

Of the six billion people alive today an estimated 1.7 billion, or 38 per cent of the world's population, live within 50 km of the coast. Nearly 45 per cent of the global population is estimated to live within 150 km of the coast. This is about the same number of people that were alive in the mid-1950s. Assuming the same proportion of coastal residents in the future, by 2050 there will be nearly 4 billion people living within 150 km of the coast, or around the same number of people alive in the mid-1970s.

Coastlines are the world's most important and intensely used of all areas settled by humans. It is this simple fact that directs special attention to the planning and management of coastlines. Coastal resources have, and will continue to be, placed under multiple, intense and often competing pressures. Techniques which attempt to assist in managing the resulting conflicts in a sustainable way will therefore become increasingly important in both developed and developing countries.

Translating sustainable development principles into tangible actions aimed at improving the long-term management of coastal areas is the main purpose of this book. We do this by providing practical guidance through the dual use of theoretical analysis and numerous examples of best practice from around the world. We draw on our personal experience and the contributions of practising coastal planners, managers and academics from four continents.

We have chosen to focus the book on coastal planning, management and the nexus between them. We believe that achieving genuine sustainable development in coastal areas will be extremely difficult, but without proper planning it will be impossible. Planning provides structured mechanisms for governments to reconcile the apparently conflicting aims of sustainable development: to promote the economic development of coastal resources while attempting to preserve their ecological, cultural and social uses. We believe a key component of coastal planning efforts is to harness the energy of coastal residents and industrial and recreational users in the day-to-day management of coastal areas. We show practical examples of stakeholder participation in coastal planning, including collaborative management and co-management approaches.

One of the biggest challenges faced by governments is to direct financial and human resources effectively to the management of coastal areas through administrative systems established on sectoral lines. Sectoral-based systems of government focus on each part of a government's operations, such as transport, employment, health and environment. These systems do not explicitly focus on the planning and management of discrete geographic areas, such as coastal areas. Governments have chosen to face this challenge through various mechanisms to coordinate and/or integrate functions within coastal areas. These mechanisms are critically analysed throughout the book.

Case studies from around the world are used to illustrate sound coastal planning practices and to show differences in approach. Four groups of case studies have been selected to provide constant themes at different planning scales and to provide links between these planning scales, listed below.

The structure of the book, outlined below, reflects our aim of emphasising the current state of best practice coastal area planning and management.

Planning scale	Case studies
International	• Millennium Assessment • Regional Seas Programme (focusing on East Africa and the Mediterranean) • European Union • OneCoast
Whole of jurisdiction	• Indonesia • United States • Sri Lanka • Western Australia • Philippines • New Zealand • South Africa
Regional (sub-national)	• South Sulawesi, Indonesia • Thames Estuary, United Kingdom • Central Coast region, Western Australia • British Columbia, Canada • Great Barrier Reef Marine Park • Strymonikos, Greece • Malta north-west coast
Local	• Hikkaduwa, Sri Lanka • Fanga'uta Lagoon, Tonga • Spermonde, Indonesia • Green Island, Australia
Site	• Apo Island, Philippines • Warnbro dunes, Western Australia • Others site level plans are used to illustrate particular techniques

Chapter 1

Coastal areas are introduced, how they are defined, and a brief history of coastal management presented; the terminology used throughout the book is discussed.

Chapter 2

The major issues facing coastal managers today are discussed, together with the emerging issues likely to be of importance in the future.

Chapter 3

Theoretical approaches of coastal planning and management are analysed with respect to their translation into the development of structured coastal management programs. The chapter emphasises principles of sustainable development, ecosystem-based management, systems analysis, pragmatism, environmentalism and adaptive management. How governments are currently attempting to work towards the implementation of sustainable coastal policies and practices through systematic and accountable approaches are outlined.

Chapter 4

In this chapter the overall theory of coastal planning and management is translated into on-the-ground actions. These actions are through a range of tools and techniques, each of which is described with reference to real-world examples.

Chapter 5

Coastal planning processes and coastal management plans are described. The mechanisms and contents of plans and strategies at a range of scales are critically examined.

Chapter 6

This chapter draws together the major findings of the book and outlines possible future directions for the management and planning of the coast.

Preface to the second edition

On the one hand, a lot has changed since we wrote the first edition of *Coastal Planning and Management*. Coastal management and planning activity has significantly increased globally with many major new international, national and regional programs. The 2002 World Summit on Sustainable Development moved the global coastal agenda forward; there have also been a number of interesting theoretical and methodological developments in the field. Information technology and in particular the Internet now play a major role in the management of much of the global coastline, particularly in the developed world. The Internet now also plays a major role in sharing lessons learned within the developing world and between the developing and developed worlds.

On the other hand, the issues and challenges facing the global coastline remain and in many cases have increased. There are a few success stories that reinforce the notion that the declining condition of the coast can be reversed and that planning and management have a key role. Nevertheless the considerable frustrations of coastal planners and managers as they strive to develop and implement meaningful tools to face these challenges appear to have increased rather than decreased. There are no new magic solutions to these newly-emerging problems, nor do there appear to be any on the horizon. Today the emphasis for coastal planners and managers is to critically use whatever tools, techniques and approaches they can in whatever way makes a difference. This places more emphasis on the individual and collective skills, knowledge and experience of coastal planners and managers than ever before.

Our lives have changed too. Both of us have challenging new work lives in the private sector. Jackie now lives in Vancouver, Canada, working on a range of consulting and academic activities including the Sea Around Us Project and the Millennium Assessment. Robert is now Principal in his own coastal zone management and information technology consulting business. Robert now has a daughter Elizabeth: a life perspective-changing event if ever there was one. These life changes are inevitably reflected in the second edition.

Writing a second edition brings with it the responsibility to critically reflect on the theory and practice of coastal planning and management. There is an obligation to reinforce what has been successful, to analyse the not so successful and to introduce emerging issues and approaches. While this is a heavy burden for just two people, we know from the encouragement and feedback from our contributors and many others across the coastal management and planning spectrum around the world, that it is a burden shared. We hope that your faith in us is rewarded in this second edition.

Robert Kay, Perth, Western Australia
Jackie Alder, Vancouver, Canada
May 2004

Credits

The authors and publishers would like to thank the following for permission to reproduce material:

Allen & Unwin (J.M. Owen (1993) *Program Evaluation: Forms and Approaches*, Table 2.1); John De Campo (R. Zigterman and J. De Campo (1993) *Green Island and Reef Management Plan*, Queensland Department of Environment and Heritage, Great Barrier Reef Marine Park Authority, Cairns City Council, Cairns Port Authority and Department of Lands, Cairns, Australia); Canadian Association of Geographers (R.W. Butler (1980) 'The Concept of a Tourist Area Cycle of Evolution', *The Canadian Geographer*, 24(1), Fig. 1); Delaware Coastal Management Program, Department of Natural Resources and Environmental Control (Delaware coastal strip definition); Earthscan (D. Pearce, R.K. Turner, R. Durborg and G. Atkinson (1993) *The Conditions for Sustainable Development. Blueprint 3: Measuring Sustainable Development*, London, pp. 18–19; T. O'Riordan and H. Voisey (eds) (1998) *The Transition to Sustainability: The Politics of Agenda 21 in Europe*, London, Fig. 1.1, p. 5); Herman Cesar (1996) *Economic Analysis of Indonesian Coral Reef*, Environment Department, World Bank, Washington, DC, Tables E-l, E-4); Dunwich Museum; Ian Dutton (K. Hotta and I. Dutton (1994) *Coastal Management in the Asia-Pacific Region: Issues and Approaches*, Japan International Marine Science and Technology Federation, Fig. 1.2); Elsevier Science Ltd (B. Cicin-Sain (1993) 'Sustainable Development and Integrated Coastal Zone Management', *Ocean and Coastal Management*, 21(1–3), Fig. 2, Table 2); Great Barrier Reef Marine Park Authority; HMSO (1988) *The Tolerability of Risk at Nuclear Power*, p. 9); Kluwer Academic Publishing (S. Gubbay (1995) *Marine Protected Areas*, Fig. 6.1); Department of Planning and Infrastructure, Western Australia (*Central Coast Regional Strategy*, Fig. 14); *New Scentist* (H. Gavaghan (1990) 'The Dangers Faced by Ships in Port', *New Scientist*, 128(1744)); Risk Unit, University of East Anglia (B.A. Soby, A.C.D. Simpson and D.P. Ives (1993) *Integrating Public and Scientific Judgements into a Tool Kit for managing Food-related Risks. Stage 1: Literature Review and Feasibility Study*, Centre for Environmental Risk Research Report No. 16); School for Resource and Environmental Studies, Dalhousie University, Nova Scotia (H.J. Ruitenbeek (1991) *Mangrove Management: An Economic Analysis of Management Options with a Focus on Bintuni Bay, Irian Jaya*, Figs A6.1, A6.3, 4.10, 4.12, Tables 2.1, 3.1); Taylor & Francis (R.C. Kay, I. Eliot, B. Caton, G. Morvell and P. Waterman (1996) 'A Review of the Intergovernmental Panel on Climate Change's Common Methodology for Assessing the Vulnerability of Coastal Areas to Sea-level Rise', *Coastal Management*, 24(2), Fig. 4);

John Wiley & Sons, Inc. (Harvey M. Rubinstein (1987) A *Guide to Site and Environmental Planning*, Fig. 1-2); UNEP-Global International Waters Assessment (Pollution impacts and large marine ecosystems); World Bank (J.C. Post and C.G. Lundin (eds) *Guidelines for Integrated Coastal Zone Management*, pp. 5–6).

While every effort has been made to contact and acknowledge copyright holders, any omissions should be reported to the publisher.

Acknowledgements

Our book is an effort shared with family, friends and colleagues. The effort has also been shared with the many contributors and supporters to this second edition and with the readers of the first edition. Without fail the enthusiasm and support for the book from first-edition readers has supported us to both start writing the second edition and to push us to completion. And of course thanks to Robert's partner – Caro Kay – who provided both the most immediate and lasting support.

In particular we would like to thank for their assistance with the second edition: Ellik Adler, Mike Allen, Sithara Atapattu, Roland Barkey, Michelle Borg, David Brown. Patrick Christie, Sarah Cooksey, Roger Cornforth, Andrew Crow, John de Campo, Indira Fernando, Sussan Gubbay, Marc Hershman, Adrian Kitchingman, Laure Ledoux, Derek McGlashan, John Qulity, Hamish Rennie, Vivienne Panizza, Fitri Putjuk, Yvette Rizzo, Johanna Rosier, Tim Smith, Doug Storey, Akbar Tahir, Clive Turnbull, Tonny Wagey, Simon Woodley, Maureen Woodrow, and Andrew Wright.

The editorial team at Taylor & Francis have also been great. Thanks to Tony Moore and Matthew Gibbons and before them Alice Hudson and Charlotte Friel. Gerry MacGill's editorial skills are gratefully acknowledged as are the drafting talents of Rita Willsdon-Jones.

Robert Kay would like to acknowledge the staff and clients of Coastal Zone Management Pty Ltd for their patience and support.

Disclaimer

The opinions expressed in this book are the personal opinions of the authors only. The authors accept responsibility for any accidental errors.

Chapter 1

Introduction

This chapter introduces the importance and uniqueness of the world's coastal areas, with a view to outlining the coastal issues and planning and management tools described in later chapters. Several important terms, including 'coastal area', 'planning' and 'management' are defined, and the use of the terms 'coastal area' and 'coastal zone' is discussed. The fundamentals of the approach taken in the book are described.

1.1 The language of coastal planning and management

The boundary between the land and ocean is generally not a clearly defined line on a map, but occurs through a gradual transitional region. The name given to this transitional region is usually 'coastal zone' or 'coastal area'. In common English there is little distinction between zone or area, but in coastal management there has been some debate as to the implied meanings associated with zone, as used in 'coastal zone management'. The debate has focussed on the implication that 'zone' may imply that geographically defined planning zones will be established and become the dominant part of the coastal management process. This implication is not important in many developed countries, where 'coastal zone management' is a phrase commonly used to describe a variety of coastal programs (OECD 1992), such as the United States' Coastal Zone Management Act (1972). But developing countries often equate coastal zone with land-use or marine-park zoning (Chapter 4). Although 'coastal zone' and 'zoning within the coastal zone' are clearly different, to avoid confusion many coastal management initiatives use the description 'coastal area' (e.g. UNEP OCA/PAC 1982; Chua and Pauly 1989).

Kaluwin (1996) describes the notion of delineating a zone or area as an essentially western concept, which places artificial boundaries on the geographical extent of this transition. He considers it culturally inappropriate for Pacific Islands, where the coast has traditionally been viewed as a transitional region between land and ocean; however, few coastal nations, especially in developed countries, take this enlightened traditional Pacific view of the coast. In this book we concur with Kaluwin (1996) that zone could be implied to mean a planning zone.

Case studies and in-text quotations in this book reveal other variations in the terminology used in the day-to-day practice of coastal planning and management around the world. While this is to be expected as the coastal initiatives of different cultures and language groups are translated into English, a decision had to be made about whether to standardise the use of language for the purposes of analysis in this book. For simplicity (except when quoting from original sources) we decided to use the shortest and most flexible terminology, and use 'coastal planning' and 'coastal management'. We therefore do not use the prefix 'integrated' to describe the bringing together of participants, initiatives and government sectors. Nor do we routinely insert 'zone' or 'area' to define that a broad geographic area beyond the immediate boundary between land and sea is the focus of attention in coastal planning and management. We take the pragmatic view that terms such as 'area', 'integrated', 'coordinated' and 'zone' will be used when it is useful to do so within the social, cultural and political circumstances of a coastal nation. In other words, we strongly advocate using terminology as a means to an end – the selection of a particular set of words for a particular section of coast should be based on what will best provide the optimum means of ensuring the sustainable development of that section of coast.

1.2 Defining the coastal area

Defining the boundaries of a coastal area is of more than academic interest to coastal planners and managers. Governments often create administrative systems, or set out policies to guide decision making, that operate within a defined coastal policy area. The variety of ways in which such areas may be delineated in order to serve the purposes of particular policies is outlined in this section.

1.2.1 Biophysical definitions of a coastal area

The coast is where land and ocean meet. If this line of meeting did not move, defining the coast would be easy – it would be simply a line on the map – but the natural processes that shape the coast are highly dynamic, varying in both space and time. Thus the line that joins land and ocean is constantly moving, with the rise and fall of tides and the passing of storms, creating a region of interaction between land and sea.

There are parts of the coastal environment that clearly have strong interactions between land and ocean, including beaches, coastal marshes, mangroves and fringing coral reefs; other parts may be more distant from the immediate coast (inland or out to sea) but they nevertheless play an important role in shaping it. One of the most important of these is the rivers that bring freshwater and sediment to the coastal environment. In this case, the inland limit to the coast is catchment boundaries that can be thousands of kilometres inland at the head of catchments. For example, the Ganges–Brahmaputra river system, whose sediments form much of Bangladesh, rises far inland in the Himalayas.

Therefore, the coast may be thought of as the area that shows a connection between land and ocean, and a coastal area defined as 'the band of dry land and adjacent ocean space (water and submerged land) in which terrestrial processes and land uses directly affect oceanic processes and uses, and vice versa' (Ketchum 1972).

The key element of Ketchum's definition is the interaction between oceanic and terrestrial processes and uses: coastal areas contain land, which interacts with the ocean in some way, and ocean space which interacts with the land. Thus coastal areas:

- contain both land and ocean components;
- have land and ocean boundaries that are determined by the degree of influence of the land on the ocean and the ocean on the land; and
- are constantly changing in width, depth, or height.

The three above elements are depicted in Figure 1.1, which shows, for a sandy beach coast, the strength of interaction between coastal and ocean processes and uses, termed here the 'degree of coastalness', against the distance away from the immediate coast. Figure 1.1 could be repeated for other coastal environments, such as delta coasts, beach/barrier systems and estuarine coasts, where the various physical and biological processes of these environments will determine the 'degrees of coastalness'. On deltaic coasts, for example, important determining factors would be the degree of salt water penetration into fresh surface- and ground-water systems, and the seaward distance to which sediments of terrestrial origin are moved.

As Figure 1.1 shows, the transition between land and ocean is often gradual, depending on local biophysical conditions. The issue here is not the nature of the actual transition which occurs, but what its implications are for defining a coastal area. Choosing the thresholds, which define the landward and seaward limits of a coastal area depends to a large extent on why the definition is needed. This 'need driven' approach to coastal area definition is discussed further in the next section.

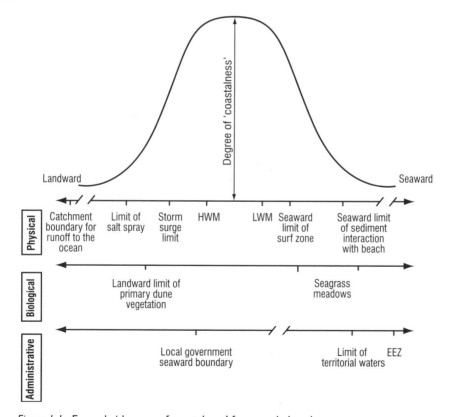

Figure 1.1   Example 'degrees of coastalness' for a sandy-beach coast

1.2.2 Policy-oriented definitions of a coastal area

In practice, the (coastal) zone (area) may include a narrowly defined area about the land–sea interface of the order of a few hundreds of metres to a few kilometres, or extend from the inland reaches of coastal watersheds to the limits of national jurisdiction in the offshore. Its definition will depend on the particular set of issues and geographic factors which are relevant to each stretch of coast (Hildebrand and Norrena 1992).

Coastal zone (area) management involves the continuous management of the use of coastal lands and waters and their resources *within some designated area*, the boundaries of which are usually politically determined by legislation or by executive order (Jones and Westmacott 1993).

At a policy level the limits of coastal areas have been defined in four possible ways:

* fixed distance definitions;
* variable distance definitions;
* definition according to use; or
* hybrid definitions.

Current or proposed examples of each of the above definitions are given in the Appendix.

Fixed distance definitions, as the name implies, specify a fixed distance away from the coast, which is considered 'coastal'. Usually this distance is calculated from some measure of the boundary between land and water at the coast, usually the high water mark. Fixed distances defined for the ocean component of a coastal area usually apply to the limit of governmental jurisdiction, for example the limits of Territorial Seas. An example of a fixed definition coastal area as used by the government of Sri Lanka is shown in Figure 1.2.

As for fixed distance definitions of coastal areas, the boundaries of variable distance definitions are set from some measure of the coast, usually the high water mark. However, their boundaries are not fixed, but vary along the coast according to a range of variables such as:

* physical features – e.g. the landward limit of Holocene dunes, or the seaward limit of submarine platforms;
* biological features – e.g. the landward limit of a coastal vegetation complex, or the seaward limit of a fringing reef;
* constructed landmarks – e.g. roads, canals, railways or well-known buildings; and
* administrative boundaries – e.g. the landward limit of local municipalities which front the ocean.

International organisations and large coastal nations often define the limits of a coastal area according to the particular coastal management issue being addressed; that is, the coastal area is defined according to the use to which that definition will be put, and the form of definition is termed 'definition according to use'. For example, tackling the issue of non-point sources of marine pollution would require the definition of an area of attention that included inland catchments and groundwater outflow regions. A coastal area defined for this purpose would be much larger than one defined

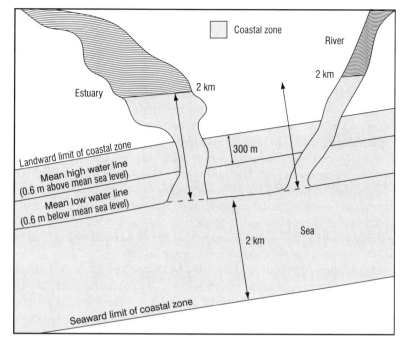

Figure 1.2 The coastal zone of Sri Lanka as defined by the Sri Lankan Coast Conservation Act (Coast Conservation Department 1996)

to manage four-wheel-drive vehicle damage of beaches and dunes. As recognised by the Coastal Committee of New South Wales (1994: 22):

> To a large extent, the definition of the coastal zone depends upon the purpose for which the definition is intended. From both management and scientific viewpoints, the extent of the coastal zone will vary according to the nature of the management issue.

Within the context of defining a coastal area according to what the purpose is, the concept of 'areal foci' used by Jones and Westmacott (1993) is useful. Areal foci include:

- an administratively designated area, in the sense that the political process or the administration will designate the responsibility to manage;
- an ecosystem area;
- a resource base area, e.g. a mineral body, oil fields, fisheries, habitats etc.; and
- a demand area, i.e. the wider area from which demands are exerted on the designated coastal area, such as for use for recreation, marine transport or waste disposal.

A good example of policy-oriented coastal zone definitions is the State of Delaware in the USA where various definitions are used. Delaware's pragmatic approach is shown in Box 1.1.

Defining a coastal area according to use has the advantage of focussing attention on particular issues. However, care needs to be taken to avoid multiple coastal area

Box 1.1

State of Delaware coastal zone definitions

A good example of pragmatic policy-oriented coastal zone definitions is the definition of coastal zones used by the Delaware State coastal program in the USA. Delaware uses two different definitions for its State's coastal zone depending on the purpose to which the definition will be applied. Delaware also uses additional specific definitions to achieve particular policy outcomes.

The first coastal zone is the state coastal zone area, known locally as the 'coastal strip', shown in the figure below. The landward boundary of the coastal strip uses state routes (road highways) 13, 113, and 1 (north–south corridors), and an area on the north and south of the Chesapeake and Delaware Canal (see Appendix). The main purpose of the definition is to define coastal lands and waters for controlling heavy industry, thus it is important that the chosen definition is clearly understood both by policy stakeholders and regulators.

The second definition of the coastal zone used by Delaware State is the federally approved coastal management area, which is used to enable the state to work with the federal government on the national coastal zone management program under the Coastal Zone Management Act (CZMA). The CZMA uses a 'definition according to use' approach. An administratively simple and effective approach under the CZMA used by Delaware has been to define the entire state as the coastal zone. However, it is not that simple because Delaware is currently in a boundary dispute with its neighbouring state New Jersey, and hence the coastal zone boundary is not fully resolved.

Delaware State coastal management policies for beach management also use additional definitions to define coastal areas in order to regulate specific activities. For example, a 'coastal line' is defined (and mapped) a fixed distance inland (either 75 or 100 feet depending on coastal sector) from a defined contour line (either 7-foot or 10-foot elevation depending on coastal sector) requiring a permit or letter of approval from the state government to undertake any activity (Delaware State Government 1999):

a   To construct, modify, repair or reconstruct any structures or facility on any beach seaward of the building line.

b   To alter, dig, mine, move, remove or deposit any substantial amount of beach or other materials, or cause the significant removal of vegetation, on any beach seaward of the building line which may affect the enhancement, preservation or protection of beaches.

In addition, the policies state that construction activities landward of the building line in Delaware on any beach, including construction of any structure or the alteration, digging, mining, moving, removal or deposition of any substantial amount of beach or other materials, shall be permitted only under a letter of approval from the state government.

Box 1.1, continued

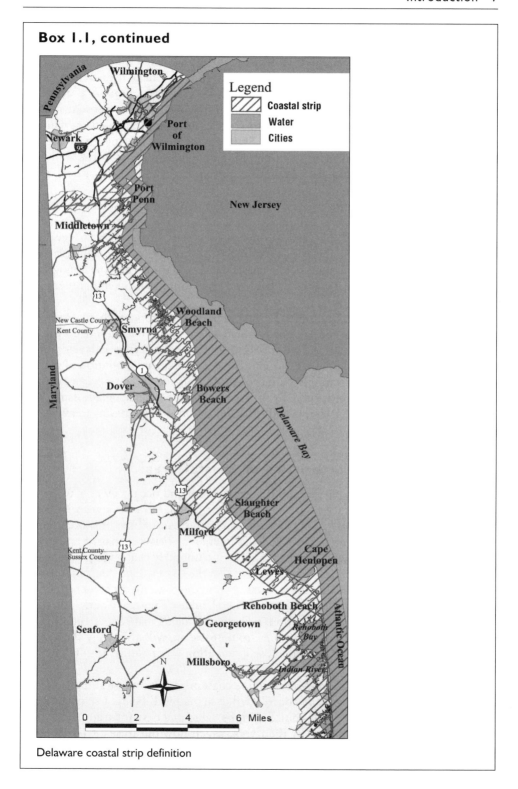

Delaware coastal strip definition

definitions being established in one region to address different coastal management issues, leading to confusion. Defining the coast according to one use only may perpetuate sectoral managerial systems and detract from an integrated management perspective.

Hybrid definitions mix one type of coastal definition for the landward limit of the coastal area and another for the seaward limit. This is relatively common practice by governments that have a fixed limit of jurisdiction over nearshore waters. Australian States, for example, have management responsibilities for coastal waters three nautical miles from the coastline. Some Australian State governments use this to define the seaward limit of their coastal areas, while choosing other means to define the landward boundary (see the Appendix). For example, the recent definitions of coastal areas adopted by the Queensland State Government are shown in Box 1.2.

The vertical dimension of any coastal area definition can also be included; that is, the depth below the surface and height above a coastal area considered to be covered by a coastal policy. Usually the vertical dimension is part of the overall legislative framework of governments, and is not explicitly covered by coast-specific policies. Examples include all mineral rights below coastal lands and waters and the atmosphere above it, which are generally covered by laws and regulations that cover all other parts of a government's jurisdiction.

In summary, a generic definition of coastal areas is not proposed here. Rather, a pragmatic view of defining a coastal area is taken, where the definition reflects the use or uses to which it will be put. If the purpose is to control certain types of development, then fixed, variable or hybrid definitions may be used. If reducing pollution of marine waters is the purpose, then variable definitions including catchment or groundwater boundaries may be more appropriate. By focussing on coastal management issues, and not on problems of definition, simple and workable definitions of coastal areas usually follow.

1.3 The unique characteristics of coastal areas

Stating that the coast is unique because it is where land and oceans meet may appear rather obvious, but it is a fact of great significance. The contrast between land and ocean may be dramatic where ocean swells crash against rock cliffs, or more gradual where tides ebb and flow over marshes. It is this interaction between marine and terrestrial environments that makes the coast unique – and uniquely challenging to manage. This unique challenge is compounded by the fact that the coast suffers the consequences of land use decisions often made a great distance from the coast in catchments draining to coastal zones. For example, seemingly detached land use changes in agriculture and forestry that cause erosion and pollution, or irrigation dams that reduce the flows of water and sediment load into coastal zones from upstream, will ultimately show up in the coast in one form or another through sedimentation (or erosion), salinity changes or increased toxic blooms.

The transition between land and ocean at the coast produces diverse and productive ecosystems, which have historically contributed to human well-being. Use of the coast for its resources has long been combined with its value as a base for trading between countries, both across oceans and by the rivers, which flow out to sea. Coastal lands and nearshore marine waters have consequently long been at a premium. As populations grow and increase their level of socio-economic development, this premium also grows.

Box 1.2

Coastal definitions used in the Queensland Coastal Protection and Management Act (1995)

Foreshore means the land lying between high water mark and low water mark as is ordinarily covered and uncovered by the flow and ebb of the tide at spring tides.

The *coast* is all areas within or neighbouring the foreshore.

Coastal management includes the protection, conservation, rehabilitation, management and ecologically sustainable development of the coastal zone.

Coastal resources means the natural and cultural resources of the coastal zone.

Coastal waters are Queensland waters to the limit of the highest astronomical tide.

Coastal wetlands include tidal wetlands, estuaries, salt marshes, melaleuca swamps (and any other coastal swamps), mangrove areas, marshes, lakes or minor coastal streams regardless of whether they are of a saline, freshwater or brackish nature.

The *coastal zone* is:
a   coastal waters; and
b   all areas to the landward side of the coastal waters in which there are physical features, ecological or natural processes or human activities that affect, or potentially affect, the coast or coastal resources.

The consequence of this intense and long-standing pressure on coastal resources is that problems with the way in which competing uses are managed within a country as a whole tend to emerge first on the coast.

To make management even more difficult, major administrative boundaries commonly follow high or low water lines, bisecting coastal areas and dividing the management of the land from that of the ocean. Coastal land is usually owned and/or managed by a multiplicity of private, communal, corporate and government bodies, whereas coastal waters are usually owned and/or managed solely by governments. Where there are significant coastal areas under private ownership, conflicts between the owners and other users emerge and present difficult problems for the managing agencies. We see this in the USA as well in countries where there are tourist resorts which marginalise locals – such as in the Caribbean and other developing countries. Furthermore, administrative boundaries can follow the centres of rivers and estuaries, splitting their management between two neighbouring authorities, in doing so dividing natural units such as bays and sediment cells.

The uniqueness of the coast is further enhanced by the value of its resources such as fish and offshore mineral reserves, and more recently aquaculture sites, considered by the populace to be common property, and in high demand by coastal dwellers for subsistence use, recreation and economic development (Berkes 1989; Feeny *et al.* 1990;

Dolak *et al.* 2003). Exploitation of such resources raises their value, with a consequential demand for equitable resource allocation. Therefore, resource planning often forms an integral part of coastal management programs.

1.4 A brief history of coastal management and planning

A brief history of the development of coastal area management and planning is presented for two main reasons. First, history provides a framework for understanding how current approaches to the planning and management of coastal resources have evolved, and the constraints these approaches are operating within. Second, by looking back at how coastal planning and management have developed, trends become evident. Projecting such trends provides an insight into the possible future development of coastal management and planning.

Humans have deliberately modified the coastal environment and exploited its resources for thousands of years. Ancient civilisations throughout the world built ports and seawalls, or diverted river water flowing into the sea; they also evolved various management systems for their fisheries, use of rich coastal soils for agriculture, trading through ports, and other coastal resources. Examples include ancient Greek and Roman port cities throughout the Mediterranean; the diversion of the Yangtze (Yellow) River, China in AD1128 (Ren 1992); or the reclamation of mangrove areas over 1,000 years ago on Pohnpei, Federated States of Micronesia (Sherwood and Howarth 1996).

Ancient interventions such as these in the coastal environment were all works of civil engineering. That is, structures were built to modify the flow of water and/or sediment. Given that such structures were all essentially hand built, the scale and intensity of their impacts on the coastal environment were limited, but over the centuries the ability of humans to influence coastal processes increased as construction techniques improved. Perhaps the most famous example of diversion of water courses and construction on the coast was the building and maintenance of the current urban form of Venice, Italy, from the seventh century AD (Frassetto 1989).

For these civilisations an informal form of resource planning was undertaken by either community consensus or by a leader who decided when, where, how and how much resources would be exploited. Resources were abundant but sparsely exploited because of limited technology. Hence resources were generally allocated on a social rather than on an economic basis.

Technological limitations were dramatically reduced as a result of the industrial revolution, which started in Europe in the mid-nineteenth century. The industrial revolution brought machines that could be used to construct grander civil engineering works. Major modifications of the coastal environment were now possible: large rivers could be dammed or diverted and vast areas of coastal wetlands could be converted to urban or agricultural land.

The industrial revolution also altered the community's view of its resources. Viewing them as tangible elements or objects of nature led to the use of the term 'natural resources', and management, including planning, now focussed on supply and demand, and the options for managing these factors. This was linked to the pervasive western cultural attitude at the time of man's dominance over other animals and natural systems.

Concentrating on economic factors, very little attention was given to the ecology (including habitats), social demands or public perceptions (O'Riordan and Vellinga 1993). The underlying objective was to maximise profits, which usually translated into maximising production. The weakness of this approach was the assumption that resources are easily valued, single purpose and static in value over time, which we now know is not valid (Chapter 4).

During the industrial age the market place began to dominate resource allocation, while social norms no longer guided resource use. Resources were perceived as limitless and there to be consumed for profit (Goldin and Winters 1995; Grigalunas and Congar 1995). It was not until the late nineteenth century that this view began to change. Resources came to be considered finite, a change in attitude attributable to:

- advances in economic theories on supply and demand;
- the developing realisation that society had the ability to destroy the environment, ultimately affecting its survival;
- social reforms; and
- studied attempts to plan for resource management.

In contrast, deliberate human intervention in the coastal environment to preserve components of its natural character or ecological integrity is a much more recent activity. Coastal ecological management grew from the national park movement of the late nineteenth century. During this era, protected areas or parks were perceived as places of significant scenic or natural value set aside for the enjoyment of visitors or for scientific pursuits (MacEwen and MacEwen 1982). The first such parks in coastal marine areas were established in the 1930s. Since then, protected areas with significant coastal components have been established throughout the world, with most being terrestrial. Currently there are an estimated 30,350 protected areas (as defined by the IUCN and interpreted by the World Conservation Monitoring Center – WCMC) extending over a total area of 13,232,275 km²; or an estimated 8.83 per cent of total land area (Green and Paine 1997). These estimates are currently being updated and included in an Internet-based tool to ensure currency (Louisa Wood, WCMC, personal communication).

Expansion of land use planning in the late nineteenth and early twentieth centuries also influenced coastal area management in developed and colonial 'new world' countries (Platt 1991). Important influences included the notion of separating conflicting land uses through zoning, planning open space areas for the public good and health, and sanitation problems which affected waste disposal into coastal waters. While the main way to effect such interventions was through the use of the engineering works described above, it is the role of land use planners in directing the expansion of urban environments into coastal areas, and their enthusiasm for embracing engineering interventions, that is important here. Urban expansion brought with it the need to develop the coast for new residential areas and industries, as well as a need to cater for increased recreational use of the coast.

Different streams of human endeavours in coastal areas, such as ecological management, resource management, engineering intervention and urban/industrial development, operated relatively independently for many years. The coastlines of developed nations had been planned and managed using land use planning and environmental management techniques, which had evolved within their various governmental and cultural settings. Each can be considered as a form of coastal area management, and their proponents as coastal managers. However, it was not until the 1960s and 1970s that these, and other disciplines, were brought together under the banner of 'coastal zone management', phrase credited to those involved in the development of the US Coastal Zone Management Act in the late 1960s and early 1970s (Godschalk 1992; Sorensen 1997).

Realisation around the world that environments were being continually degraded by a rapidly expanding human population led to the concept of sustainable development in the late 1980s and early 1990s. The basis of sustainable development is 'development that meets the needs of the present without compromising the ability of future generations to meet their own needs' (World Commission on Environment and Development 1987), a concept now central to most coastal management efforts worldwide, as will be shown in Chapter 3.

The challenges of managing for sustainability led to a realisation that sustainable resource use could not be achieved by maximising short-term economic growth and that an ecosystem approach was necessary. Ecosystem management is not easy, especially on the coast, where there is a lack of information on its biodiversity, the function and dynamics of coastal systems, and of the interconnectedness within coastal ecosystems as well as other ecosystems (Christensen *et al.* 1996). However, the elements of ecosystem management are easily incorporated into coastal planning and management processes. Given the importance of sustainability principles and ecosystem approaches in coastal management, these topics are discussed further in the next section.

Today, it is generally accepted that coastal resources can only be effectively evaluated and managed in the total context of the ecosystem and associated social and cultural environments (e.g. Ehler 1995; Agardy and Alder, in press). Hence, effective resource planning provides for decision making which allocates resources over space and time according to the needs, aspirations and desires of society, taking into account society's ability to exploit resources, its social and political institutions, and its legal and administrative arrangements.

O'Riordan and Vellinga (1993), in reviewing the history of coastal area management up to the early 1990s summarised its development over the past forty years as a professional activity into four phases. We have extended their analysis to include developments during the last ten years (Table 1.1).

Chapters 2–5 discuss and analyse the elements of recent developments in coastal management shown in phase IV of Table 1.1. Chapter 6 provides the justification for the inclusion of the potential future development of coastal management shown in phases IV and V in Table 1.1.

1.4.1 Sustainability – the dominant paradigm in coastal planning and management

> Sustainable development requires a broader view of both economics and ecology than most practitioners in either discipline are prepared to admit, together with a political commitment to ensure that development is 'sustainable'.
>
> (Redclift 1987: 33)

Table 1.1   Phases in the development of coastal management

Phase	Period	Key features
I	1950–1970	• Sectoral approach • Man-against-nature ethos • Public participation low • Limited ecological considerations • Reactive focus
II	1970–1990	• Increase in environmental assessment • Greater integration and coordination between sectors • Increased public participation • Heightened ecological awareness • Maintenance of engineering dominance • Combined proactive and reactive focus
III	1990–2000	• Focus on sustainable development • Increased focus on comprehensive environmental management • Environmental restoration • Emphasis on public participation
IV	2000–2010	• Focus on tangible implementation of sustainable development principles • Ecosystem-based management becoming embedded in national legislation • Shared governance emerging • Exploration of new coastal management approaches, including learning networks and adaptive management systems • Increased impact of globalisation and the Internet on management approaches and impacts • Emerging re-analysis of the basic tenets of coastal management
V	Future	• Integrated suite of theories and tools applicable with confidence over all scales, timeframes, locations and issues • Comprehensive ecosystem-based management • Connected coastal management communities of practice • Verified set of governance models

Source: adapted from O'Riordan and Vellinga (1993).

Sustainability has emerged as the dominant paradigm of the world's coastal management programs in the late twentieth century – and remains valid today albeit with continued debate over the tangible measures required for sustainable coastal management. The historical context of this emergence is described in the previous section; here we describe the concept of sustainability and discuss its influence on coastal programs, from broad scale strategic planning to day-to-day management regimes. This discussion forms the basis for the more detailed treatment in Chapters 3, 4 and 5 of tools and techniques to help achieve the sustainable development of coastal areas.

The concept of sustainability came into prominence with the publication of the World Commission on Environment and Development (WCED) report called 'Our Common Future' (World Commission on Environment and Development 1987). The WCED group was chaired by Gro Harlem Brundtland, hence the report came to be known as the Brundtland Report. The message of the Brundtland Report was that it is possible to achieve a path of economic development for the global economy 'which meets the needs of the present generation without compromising the ability of future generations to meet their own needs' (WCED 1987: 8). This quotation from the Brundtland Report is often seen as its 'punch line'. However, reading directly on from the above definition of sustainability Brundtland continues:

It (sustainable development) contains within it two key concepts:

- the concept of 'needs', in particular the needs of the world's poor, to which overriding priority should be given; and
- the idea of limitations imposed by the state of technology and social organisation on the environment's ability to meet present and future needs.

Thus the goals of economic and social development must be defined in terms of sustainability in all countries – developed or developing, market-oriented or centrally planned. Interpretations will vary, but must have certain general features and must flow from a consensus on the basic concept of sustainable development and on a broad strategic framework for achieving it.

Development involves a progressive transformation of economy and society. A development path that is sustainable in a physical sense could theoretically be pursued even in a rigid social and political setting. But physical sustainability cannot be secured unless development policies pay attention to such considerations as change in access to resources and the distribution of costs and benefits. Even the narrow notion of physical sustainability implies a concern for social equity between generations, a concern that must logically be extended to equity within each generation.

Dresner (2002) views this extended quotation from the Brundtland Report as essentially identifying the important components of sustainable development, namely:

- meeting basic needs;
- reorganising environmental limits; and
- the principles of intergenerational and intragenerational equity.

In the above sense Dresner (2002: 68) believes:

... sustainable development is not such a vague idea as it is sometimes accused of being. The problem of actually operationalizing sustainable development remains, however. The difficulty in giving an operational definition of sustainable development, or even reaching agreement on what are the key elements of the idea, lies in the fusion of two concerns that pull in somewhat different directions: the environmental and the social.

O'Riordan (1998) gives a useful way of thinking about the 'domains of sustainability' through the now commonly adopted notion of the 'triple bottom line' of (in O'Riordan and Voisey's order of priority):

- planetary maintenance (ecological sustainability);
- social equity (ethically laded sustainability); and
- economic enterprise (livelihood sustainability).

Indeed, Figure 1.3 provides a useful starting point for consideration of balanced, tangible management responses to operationalising sustainable development. The notion of balanced triple bottom line sustainability-oriented accountability reporting by business has been an important spin-off of the above debate on sustainability definition and its operationalisation (Elkington 1997).

It is the debate over the sustainability of urban systems that has drawn out many of the issues surrounding operationalision of sustainable development principles. Given the increasingly urban nature of the global coastal zone this debate is directly relevant to coastal managers as shown in Box 1.3.

Importantly, the principles of sustainability were recently re-validated at the 2002 World Summit on Sustainable Development (WSSD). The Political Declaration from the WSSD stated (UN 2002):

Accordingly, we assume a collective responsibility to advance and strengthen the interdependent and mutually reinforcing pillars of sustainable development – economic development, social development and environmental protection – at local, national, regional and global levels.

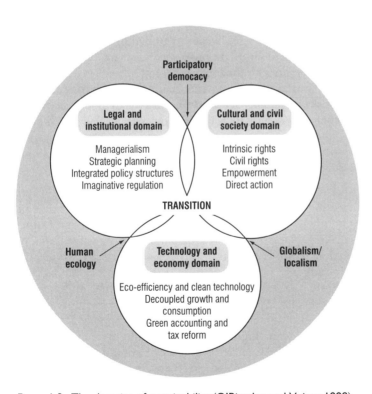

Figure 1.3 The domains of sustainability (O'Riordan and Voisey 1998)

Box 1.3

Urban sustainability issues

Traditionally, sustainable development has been described as the balance between social, environmental and economic goals. The problem with this model is that it offers relatively little understanding of the inherent trade-offs found in the simultaneous pursuit of these goals. Furthermore, the picture it provides is too abstract to appreciate how sustainable development unfolds at the urban level, and fails to acknowledge the political dimension of the process. By definition, cities are not sustainable in themselves, as urban dwellers and economic activities inevitably depend on environmental resources and services from outside the built up area. So what does urban sustainability mean and how can the sustainability of urbanisation and urban development be appraised?

These questions require a more encompassing vision of the concept, one that adequately defines the goals and means of the process. Quite rightly the environmental, economic and social goals still apply. However, in an increasingly urbanised world, the built environment needs to be recognised as a central component of sustainability. Furthermore, the search for more sustainable forms of urbanisation also depends on political and institutional processes prompting competition or cooperation between different agents. Thus to assess whether any given practice, policy or trend is moving towards or against urban sustainability, it is necessary to consider two additional dimensions.

• Physical sustainability that addresses liveability of buildings and urban infrastructures and their relation to the urban-region environment. It also considers building efficiency in supporting the local economy.
• Political sustainability that focuses on the quality of governance systems guiding the relationships between actors. It implies the democratisation and participation of civil society and areas of decision-making.

(quoted from Allen and You 2002: 16)

Though a precise operational definition of sustainability may be rather elusive, it is clearly not a set of prescriptive actions; rather it is the basis for a fundamental reassessment of the way in which resource, environment, social and equity issues are considered in decision making. Nevertheless, the articulation of sustainable development principles into policy and management practice has been a major field of endeavour in recent years. The profoundness of its implications has caused sustainability to be compared with such basic societal values as freedom, justice and democracy (Buckingham-Hatfield and Evans 1996). Seen in this light, sustainability becomes a 'way of thinking', helping to modify the context to which it is applied (Turner 1991). Thus, sustainability principles can 'highlight unsustainable systems and resource management practices' (Turner 1991: 209). The tests of sustainability having been applied and unsustainable practices revealed, the way opens for new, sustainable management approaches to coastal area management to be devised and adopted (Figure 1.4). This view is supported by Dresner (2002) who referring to Nitin Desai, the

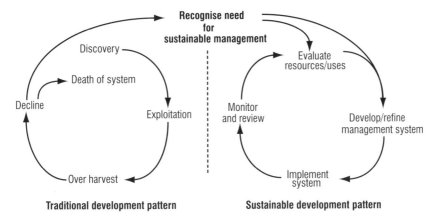

Traditional development pattern   Sustainable development pattern

Figure 1.4 Sustainable and unsustainable approaches to coastal resource use (Dutton and Saenger 1994)

economic advisor to the Brundtland Commission, believes that it is the values underlying the concept of sustainability that are critical.

As a way of thinking, sustainability has not only become part of the mainstream of decision-making processes, it has also in many nations become a political reality (Buckingham-Hatfield and Evans 1996; O'Riordan and Voisey 1998) – though remaining elusive in many others (Kirkby *et al*. 1991). However, the idea that the present generation can through the application of sustainability principles act as stewards of the earth for future generations is as much an act of faith as it is one based on technical or scientific evidence (Buckingham-Hatfield and Evans 1996; Dresner 2002). This raises two important issues: the weight to be given to technical information, and the time-dependence of decision making.

Sustainability has acted as the catalyst for a new mix in the information sources on which decisions are based. It has seen the 'hard science' emphasis of the 1970s and 1980s evolve into a more balanced appreciation of scientific and non-scientific inputs into decisions. This balancing has manifested itself in various ways, for example the Best Practicable Environmental Option system in the United Kingdom; but its most pervasive expression is the 'precautionary principle', commonly defined in the language of Principle 15 of the Rio Declaration:

> In order to protect the environment, the precautionary approach shall be widely accepted by the States according to their capabilities. Where there are threats of serious or irreversible damage, lack of full scientific certainty shall not be used as a reason for postponing cost effective measures to prevent environmental degradation.

This principle is now incorporated into the London and Hague Declarations dealing with marine pollution.

Precautionary action has three central components. First, there is an economic dimension of cost-effectiveness; second, decisions which may have irreversible impacts, so providing a legacy for future generations, gain heightened importance in the decision-making process; and third, the lack of a requirement for complete scientific information in the face of economically inefficient and/or irreversible impacts – a substantial shift from a rational-comprehensive view of decision making, as will be shown in Chapter 3. It is important to note that a precautionary approach to guiding decision making is a very recent phenomenon and its use is not uniform around the world (O'Riordan and Cameron 1994; Tickner 2003). However, its current use in some coastal nations, and probable spread to many more, is likely to see precaution entering the lexicon of most coastal managers in the next few years.

A central part of the 'way of thinking' introduced in this section is the consideration of the time dependence in decision making; that is consideration of the effects of present-day activities on future generations (Young 1992). Sustainability thinking requires that future effects and impacts of decisions, and not simply those in the present day, be considered. Relating to this concept, many planners have seized upon sustainability with the notion that planning and sustainability principles are similar, and that a convergence of planning and sustainable development is emerging under the banner of environmental planning (Blowers 1993; van Lier *et al.* 1994; Buckingham-Hatfield and Evans 1996).

Having looked at the general principles of sustainability, three specific effects of sustainable thinking on coastal management can now be briefly considered. First is the effect on the use of economics and economic instruments in decision making. Environmental Economics, as we shall demonstrate in Chapter 4, is rapidly becoming one of the mainstays of the practical use of sustainable development in decision making. Sustainability has provided many economists with a basis for implicitly including equity, environmental considerations and a long-term view into the cost–benefit and other economic analyses (Jacobs 1991). Likewise, sustainability has also allowed environmental issues, such as conservation of biodiversity, to become a central part of decision making through techniques such as ecosystem-based management (Christensen *et al.* 1996) and consideration of large marine ecosystems (Wang 2004), especially in those areas previously the exclusive domain of economists: most notably economic development. Finally, sustainable development explicitly recognises the quality of human life of both current and future generations. Thus, social and cultural equity is recognised as an equal partner with economic and environmental considerations.

In summary, sustainable development principles have had four main effects on the way the coast is managed, one general and three specific. The general effect is the influence 'sustainability thinking' or 'sustainability values' have on the overall decision-making context. The mixture of equity, environmental and economic concepts moves the decision-making paradigm away from considering economic, environmental or social-oriented decisions in isolation from each other. The three specific impacts are in the fields of economics, environmental resource management and social and cultural development, summarised by Reid (1995) as requiring the following characteristics:

- integration of conservation and development;
- satisfaction of basic human needs;
- opportunities to fulfil other non-material human needs;
- progress towards equity and social justice;
- respect and support for cultural diversity;
- provision for social self-determination and the nurturing of self-reliance; and
- maintenance of ecological integrity.

Clearly, these are major issues which go to the heart of the human cultural and spiritual and developmental aspirations as well as fundamental issues of governance, democracy and the relationship of humans and the environment. They are weighty issues but, nevertheless, ones which must be confronted to ensure a viable future for the world's coastal regions. Sustainability then is 'not just about managing and allocating natural capital. It is also about deciding who has the power both to do this and to institute whatever social, economic and political reforms are considered necessary' (Reid 1995: 231). Any discussion of approaches to the sustainable development of coastal areas must, as a result, analyse techniques for environmental management, systems of governance and the role of individuals in decision-making and planning processes. It is the interplay between these factors, which is explored at length in the coming chapters.

1.5 Issues to actions

Coastal management programs have generally developed in response to problems experienced in the use and allocation of coastal resources. Development of a coastal program usually follows a period of mounting public, political and scientific pressure on governments to tackle problems, usually resulting in a time lag between the identification of problems and the development of responses. Development of the US Coastal Zone Management Act in 1972, for example, followed a period of intense pressure for improvement in coastal land and water management which started more than twelve years earlier (Godschalk 1992). A similar pattern was followed in the United Kingdom during the late 1980s to the early 1990s (King and Bridge 1994) and recently more broadly in the European Union. Much of the history of coastal management and planning illustrates similar reactions to problems experienced in coastal regions. Many other national and international initiatives can be traced to the time when the problems could no longer be ignored.

A further stimulus to the development of coastal programs was the realisation that coastal area management programs could be used to avoid future problems. However, unlike the development of programs which respond to existing problems, it is unclear when this proactive approach became important. It may be inferred that although there were some important forward-looking parts of the US Coastal Zone Management Act, this is not formally reflected in its aims (Godschalk 1992). Proactive management, through the use of various coastal planning approaches, is now one of the most important components of coastal area management around the world, with modern programs blending proactive and reactive elements to address current problems, such as ecosystem degradation, and to avoid future problems.

1.6   Chapter summary

It is worth re-emphasising at this point that the deliberate actions of humans to influence the natural processes of the coast have been occurring for thousands of years. Coastal management choices made during this time reflected the cultural and spiritual relationship between people and the coastal environment. Historically, it is the perception of how the coast should be managed, and for what purpose coastal resources will be used, that has shaped management of coastal areas. These perceptions are culturally and politically influenced; they have clearly changed over time, will continue to change, and are demonstrably different around the world. The diversity of coastal area management approaches reflects these differences.

The documented development of coastal planning and management described above is largely a history of western nations, or those countries colonised by western nations. In this group of nations, the evolution of coastal programs, as they reflect changes in cultural values, has been well described (Table 1.1). In contrast, the traditional coastal management systems of indigenous cultures in other parts of the world are relatively poorly documented. Although culturally appropriate coastal management programs are making something of a resurgence in many developing countries (Chapter 3), there remains much to be done in understanding how these traditional practices evolved and how they have been extended into modern and postmodern ages. This is especially so since their integration with western management practices is becoming increasingly important with the pervasive spread of western technologies and management approaches in an increasingly globalised world. These are recurring themes of the following chapters.

In summary, the early development of coastal area planning and management programs in the early 1960s and 1970s was generally in response to urgent problems on the coast. As these reactive coastal programs became more established they gradually evolved into a combination of reactive and proactive programs during the 1980s and 1990s and the more recent probing of theoretically rigorous approaches to collaborative coastal management. This evolution may reflect the heightened influence of planning on the management process, or the increasing evidence of long-term systemic environmental degradation, or it may reflect the need to manage existing problems by addressing possible future pressures. However, perhaps the key lesson to be drawn from this brief history is the need to combine present and future perspectives; that is, attempting to address present day problems whilst preventing new ones, an aim which fits well within the techniques described in Chapters 3, 4 and 5.

Chapter 2

Coastal management issues

> Man has only recently come to realize the finite limitations of the coast as a place to live, work, and play and as a source of valuable resources. This realization has come along with overcrowding, overdevelopment in some areas, and destruction of valuable resources by his misuse of this unique environment.
>
> (Ketchum 1972: 10)

This chapter provides an overview of the major issues, problems and opportunities in coastal management. The chapter does not attempt to analyse and describe every issue at length. Consistent with the general focus of this book, particular emphasis is placed on describing and analysing management tools and planning techniques to assist in dealing with the issues.

There is a strengthening consensus among scientists that many ecosystems, including coasts, continue to degrade (Agardy and Alder, in press). The Millennium Ecosystem Assessment, modelled on the Intergovernmental Panel on Climate Change, provides a detailed up-to-date picture of the state of the world's coastal ecosystems. It is based on a consensual approach of leading experts in the field on coastal systems and it has an extensive peer review process. The assessment highlights the lack of progress in addressing many of the challenges faced in managing coastal ecosystems and in stemming the decline of many coastal ecosystems, including coral reefs, mangroves and temperate marshes.

Globally, coasts are under a number of threats or face issues with a number of underlying social and economic drivers, resulting in many complex and inter-related problems (Table 2.1). While considerable progress has been made over the last 30 years, these threats remain, along with the limited development of tools and approaches with which to address them. Works such as Clark (1996) and Beukenkamp *et al.* (1993) also provide a useful treatment of the issues, as well as numerous conference and workshop proceedings that contain specific examples of coastal problems from around the world. Further information on the range and depth of coastal issues can be obtained through reference to the sources of the many case studies listed throughout the book.

Coastal management initiatives are usually a response to a demand to resolve problems such as conflicting uses of coastal resources, urbanisation, access, pollution and environmental degradation. Problems may also be related to poor liaison or inefficient coordination between those responsible for making decisions on the allocation of coastal resources; or they may even be a perception among decision-makers that a problem does not exist. A sound understanding of such issues is integral to planning an effective approach to coastal management. It is important that issues be addressed in a

Table 2.1   Threats to coastal ecosystems and key drivers

Type of threat	Drivers
Habitat loss or conversion	
Coastal development (ports, urbanisation, tourism-related development, industrial sites)	Population growth, poor siting due to under-valuing and lack of knowledge, poor develop-ment policies for industry, and tourism, environmental refugees and internal migration
Destructive fisheries (dynamite, cyanide, bottom trawling)	Shift to market economies, demand for aquaria fish and live food fish, increasing competition in light of diminishing resources
Coastal deforestation (esp. mangrove deforestation)	Lack of alternative materials, poor national policies, increased competition
Mining (coral, sand, minerals, dredging)	Lack of alternative materials, global common perceptions
Civil engineering works	Transport and energy demands, poor public policy, lack of knowledge about impacts and costs
Environmental change brought about by war and conflict	Increased competition for scarce resources, political instability, inequality in wealth distribution
Aquaculture-related habitat conversion	International demand for luxury items, regional food security needs, declining wild stocks, loss of property rights in fisheries, inability to compete
Habitat degradation	
Eutrophication from land-based sources (agricultural waste, sewage, fertilisers)	Urbanisation, lack of waste water and sewage treatment systems, poor agricultural practices, loss of wetlands and other natural controls
Pollution: toxics and pathogens from land-based sources	Lack of awareness, increasing pesticide and fertiliser use (especially as soil quality diminishes), unregulated industry
Pollution: dumping and dredge spoils	Lack of alternative disposal methods, increasing costs for land disposal, belief in unlimited assimilative capacities, waste as a commodity
Pollution: shipping-related	Substandard shipping regulations, no investment in safety, policies promoting flags of convenience, increases in ship-based trade
Salinisation of estuaries due to decreased freshwater inflow	Demand for electricity and water, territorial disputes
Alien species invasions	Ballast discharge regulations lacking, increased aquaculture-related escapes, lack of international agreements on deliberate introductions
Global warming and sea level rise	Emissions controls lacking, poorly planned development (vulnerable development), stressed ecosystems less able to cope
Over-exploitation	
Directed take of low value species at high volumes exceeding sustainable levels	Subsistence and market demands (food and medicinal), industrialisation of fisheries, improved fish-finding technology, poor regional agreements, lack of enforcement, breakdown of traditional regulation systems, subsidies

continued...

Table 2.1   Threats to coastal ecosystems and key drivers, continued

Type of threat	Drivers
Over-exploitation	
Directed take for luxury markets (high value, low volume) exceeding sustainable levels	Demand for specialty foods and medicines, aquarium fish, and curios, lack of awareness or concern about impacts, technological advances, commodification
Incidental take or by-catch	Subsidies, by-catch has no cost
Subsistence and artisanal effort increasing as food security declines	Unempowered local peoples, breakdown of traditional structures

Source: Agardy and Alder (in press).

coordinated and integrated framework – a feature of good coastal planning. Such a framework ensures that individual issues are recognised, yet enables solutions to be developed that cut across as many issues as possible while avoiding the creation future issues.

The issues described in this chapter are those common to many coastal areas around the world. Inevitably, they are more critical in some places than in others, and hence will be of differing levels of interest to managers in different places. Nevertheless, they are all relevant to the development of an understanding of coastal problems and the approaches to avoiding or mitigating their impacts.

Issues are discussed under the broad groupings of population growth and urbanisation, coastal use, the impacts of coastal use and impacts on coastal uses, and administrative issues. The groupings are not mutually exclusive, but are designed to give a general feel for the major challenges facing coastal managers today.

A useful introduction to the range of typical issues for coastal nations is provided from the Philippines (Figure 2.1) (Milne *et al.* 2003) that builds on descriptions of coastal issues facing Bangladesh (LGED 1992 cited in Clark 1996). Indeed the Philippines example provides a useful framework for other coastal locations to consider both the source of impacts derived from human activities upland of the coast and those from inshore and offshore marine activities. Together with the topics addressed by the Thames Estuary Management Plan (Box 2.1), discussed further in Chapters 3 and 5, they provide a concise introduction to the issues outlined in the following sections.

2.1 Population growth and urbanisation

Population growth is the driver behind many, if not most, coastal problems. Increasingly accurate estimates of population growth in the coast are now possible with the development of GIS tools and global efforts to improve the resolution of demographic information. The population of the coast was first estimated as 60 per cent living within 100 km of the coast (Hinrichsen 1990). Cohen *et al.* (1997) corrected this estimate to 38 per cent using the 1995 global population and the digital World Vector Shoreline. The proportion of the world's population living within 100 km of the coast has remained at around 38 per cent since 1990, the year when spatially referenced global population information became available.

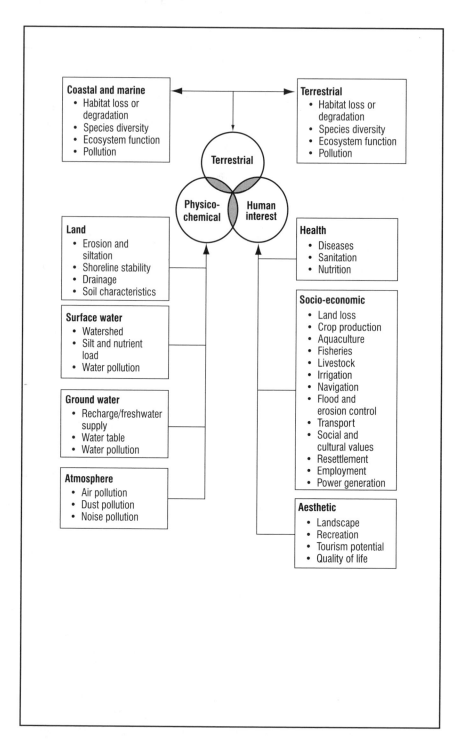

Figure 2.1 Examples of impacts on coastal systems in the Philippines (adapted from LGED 1992 cited in Clark 1996; Milne *et al.* 2003)

Box 2.1

Issues and topics addressed by the Thames Estuary Management Plan

The Thames, one of the world's most famous estuaries, has multiple management conflicts. It is the United Kingdom's busiest and most commercially significant tideway; 12 million people live within easy reach of it and the port alone supports 37,000 jobs. Nowhere in the country are environmental pressures and competing demands for space and resources greater than on Thames-side. Despite the enormous pressure, the Thames is also internationally important for wildlife. Following massive cleanup operations that started in the 1960s it is now one of Europe's cleanest estuaries. It supports 121 different species of fish, and its mudflats and marshes are home to an estimated 170,000 birds.

In recognition of the need to continually plan for the future, many of the users of the Thames worked together to produce an estuary management plan, described in Chapters 3 and 5 (Boxes 3.5, 3.10 and 5.26). The general issues and specific topics addressed by the Thames Estuary Management Plan are listed below.

General issues	*Specific topics*
• Facilitating communication between different sectors	• Agriculture
• Fostering understanding of different organisational cultures	• Coastal processes
	• Commercial use of the estuary
• Sharing technical information of agreed standards	• Fisheries
• Overcoming administrative fragmentation	• Flood defence
• Maintaining a shared vision among stakeholders to deal with problems as they emerge	• Historical and cultural resources
	• Landscape
	• Nature conservation
	• Recreation
	• Waste transfer and disposal
	• Water management
	• Public awareness
	• Enhancement opportunities
	• Targets and monitoring

Global population growth from the turn of the twentieth century to recent years has been staggering (Haub 1996). Broad estimates put the world's early 1990s population in coastal areas as equal to that of the entire global population of the 1950s (Edgren 1993). More recently coastal populations have been quantified with absolute numbers, revealing that the number of people living within 100 km of the coast has increased from 2 billion in 1990 to 2.3 billion in 2000; if current trends continue this will grow to 3.1 billion in 2025 (UN Population Division 2000). In terms of population density, there were 88.5 people per km^2 in 1990, increasing to 99.6 people per km^2 by 2000. The comparative population density figures for inland areas were 33 per km^2 in 1990 and 37.9 per km^2 in 2000 (Figure 2.2). More than 50 per cent of the habitable

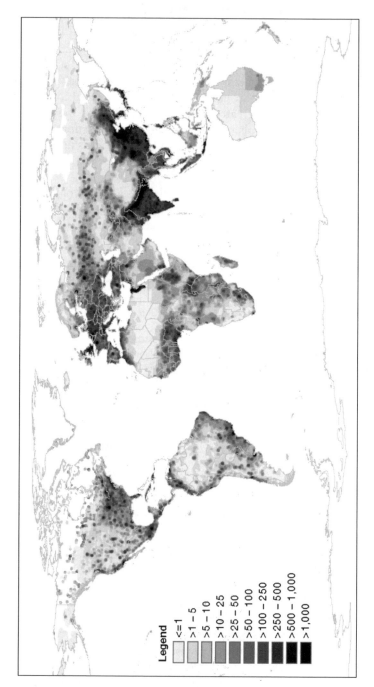

Figure 2.2 Global population densities 2000 (CIESIN 2004)

coast is sparsely populated (holding less than 3 per cent of the coastal population) showing clearly that most people live in urban centres along the coast (Small and Nicholls 2003).

Most of the population increase in the coast has been in major urban areas where the coast supports 50 per cent of the population that lives in large cities globally. The explosive growth in urban populations since 1960 is marginally higher in coastal areas (333 per cent) compared to inland areas (313 per cent). In 1960 there were only 119 large urban centres (population over 500,000) including four megacities (population exceeding 8 million people) within 100 km of the coast. In 2000 the number of large urban areas nearly doubled to 216 and of these seventeen were megacities. Currently eight of the ten largest cities in the world are in the coast (UNEP 2002).

Population growth in coastal areas has two main causes. First, it reflects the general trend of population growth in developing countries, linked to rural-urban migration; and second, the migration from inland areas to the coast, which often offers people more economic, social and recreational opportunities than inland areas (Goldberg 1994). Example of coastal population growths and their impacts in Florida and California (USA), and in the Indonesian province of Sulawesi Selatan, are shown in Box 2.2 and Box 2.3.

The clearest result of population growth in the coast is the accelerating rate of urbanisation. Between 1950 and 1990 the coastal population density of the US increased from 275 to nearly 400 people per km². In 1990 the population density in the coastal area from Boston to Washington DC was 2,500 people per km² (Hinrichsen 1998). By 2010, 320 million people will live in megacities; most of these cities are in the coast

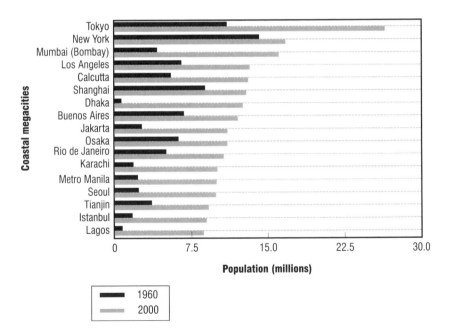

Figure 2.3 Population change in coastal megacities 1960–2000 (based on UN Population Division 2004)

Box 2.2

Coastal issues in Florida and California

The 673 coastal counties of the contiguous USA make up 17 per cent of its land area, yet as of 1994[1] accounted for 53 per cent of its population and housing supply. This coastal population is estimated to be increasing by 3,600 people per day. The result is a projected increase of 27 million people between 1998 and 2015 (Culliton 1998).

California and Florida are two states in the USA carrying the brunt of this population growth. Between 1990 and 2000 the coastal population of California increased by 11.1 per cent and that of Florida by 24.8 per cent (National Atlas of the United States). Their warm and sunny climate and resulting outdoor lifestyles have attracted migrants from northern states – from the 'frostbelt' to the 'sunbelt'. Many settle on the coast, creating coastal development and management issues which have required concerted efforts for many years. These migration trends continue as many Americans, with improved access to technology, greater personal income and transport routes are able to chose their lifestyle (Ullmann *et al.* 2000).

California

California's 3,427 miles of coast are some of the longest of all the states of the USA, made up of spectacular sea cliffs, rocky shores and beaches. The coastal area contains abundant living and non-living resources as well as one of the largest bay-estuary systems in the world – San Francisco Bay. The major impacts on the California coast include increased residential and commercial development, the effects of relative sea-level rise on coastal structures, and degraded coastal water quality from urban and industrial runoff.

Florida

Florida's tropical and sub-tropical coastal area contains the most extensive mangrove and wetland areas in the USA as well as the nation's greatest concentration of coral reefs, found around the Florida Keys. Major impacts to the Florida coastal areas have centred on the rapidly expanding commercial and residential construction resulting from the 700 per cent population increase since 1945 (Culliton 1998). These impacts include primary sand dune removal, wetlandfilling, channel dredging, coastal erosion, pollution runoff and threats to the preservation of Florida's unique wetland and coral reef areas enhanced by the impact of tropical storms and sea-level rise.

The administrative mechanisms for organising coastal management programs to tackle the above problems in California and Florida are described in Box 3.9.

Box 2.3

Coastal pressures in Sulawesi Selatan province, Indonesia

Indonesia is a rapidly developing country. Like many Asian nations it experienced strong economic growth during the early 1990s, which was halted with the Asian crisis towards the end of that decade. The Indonesian economy is now growing much slower, at approximately 3.7 per cent in 2002 (World Bank 2003). This continuing economic and population growth (1.3 per cent annually) and urban expansion have strained coastal environments. Politically, Indonesia, like its neighbours Thailand and the Philippines, is attempting to decentralise much of its authority, including the management of coastal areas to provincial and local government authorities (Patlis *et al*. 2001). The wording of the Regional Autonomy Act No. 22/1999 suggests that local governments could have jurisdiction 4 nautical miles seaward and provincial governments jurisdiction from 4 to 12 nautical miles offshore. This has led to widespread ambiguity in interpreting of the law. The government and the House of Representatives are revising the law so that it is clear that central government has jurisdiction from the coastline to the edge of the EEZ. The provincial government will have the role of representing the federal government in the regions. Regional and local governments will only have jurisdiction for a few selected activities such as aquaculture.

The national government is also responsible for setting policies, criteria and standards for a host of issues nationally (Patlis *et al*. 2001). In an archipelagic nation such as Indonesia, this new law impacts on thirty-two coastal provinces and 198 coastal districts and creates considerable challenges in developing integrated coastal management plans and policies within the newly created Ministry of Marine Affairs and Fisheries (Patlis *et al*. 2001).

Eastern Indonesia has been the focus of many economic initiatives and rapid urban development and it has embraced the notion of local autonomy. One area which has experienced these changes is the province of South Sulawesi (locally called Sulawesi Selatan or SulSel).

More than 80 per cent of Sulawesi Seletan residents live in coastal settlements, most of which are located on the fertile coastal plan adjacent to the Makassar Straits (Hasanuddin University 1999). Many of these residents are economically dependent on fisheries resources, especially the Spermonde Archipelago and Taka Bone Rate reef systems (see Box 5.17). These coral reef systems are considered to contain some of world's highest marine biodiversity. The highest number of coral reef species are found in South Sulawesi: they also support one of the world's most intensive reef fisheries. The importance of these reefs and the threats they face are recognised in the National Coral Reef Rehabilitation and Management Project, where Taka Bone Rate has been the focus of community based management interventions since 1997. More recently (2004) the Spermonde Archipelago has been included in these interventions.

These rich waters have enabled coastal communities in Sulawesi Seletan to develop a strong marine and coastal culture. Many communities rely on coastal

continued...

Box 2.3, continued

and marine resources for subsistence and income generation. As a consequence, the demands for access and use of coastal and marine resources have increased, with significant costs to the environment. Fifty-one percent of the province's mangroves have been destroyed since 1982. Many of the mangroves have been converted to aquaculture ponds which operate with no environmental controls. Other marine environments have been destroyed due to destructive fishing practices such as blasting and cyaniding.

Shipping within the Makassar Strait has grown and is expected to continue expanding now that the strait is an international shipping lane. The demand for access to the coast and islands for tourist developments has increased: many developments will displace local residents and place a burden on existing water supplies. In addition, many developments are not required to provide sewage treatment facilities.

To address these pressing issues, the Indonesian governments are working to develop a coastal planning and management framework including national guidelines and regional and local plans described in Boxes 3.6, 5.9, 5.13 and 5.17.

(GESAMP 1999). It is predicted that most of the growth of urban centres will be in developing countries where birth rates are high. The population in the urban area between São Paulo and Rio de Janeiro is currently 30 million; it is expected to increase to 40 million by 2010 (Hinrichsen 1998). In developed countries, growth is predicted to be much slower due to declining birth rates.

Cities on the coast are often associated with major ports, which facilitate cheap sea transport of goods, which in turn attracts major industries. Economic growth provides employment and investment opportunities, coastal cities acting as a magnet for people looking to improve their economic status (Ehler 1995). The coast's attractiveness also draws people for holidays, retirement and those seeking coastal lifestyles. In response, many urban areas are being developed or expanded to meet the needs of new coastal residents for housing, sanitation and transport. Since 1970, houses have been built at the rate of 2000 per day in coastal areas of the US (UNEP 2002a). If tourism continues to develop in the Mediterranean, the area could host up to 350 million seasonal tourists every year by 2025 (Hinrichsen 1998).

Many specific resource allocation and planning issues are raised by the urbanisation debate: urban residential densities, the development of high rise buildings, and public versus private access to beaches and foreshores are among the more prominent. These in turn impact on the visual landscape, and create increased pressure on coastal resources and the use of facilities such as transport, landfill and sewerage.

Management of urban areas expanding along the coast can be one of the most difficult tasks of coastal planning. The often enormous values of coastal land which can be developed for residential and tourist developments can see the widespread conversion of agricultural, forestry and other low intensity land uses to urban. A result can be urban 'strip development' as tentacles of urban sprawl spread monotonously up and down the coast from urban centres. Ultimately, cities hundreds of kilometres apart can become joined, effectively becoming one coastal 'megacity' (e.g. Toyko–Osaka and São Paulo–Rio).

These trends in population growth are driven in developed countries largely by lifestyle choices through both 'baby boomers' and those working in sectors of the economy where location can be chosen. Most notably in economic sectors where high-speed connection to the Internet is the primary employment requirement, there are increasing pressures on coastal towns and resorts. Urban and regional planning attempts to resolve these competing demands (Box 2.4). Techniques for consideration of such issues are presented in Chapters 4 and 5.

Box 2.4

Coastal urban expansion issues north of Perth, Western Australia

The 1.96 million population of Western Australia is concentrated in the State's south-west, with 1.43 million people living in the capital, Perth (Australian Bureau of Statistics 2004). The state's economy has experienced a continued annual gross state product (current prices) growth rate, ranging from 7.8 per cent in 1994–5 to 5.5 per cent in 2002–3 (Australian Bureau of Statistics 2003). Perth's population is expected to continue to grow. Projections are for a total population of 2.76 million by 2026, of which it is predicted 1.98 million will live in Perth (Western Australian Planning Commission 2000).

The Central Coast region, immediately north of Perth, is currently sparsely populated. A risk for this area as Perth expands is an unplanned urban sprawl northwards along the coast. The Central Coast Regional Strategy was developed for this 250 km of coastline with the aim of balancing urban expansion pressures with conservation, recreation and tourism opportunities (Western Australian Planning Commission 1996). Four major issues prompted the strategy:

- access, protection and use of the coastline;
- the need for new road connections;
- the future use and management of the large amount of public land; and
- the impact of metropolitan development on the future of the region.

Coastal management issues and values addressed by this study were:

- the scenic attractions and natural recreation opportunities of the coast which are valuable to the region and make it a desirable place to live and visit;
- the illegal squatter developments causing significant land management problems and jeopardising recreational and conservation opportunities;
- development associated with settlements occurring too close to the coast;
- loss of seagrass could affect marine environments;

continued…

Box 2.4, continued

- the multipurpose nature of coastal activities, requiring different facilities and access considerations;
- the attractions of the coast for recreation and tourism, necessitating low key, low impact development, taking into account environmental and social considerations; and
- the potential, without adequate rehabilitation and planning, of mining and extraction of basic raw materials to damage the coastal environment.

The outcomes of the strategy and its implementation activities are discussed in Chapter 5.

2.2 Coastal use

Coastal uses are considered here under four main categories: resource exploitation (including fisheries, forestry, gas and oil, and mining); infrastructure (including transportation, ports, harbours shoreline protection works and defence); tourism and recreation; and the conservation and protection of biodiversity. Each category is described in turn. The use of land for residential purposes was outlined in the previous section, and is not considered further in this section.

2.2.1 Resource exploitation – fisheries, forestry, gas and oil, and mining

Renewable coastal resources are primarily exploited in the fisheries sector by commercial, subsistence and recreational fishers and the aquaculture industry. Worldwide attention has been focused on the sustainability of today's fisheries. Industry, resource managers and conservation groups are concerned with over-fishing of most stocks, especially inshore fisheries, and the long-term sustainability of these fish stocks. In the mid-1990s it was thought that an estimated 70 per cent of the world's commercially important marine fish stocks were either fully fished, overexploited, depleted or slowly recovering (Mace 1996; World Wide Fund for Nature 1996). Recent studies of global fisheries revealed that fish landings are in fact in decline, and not stable as previously thought, due to misreporting of catches by some countries such as China (Watson *et al*. 2001). The United Nations Food and Agriculture Organisation (FAO) (2002) estimates that 75 per cent of the major fish stocks are fully exploited or worse.

Current trends in the development of new fisheries such as the live fish trade, which has been responsible for the collapse of a number of reef fisheries throughout Asia and the South Pacific, are also of concern. The fishery is now worth between US$500 million and US$1 billion annually (World Resources Institute 2003). Coastal management has a critical role to play in managing fisheries since many coastal habitats such as mangroves and seagrass beds are part of the life cycles of many commercially important species.

Aquaculture, pond and cage culturing, have been practiced in Asia for centuries. The last 50 years has seen an exponential expansion of this industry, not just for fisheries, but for other emerging marine resources such as seaweed, prawns and sea cucumbers. Sea cage culturing has also developed in a number of areas. There are a number of issues associated with both forms of culturing. The conversion of land to ponds and the consequential loss of productive agricultural land is a major concern amongst coastal managers (Figure 2.4) (especially as in some areas pond production is sustainable for less than 20 years and the conversion of coastal habitats such as mangroves leads to a loss of fish habitats (Hay *et al.* 1994)). The World Resources Institute (1992) estimates that around 45 per cent of the world's mangroves have been lost due to human use. Pond systems produce high nutrient levels which ultimately enter coastal waters, a problem which is compounded when antibiotics, algicides and other chemicals are used. Cage culturing in marine areas causes local pollution and can introduce diseases into wild populations. The introduction of exotic species and the consequential displacement of native species is a potential problem with all forms of culturing. The repeated episodes of mass mortalities of Australian pilchards across southern Australian coasts are considered to be linked to the importation of frozen pilchards from California, Chile, Peru, and Japan to feed caged southern bluefin tuna in marine aquaculture operations (Gaughan *et al.* 2004). Other emerging issues for aquaculture include the use of genetically modified fish and the potential for genetic dilution of wild stocks.

Coastal forestry focusses on the commercial and subsistence exploitation of mangrove stands. Historically, exploitation of mangroves for charcoal, furniture and other uses was sustainable, but current demand for fuel far exceeds supply in many parts of the developing world. The result is that mangrove stands are commonly no longer a sustainable supply of cooking fuel. These issues are evident in Indonesia, as shown in Box 2.5. Clearly the loss of mangrove forests is a loss in biodiversity and habitat with potential impacts on adjacent commercial fisheries. When mangroves are cut, sediments from

Figure 2.4 Aquaculture ponds, South Sulawesi, Indonesia (Source: Reg Watson)

Box 2.5

Mangrove conversion to prawn aquaculture issues – South Sulawesi, Indonesia

Mangroves are an important coastal resource and serve a number of functions. They are critical to maintaining foreshore stability and trapping sediments from river runoff. Many commercially important fish spend a part of their early life cycle in mangrove areas (Mumby *et al.* 2004). Mangroves are also important habitats or sources for other marine products. For many people, mangroves are a source of cooking fuel, subsistence and income generation (Table 4.13).

In Indonesia, as in many areas of the world, the maintenance of mangroves is threatened, mostly by competing resource uses. The harvesting of mangroves for charcoal as a cooking fuel, their conversion to ponds for aquaculture production, or their infilling for development, industrial or urban, are just a few examples of the competing uses facing coastal managers.

Many competing uses limit the production of mangroves to a single activity: the harvesting of mangroves for charcoal cannot be maintained if the forest is converted to a port. Uses which convert mangroves to other forms of land use such as pond aquaculture, urban expansion or industrial estate development are permanent. There are no options to rehabilitate the area back to a mangrove, with the result that biodiversity is lost, and a source of food production and cooking fuel is reduced, shifting and exacerbating the problem in another area. In addition, a source of income is eliminated for a group who are already considered socially and economically the worst off in Indonesia.

In the past, decisions to convert mangroves were made without due consideration of the long term impacts. In the province of South Sulawesi the area of mangroves has been reduced by 51 per cent, with conversion to pond aquaculture systems being the primary reason.

Measures such as maintaining a buffer zone of mangroves between the converted land and open water, selective cutting and encouraging replanting have been promoted to address the loss of mangroves throughout the country. The implementation of these measures, however, has been variable (Ruitenbeek 1991).

---

upland areas entering coastal areas are no longer trapped, and shoreline stability can be adversely affected. The importance of mangroves in the community structure of coral reef fisheries has been demonstrated in the Caribbean. The biomass of several commercially important species is more than doubled when adult habitat is connected to mangroves (Mumby *et al.* 2004). This shows an emerging understanding of the complexity of linkages between coastal ecosystems.

Inland forestry practices in many developing countries can have indirect impacts such as increased sedimentation due to soil loss, especially in poorly managed rainforest extractions. Agricultural land-uses in both the developing and developed world can have similar effects, as well as the potential impacts of herbicides and pesticides.

Oil and gas are the major non-renewable resources exploited in many coastal areas, and are a major source of revenue for many coastal nations. Ancient coastal deposits and sedimentary basins adjoining continents commonly favour oil and gas accumulation. Examples include deposits found under, or adjacent to modern deltas, such as the Mississippi, Niger and Nile.

The siting of oil and gas facilities on the coast requires careful planning and management. The facilities themselves can conflict with commercial and recreational fishing areas, and can affect visual amenity and reduce recreational potential. Access roads and shipping channels to facilities dug through deltas and other sensitive coastal environments can significantly alter ecosystems and sediment balances. The risk of blow-outs and oil spills is a major environmental issue associated with this industry. There are, unfortunately, numerous examples of spills associated with both the production and transportation of oil and gas products. Other issues include the impacts of seismic surveys on marine communities. A longer-term problem of oil and gas production can be the subsidence of land due to the collapse of sub-surface reservoirs (Dolan and Goodell 1986). In response to these concerns, the oil and gas industry has been active in monitoring various marine and coastal parameters, providing much needed information for managing the coast.

Other resources such as mineral sands, coral and salt are exploited at the coast and can result in major environmental impacts when improperly managed. Again there are conflicting uses when land is used in conjunction with these activities. Waste products from mining operations can enter the system either through runoff or leakage from settling ponds or tailings sites.

In many tropical nations, coral is a cheap source of building and road making material. Many of the coastal erosion problems in the developing world are due to unmanaged mining of fringing reefs. Mined coral reefs lose their ability to stabilise the coast, since wave energy is no longer dissipated by the reefs but acts directly at the beach edge, causing the redistribution of vast amounts of sand from reef flat areas to deeper waters. Indeed, this was one of the key drivers for the development of a coordinated approach to coastal management in Sri Lanka (Box 2.6).

Box 2.6

Critical coastal management issues in Sri Lanka

Sri Lanka, like many developing countries, has a range of coastal management issues centred around the mix of subsistence uses of the coast combined with increased industrial and tourist developments (Kahawita 1993). An estimated 32 per cent of Sri Lanka's population and 65 per cent of urban areas, and 67 per cent of industrial areas are within coastal zones (IUCN 2002). An estimated 80 per cent of the country's tourist infrastructure is on the coast to capitalise on its beaches, marine waters and coral reefs. Poorly planned tourist developments have aggravated pre-existing natural coastal erosion problems, especially on the south coast which faces the Indian Ocean. This erosion problem has been found to be very sensitive to sand and coral mining, improperly sited coastal protection structures and loss of coastal vegetation. Other critical coastal management issues in Sri Lanka include (Kahawita 1993; Asian Development Bank 1999; Samaranayake 2000):

Box 2.6, continued

- degradation and depletion of natural habitats caused by physical impacts of fishing and tourism on coral reefs, over-exploitation of resources, some land reclamation, pollution, dredging and other causes;
- loss and degradation of historic, cultural and archaeological sites and monuments due to building construction; and
- loss of physical and visual access to the ocean caused by siting of hotels and other facilities impeding access.

These issues prompted the development of long-standing interventions to develop systematic coastal management approaches described in Boxes 3.11, 3.14, 5.12 and 5.18. These issues have also contributed to the impacts of the 26 December 2004 Indian Ocean tsunami as outlined in Box 4.16.

Conflicting uses can be effectively managed within a planning framework. Planning can be at the strategic level if conflicts apply on a wide geographic scale, or at the site level if issues are local in nature. Chapters 3 and 5 describe these planning approaches.

2.2.2 Infrastructure – transportation, ports, harbours, shoreline protection works and defence

Major infrastructure developments on the coast include:

- ports and harbours;
- support facilities for and operation of various transport systems;
- roads, bridges and causeways;
- wind farms, and
- defence installations.

Ports have historically been the link between inland and marine transport. As transportation technology has evolved with larger ships and advanced cargo transfer capabilities (e.g. containers, bulk handling), ports have expanded from the natural sheltered waters of estuaries and inlets to the open ocean, and in some cases new artificial offshore islands (Couper 1983).

The thousands of ports around the world can be multi-functional or used for a single commodity such as mineral exports or containers. Irrespective of the type, port development results in a number of environmental and social impacts. Generally, port developments involve the manipulation of coastal areas by dredging, land reclamation and clearing of coastal forests. Socially, port development can displace pre-existing coastal inhabitants, limiting areas for subsistence and recreation, and creating increased local traffic (road and rail). In worst-case scenarios, port development constrains dwellers from using the area for subsistence and income generation.

Port development can act as a driver for regional economic growth and employment opportunity, mainly for skilled workers. Once a port and associated infrastructure is established, port-related industries develop, which in turn enhances trade through the port, fuelling more industrial development and job growth. This feedback mechanism has been one of the most important drivers of coastal urban grown for thousands of years. The benefits of ports must be balanced with natural habitat loss, pollution, changes to visual amenity, increased road and rail traffic, and loss of recreation sites.

Maintenance dredging and channelling of ports and harbours, and the dumping of the dredged material which affects water quality, can raise environmental issues and associated pollution concerns such as oil spills, hazardous cargo, and dumping of ballast water. Port traffic can also conflict with recreational boating. An example of environmental issues in the planning and management of ports is shown by the Port of Vancouver, Canada (Box 2.7).

Transportation within coastal areas consists of domestic and international shipping, and passenger ferry services. Efficient and safe ships combined with state of the art navigation systems have the potential to ensure the industry has minimal environmental impact. Unfortunately collisions and sinking of ships do occur, especially those under 'flags of convenience'. In 2000, 199 ships with greater than 100 Gross Registered Tonnage were lost in a total fleet of 87,546 ships (Lloyd's Register). The number of tankers lost at sea declined from twenty-four in 1997 to fourteen in 2000.

Management of oil spills and other pollution problems associated with transportation is addressed in the MARPOL Convention. For example, in the Great Barrier Reef, which is an Environmentally Sensitive Area in MARPOL, pilotage of international ships (cargo and passenger) is compulsory. Other cases where pilotage is compulsory is in the approaches to ports which are inherently dangerous, or where shipping lanes conflict with other users in the area. Pollution associated with transportation is discussed in Section 2.3.1. Since 1992 the International Maritime Organisation members have been implementing a program to phase out single hulled tankers for double hulled

Figure 2.5 Container port, Yokohama, Japan

Box 2.7

Issues in the Port of Vancouver

Vancouver is the largest city on Canada's Pacific coast, contains the nation's busiest port and is ranked second in foreign exports in North America. The port includes an outer harbour and Burrard Inlet in the north end of the city. It extends offshore to the Canada–USA border in the south. The Fraser River, one of the largest salmon-producing rivers in the world, discharges into the port in the south. The port is composed of container terminals, bulk and break bulk cargo (coal, grain, sulphur, wood pulp, etc.) facilities, cruise ship terminals, port related infrastructure and non-port industrial and commercial developments. These facilities are found along approximately 153 kilometres of shoreline.

The port encompasses a large area of shoreline, which is adjacent to industrial areas, to one of the fastest growing urban areas in Canada and to wilderness settings. The Vancouver Port Authority (VPA) manages the port and its mandate is to oversee all developments and activities on the land and water that it manages as a cost effective corporation (Yarnell 1999). The urban growth of Vancouver and growing tourism place increasing demands on the VPA to improve non-port access to the waterfront. Balancing industrial, commercial, urban and recreational demands within a sustainable development framework is a challenge for the VPA. The development of the port has been guided since 1994 by its Port 2010 strategic document, which is currently under review. Port 2010 articulates policies regarding the industrial as well as the commercial and recreational use of the port. There is no explicit mandate to protect public access to the waterfront; the VPA does so through policies in Port 2010.

---

vessels. In 2003 phasing out was accelerated with all tankers decommissioned or replaced by 2010 instead of 2015 (IMO 2003).

The operation of seaplanes, helicopters, hydrofoils, jet foils and other ferry services within coastal areas can be a source of conflicts between users. Some services are visually disruptive and noisy, while others can be hazardous. Environmental concerns regarding the operation of vessels, especially hydrofoils and jet foils, may include disruption to whale and dugong populations, both of which can be a focus for marine tourism. The operation of the foils may also damage fragile benthic communities and sensitive areas such as those used for recreation.

The location of scenic drives, bridges and causeways at the coast can also raise environmental and amenity concerns. Development of road works provides easier access to the coastal area and consequently the natural features or wilderness setting of the area may be diminished. Improved access may also result in demands for amenities in the area and a consequential loss in the area's scenic value. Increased access to coastal areas, especially those in sensitive areas, raise environmental issues such as dune erosion.

An emerging issue is coastal wind farms – either on land or immediately offshore. In Europe, construction of farms began in the 1990s as a response to solving the environmental problems of global warming. They have proved to be economically viable and it is anticipated that development of this form of energy will continue to expand into the foreseeable future. Countries such as Holland, Spain and the United Kingdom are in the initial stages of developing their offshore wind resources (ETSO 2003). The development of these farms has raised public concern over issues such as marine mammals, birds and visual amenity (Metoc PLC 2000). In the United States, for example, the introduction of a wind farm at Cape Cod has generated considerable public concern primarily over the issue of visual amenity (Burkett 2003). There is also in Europe a growing public concern about the effects on human health. Further information on the steps taken to integrate coastal wind farm management into broader coastal amenity assessment are provided in Chapter 4.

Finally, the infrastructure associated with military and defence uses of the coast can be significant. Defence infrastructure on the coast includes ports and harbours, repair yards, surveillance and communications facilities, and training grounds. The continuing provision of such facilities has become more important in many countries in recent years due to changing geopolitical circumstances, particularly an increase in terrorism. In contrast, reduction in military facilities in coastal areas , e.g. Portland, England, has allowed development of recreational facilities within the sheltered waters of the artificial harbour.

2.2.3 Tourism and recreation

International and domestic tourism is recognised as a growth industry, and much of it is focussed on the coast. World tourism peaked in 2000 with a growth rate of 7.4 per cent and 687 million international arrivals. Growth in the industry declined after the 11 September 2001 terrorist attack and ranged between −1.2 per cent and 2.7 per cent annually between 2001 and 2003 (WTO 2004). Nevertheless, tourism is projected to double by 2020. Coastal systems are important to the tourism sector. For example coral reefs support a tourism industry worth US$1.2 billion in the Florida Keys (Leeworthy and Bowker 1997). In Canada, marine tourism was worth US$421 million and recreational fishing US$774 million (Department of Fisheries and Oceans 1999).

Many developing nations see tourism as a potential source of foreign revenue, but lack the expertise to plan for a sustainable and well managed industry. Many have embraced tourism, especially on the coast, to meet the Northern Hemisphere's demand for tropical destinations close to the coast. The tourism industry in the Red Sea region has, for example, expanded rapidly as European holidaymakers seek an alternative destination to the Mediterranean (see Box 4.13). Moreover, Israel plans to increase coastal tourism by 43 per cent, Jordan by 100 per cent, and Egypt expects to expand tourism eleven-fold in the coast (Hall 2001). This increase is a result of a strategy to develop the Red Sea as a low cost destination for coastal tourists, since land and labour are much cheaper than in European coastal tourist destinations such as Spain.

Tourism can be an environmentally appropriate industry if managed correctly. There are many examples of where tourism has not been well managed, and not only have the natural resources of the area diminished, but local communities and economies have suffered (Chapter 4). But there are successes in developing sustainable tourism, which also benefits local communities. An example of planning for sustainable coastal tourism in Sri Lanka is given in Chapter 5.

Most of the issues associated with tourism development fall into two categories: environmental and social. Environmental issues include the impacts of developing tourist facilities such as resorts, caravan parks, golf courses, marinas and offshore structures (Huttche *et al.* 2002). Tourist facilities alter the natural landscape, disturb natural areas, and if they are not properly managed become a source of pollution. Throughout the developing world, coastal resorts are often established with little consideration of environmental issues such as sewage disposal (Chon 2000; Wong 2003). In areas where there are a number of resorts without some form of treatment and poor flushing of system, sewage can be a public health hazard. Some countries have standards as part of the permit/licence but do not enforce them for fear of losing investors to other destinations that have lower or no enforced standards; others have voluntary codes that are not enforceable but operators follow them so that they are seen as good corporate citizens.

Other environmental impacts of increased use of coastal and marine resources by recreationalists include anchor and mooring damage to benthic communities, overfishing and littering (Figure 2.6 and Figure 2.7).

Social issues related to coastal tourism development and recreational activities include: the displacement of indigenous residents, restricted access to coastal resources for income generation and subsistence, loss of wilderness opportunities, conflicts between users, changes to the area's amenity and possible lifestyle changes. There is also typically considerable leakage of the economic benefits out of developing countries to the developed countries and the globalised tourism providers, so that the much-lauded benefits often do not accrue to the host country or community.

2.2.4 Conservation reserves and protection of biodiversity

Only a small proportion (0.5 per cent) of the biodiversity of coastal areas is held in parks and reserves which aim to protect flora and fauna (Spalding and Chape, in press). Despite these small percentages, current and proposed future parks and reserves have the potential to meet the conservation objectives set out in Agenda 21. More recently an urgent commitment was made at the World Summit on Sustainable Development to apply an ecosystem approach, reduce the current rate of biodiversity loss by 2010, and develop an ecologically representative network of marine protected areas by 2012 in marine and coastal ecosystems. How to capitalise on such reserves is the subject of current research efforts, especially on how protected areas can be linked to the conservation values of coastal areas lacking specific habitat protection. The level of protection of natural coastal systems versus the level of human development and use of such systems is an ongoing debate with any coastal project. Often a coastal development will be required to include a foreshore reserve/buffer zone, the purpose of which is to act as a buffer for physical processes, provide recreation for local residents and to meet conservation requirements.

The ability of reserves to meet the multiple-use demands of coastal users and provide for conservation is questioned by environmental preservationists who see multiple-use as a trade-off between economic development and preservation. The introduction of the revised Great Barrier Reef Marine Park (GBRMP) Zoning Plan, which incorporates a representative area strategy, may address this question (GBRMPA 2003a). The new zoning plan protects the seventy different bioregions within the GBRMP and increases the no-extraction area from less than 5 per cent to more than 33 per cent. The new

Figure 2.6 Seagrass damage from recreational boating, Florida (Source: Curtis Kruer)

Figure 2.7 Anchor damage, Great Barrier Reef Marine Park

zoning places an emphasis on accommodating current and future uses as long as they do not degrade the Park's resources (GBRMPA 2003a). Multiple use plans have been effective for broadly managing large marine areas, where maintaining the self-interests of existing users is possible and where localised problems can be masked. However, new uses and users who are entitled to access rights can be difficult to absorb in such a system since users with long-term historical rights are often reluctant to reduce their access to the resources. Transferring these types of plans into coastal systems, which need more detailed planning, has not been tested.

2.3 Impacts of human use

As shown in Figure 2.1, a number of problems can result from the coastal uses listed in the previous section. In this section these problems considered under the headings of pollution (including industrial, sewage and runoff) and coastal hazards (climate change and liability).

2.3.1 Pollution – industrial, sewage and runoff

Major coastal pollution issues are:

• diminished water quality from urban and industrial sources;
• diminished water quality from the runoff from intensively farmed agricultural/aquaculture areas;
• oil pollution, including the risk of oil spills;
• transport of hazardous goods and wastes;
• dumping at sea; and
• ballast water and hull fouling.

Monitoring in coastal areas throughout the world has detected declining water quality, especially in proximity to urban areas (GIWA 2004) (Figure 2.8). It is estimated that the global economic burden due to ill-health, disease and death related to the pollution of coastal waters is running at US$16 billion a year (UNEP 2002). The Global International Waters Assessment has summarised the current and potential future conditions for various marine and coastal areas globally (GIWA 2004). Water quality in the coasts of Europe and North America improved or remained unchanged over the last decade. However, in the developing world coastal water continues to decline as population growth overwhelms efforts to provide sanitation and waste water facilities.

In the period between 1990 and 2000, 220 million people in the South Asia Seas Region benefited from improved sanitation. But during that period the population grew by 222 million, leaving 825 million still without access to acceptable sanitation systems, and thousands of miles of coastline vulnerable to pollution (UNEP 2003). In Indonesia, Tomasick *et al.* (1993) have demonstrated a decline in water quality and consequential loss in reef habitats offshore of the nation's capital, Jakarta. Australia, which is noted for its clean marine and coastal environments, also concedes that water quality around major urban centres has declined over recent years (Zann 1995).

Changes in water quality can be attributed to several sources: sewage outfall from primary and secondary treatment directly into the oceans or via river systems, storm

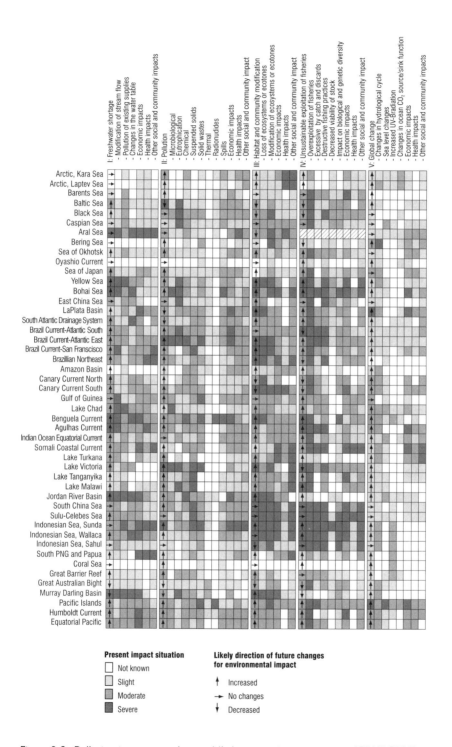

Figure 2.8 Pollution impacts on the world's large marine ecosystems (GIWA 2004)

water drainage, industrial wastes, runoff from pastoral lands and groundwater inputs (Box 2.8). Diminished water quality can lead to a loss of important coastal habitats, such as seagrasses, or an increase in unwanted species such as toxic algae, with a corresponding decrease in fish populations and resultant loss or coastal values for human recreational and amenity values. Concern has also been raised regarding the pumping of sewage from vessels, especially in sheltered embayments and estuaries. The disposal of garbage from ships, cargoes and ferries is a major source of litter washing up on beaches. Siting of landfill sites in close proximity to coastal areas, where leachates can be a source of pollution, exacerbates this problem.

Box 2.8

Pollution of urban coastal waters – the case of Jakarta, Indonesia

Pollution of nearshore waters adjacent to coastal cities has long been a problem. Since the 1960s, when critical pollution levels were reached in the developed world, a number of concerted efforts have been made to improve urban coastal water quality and to remediate polluted bottom sediments.

Like many capital cities in developing countries (see Figure 2.3), Jakarta has experienced rapid growth in its area, population and industry over the last fifty years (see figure below). This growth, however, has been at a cost to the coastal environment of Jakarta Bay and adjacent coral reefs (Kepulauan Seribu), primarily from pollution.

A number of studies have demonstrated that human impacts have severely degraded coral reefs in the Jakarta Bay area (Harger 1986; Moll and Suharsono 1986; Tomasick et al. 1993; UNESCO 2000). The studies have shown:

• water transparency increases with increasing distance from Jakarta Bay, which also corresponds with the maximum depths where corals are found;
• low water transparency reduces the maximum depth at which coral communities can survive;
• algal blooms are spreading further offshore; in 1986 blooms were only reported within 2 km of Jakarta's port, whereas in 1991 blooms were reported 12 km offshore; and
• a decline in fish landings from the Muro-ami reef fishery.

Causes of these impacts include the lack of sewage facilities throughout most of Jakarta and surrounding urban areas, where a series of canals above and below the ground carry raw sewage to the bay. A city of at least 9.5 million without a sewage treatment system is clearly a significant source of nutrient input into the bay. Existing waste disposal facilities, where much of the waste ends up as coastal landfill or in the rivers emptying into the bay, are inadequate for the city. Port activities including dredging and dry-docking have also contributed to the decline in water quality.

continued…

Box 2.8, continued

The impact of adjacent land use has been analysed by Tomasick *et al.* (1993) and UNESCO (2000). They found that nutrient runoff from land contributed to coral growth, but that wastes from industrial, agricultural and urban land uses impacted detrimentally on corals. Until recently, coral reefs were a source of building material and road construction in Jakarta. Coral extraction from shallow reef flats in 1982 totalled 840,000 m³ (Ongkosongko and Sukarno 1986 in Tomasick *et al.* 1993) and continues today.

The price paid by the environment for the rapid development of Jakarta Bay and Kepulauan Seribu is typical of many coastal areas throughout the world. Considerable resources will be needed to reduce these impacts let alone rehabilitate areas. Impact mitigation measures required include Environmental Impact Assessment and Strategic Environmental Assessment which are discussed in Chapter 4, while Chapter 5 highlights how integrated coastal planning at the local and regional level can also contribute.

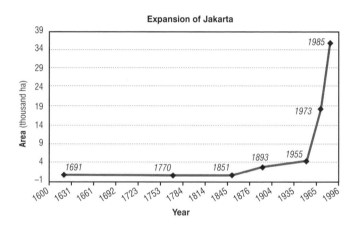

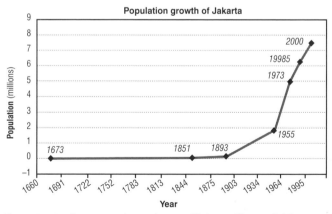

Expansion in the area and population of Jakarta (Dunwich Museum, from Setyawan 2002)

The potential impact of oil spills is a major pollution issue in coastal areas. Emergency oil spill response plans are in place in several countries, and when implemented they can reduce the impact of most spills. Oil pollution also occurs from other sources – shipwrecked vessels, oil exploration, bilge pumping and recreational craft. These sources are much more difficult to manage in a planning framework.

In nations where landfill sites are limited or the community is opposed to disposal of particular wastes (e.g. radioactive), the sea is often viewed as an easy and cheap dumping ground. Clearly this is not an acceptable practice except under very strict controls. International agreements such as MARPOL prohibit dumping at sea, and many nations have also enacted national legislation banning disposal at sea. Disposal of toxic substances such as radioactive wastes carries considerable risks since our knowledge of the long-term storage of such materials in marine environments is very limited. Disposal of landfill waste or dangerous wastes is prohibited in many nations; however, enforcement of regulations is difficult. The difficulty of waste disposal in coastal environments is shown by the example of the waste disposal issue in Tonga (Box 2.9 and Figure 2.9).

Box 2.9

Waste disposal on a coral atoll – Tonga

The disposal of waste generated by residents of coral atolls and small islands is often a major problem. Disposal options are limited, and with an increased use of consumer goods and packaged foods in island nations the problem is likely to remain in the foreseeable future. Disposal options and their problems include:

Disposal option	Problem
Deep landfill	Shallow freshwater lenses would be readily polluted
Shallow landfill	Potential to contaminate groundwater, unsightly, odour problems; vulnerable if there is a potential coastal erosion problem
Dumping at sea	Pollution risks; ecosystem damage
Removal to mainlands	Expensive; can just shift the problem elsewhere
Recycling	Not practical for all wastes and expensive for others

Waste disposal options and problems on small islands

The waste management system adopted on Tongatapu Island in Tonga highlights the problems facing many small island states. In the 1970s a central waste disposal site was established adjacent to the island's lagoon system and only a few metres from the coast. The economy of Tongatapu expanded during the 1970s and 1980s resulting in many people migrating to the island and generating much more waste as their standard of living improved. The waste facility reached its capacity in the 1990s and posed a health risk to residents in the area. Recycling programs have met with limited success, exacerbating waste disposal issues. The task of finding a new site has been problematic since there is limited public land available for non-productive use. Larger items such as cars and stoves are littered throughout the island since there are no facilities to transport them to the landfill and the cost of transporting such wastes off island for recycling is prohibitive.

Figure 2.9 Coastal landfill, Tonga

The introduction of exotic species through the pumping of ballast water in ports is a major environmental issue since in many places these exotic species have virtually destroyed the native fauna, reduced the biodiversity and altered the port's ecosystems (and subsequently adjacent marine ecosystems). The International Convention for the Control and Management of Ships' Ballast Water and Sediments was adopted by IMO members in 2004 to tackle this issue. The Convention requires all ships to implement a Ballast Water and Sediments Management Plan, carry a Ballast Water Record Book, and undertake ballast water management procedures to a given standard. Eradication of introduced pests is impossible and in many cases it is difficult even to control populations. Various countries have (mainly voluntary) guidelines which require mid-ocean exchange of water, where there is a greater chance that conditions will not favour survival of the exotic species, and taking relatively clean water onboard for disposal close to port. Similar voluntary guidelines also exist for the management of the impacts of toxic anti-fouling paints used on the underside of vessels.

2.3.2 Coastal hazards and climate change

The coast's highly dynamic nature has potential to damage property and threaten public safety. For those living on the coast, cyclones, storm surges and tsunami hazards are inherent and damaging natural events. For example, a tsunami killed 2,200 people in Papua New Guinea in 1999 and the damage from hurricane Mitch (1999) in Florida was estimated to be US$40 million (Lawrence and Guiney 2000). Hazards like these are difficult to manage and clearly pose risks to people and the environment. Increasingly, managing the economic impacts and liability issues emerging from these hazards is problematic in developing countries where insurance institutions are lacking. The question managers need to discuss with the community is who pays to manage these natural events. In developing nations, often there is no compensation for coastal dwellers who lose their property from cyclones and similar events. Depending on the nature of the event, in some developed countries compensation is provided at great cost to the community, which ultimately pays through higher insurance premiums or property taxes. In these cases the question arises as to whether the wider community should subsidise those who choose to live close to the coast and therefore risk damage. The answer to this question is not easy to formulate since it will depend on the social, economic and political culture of each country. Indeed, this issue is one that the global insurance industry is acutely aware of. The IPCC recently estimated that the global economic losses from catastrophic weather events increased from US$ 3.9 billion per year in the 1950s to US$ 40 billion per year in the 1990s (IPCC 2001).

Other coastal hazards can either be permanent, such as cliffs and headland erosion caused by long-term tectonic or isostatic land subsidence (Figure 2.10), or intermittent, such as rip currents on sandy ocean beaches. Human changes affecting coastal systems such as beach sediment budgets or interfering with riverine inputs can also create coastal hazards. In all cases, they pose serious risks to public safety and property. Public liability needs careful consideration when access to hazardous areas is provided by managing agencies, and when rescue aids, signage and other safety management elements are provided (Short, in press).

As shown above, planners and decision-makers face many hazards in coastal areas in the here and now. On the horizon, though, is the increasingly likely possibility that, over coming decades, the scale of the threats faced on the interface between sea and land may escalate as a result of global environmental change.

The scientific consensus is confirming that pollution of the atmosphere by greenhouse gases such as carbon dioxide and methane will bring about a significant change in the earth's climate – global warming – which could have widespread consequences (Houghton and Ding 2001). The understanding of the potential impacts of sea-level rise on climate change has progressed in recent years. It is important to note that as the scientific basis for greenhouse-induced climate change strengthens, the certainty of future predictions also strengthens. This is due to both a higher degree of independent understanding of global climate systems and ocean circulations, and an enhanced understanding of climate–ocean coupling. In turn, these insights are enhanced by greater accuracy of computer modelling.

The net result is a significant change in language from vague to clear statements by the Intergovernmental Panel on Climate Change (IPCC) in their most recent third (2001) assessment, compared to their 1996 second assessment. For example:

> ... most of the observed warming over the last 50 years is likely to have been due to the increase in greenhouse gas concentrations. Furthermore, it is very likely that the twentieth century warming has contributed significantly to the observed sea level rise ...

(IPCC 2001: 10)

Figure 2.10 Erosion of Dunwich, United Kingdom, 1886–1919 (Credit: Dunwich Museum)

Box 2.10

Bangladesh cyclone hazards (Kausher et al. 1996)

The Bangladesh coastal zone could be termed a geographical 'death trap' due to its extreme vulnerability to cyclones and storm surges. The massive loss of life from cyclones is due to the large number of coastal people living in poverty within poorly constructed houses, the inadequate number of cyclone shelters, the poor cyclone forecasting and warning systems, and the extremely low-lying land of the coastal zone. Approximately 5.2 million people live within coastal areas of high risk from cyclone and storm flooding within an area of 9,000 km².

(Kausher et al. 1996)

Nearly 1 million people have been killed in Bangladesh by cyclones since 1820 (Talukder et al. 1992) due to the estimated 10 per cent of the world's cyclones which develop in the Indian Ocean (Gray 1968) (an average of just under two (1.77) cyclones occurring each year) (Talukder et al. 1992). Once the cyclones have formed they generally move in a direction between north-west to north-east and can cross the coast in Burma, Bangladesh or India.

A particularly devastating cyclone to hit Bangladesh occurred on 29 April 1991. By the time this book is published there may have been another one! An estimated 131,000 to 139,000 people died, with the majority of victims below the age of ten, and a third of them below the age of 5; also, more women than men died (Talukder et al. 1992). An estimated 1 million homes were completely destroyed, and a further 1 million damaged. Up to 60 per cent of cattle and 80 per cent of poultry stocks were destroyed, and up to 280,000 acres of standing crops destroyed. Four hundred and seventy kilometres of flood embankments were destroyed or badly damaged, exposing 72,000 ha of rice paddy to salt water intrusion. Coastal industries and salt and shrimp fields were also badly damaged. The flood waters brought disease and hunger to the survivors. The total economic impact of the cyclone was of the order of US$2.4–4.0 billion (Kausher et al. 1996).

How the government of Bangladesh is attempting to plan for the impacts of future cyclones within their coastal management program is described in Chapter 4.

Of course, mean sea-level is just one of the outcomes of greenhouse-induced climate change: sea-surface temperature rises, changes in ocean current circulations, changes to the location, intensity and duration of extreme climate events, and changes to freshwater flows (and sediment input) will also play a part.

Importantly, too, the IPCC has also recently strengthened its assessment of the long-term impacts of climate change, particularly with respect to sea-level rise. The impacts are expected to continue for hundreds of years after stabilisation of greenhouse gas concentrations (IPCC 2001). Clearly, the world's coasts will be impacted for many,

many years to come. But equally clearly many of the world's immediate coastal management problems exist because of previous poor management decisions and present practices unconnected with, but made more urgent in the medium term, because of the likelihood of climate change impacts.

Coastal areas may face primary impacts as a result of, for example, a change in the risk of storm impacts, changes in ocean temperatures or rising sea level; and also the impact of secondary effects, since regional changes in climate influence economic performance and other aspects of human well-being (McLean and Tsyban 2001) (see Table 2.2).

While the global-scale consensus on greenhouse effects is firming there remain considerable uncertainties at a regional level as well as uncertainty of the climate predictions. The community of climate scientists has therefore recommended that a 'precautionary' approach be taken to the global warming problem at this time (Chapter 1). For example, Tri *et al.* (1996) have demonstrated through benefit-cost analysis that rehabilitating mangroves in northern Vietnam represents a sensible precautionary response to the threat of global warming as it is a 'win–win' strategy, providing additional storm protection, reduced dyke maintenance costs over time and, managed sustainably, providing an immediate boost to local incomes through the provision of extractable resources such as fish, crabs, fuelwood, and honey (Chapter 4).

The broader lesson here is that many measures which might be taken to protect the coastline against long-term climate impacts are precisely those which should be adopted on the basis of more immediate priorities. At this precautionary stage, there need be no contradiction between present-day goals and the longer-term aim of 'climate-proofing' management plans (Kelly *et al.* 1994). This approach to management-oriented adaptation strategies has recently been adopted by the IPCC, replacing its previous

Table 2.2 Potential impacts of climate change and sea-level rise on coastal systems

Biophysical impacts can include the following:

- Increased coastal erosion
- Inhabitation of primary production processes
- More extensive coastal inundation
- Higher storm surge flooding
- Landward intrusion of seawater in estuaries and aquifers
- Changed in surface water quality and groundwater characteristics
- Changes in the distribution of pathogenic micro-organisms
- Higher sea-surface temperatures
- Reduced sea-ice cover

Related socio-economic impacts can include the following:

- Increased loss of property and coastal habitats
- Increased flood risk and potential loss of life
- Damage to coastal protection works and other infrastructure
- Increased disease risk
- Loss of renewable and subsistence resources
- Loss of tourism, recreation and transportation functions
- Loss of non-monetary cultural resources and values
- Impacts on agriculture and aquaculture through decline in soil and water quality

Source: McLean and Tayban (2001) adapted by Choke *et al.*.

simplistic options of protection, accommodation or retreat (IPCC 1990, 1992). There is now a more concerted effort to consider the complexities of socio-cultural-economic impacts on coasts and their inhabitants (IPCC 2001) by considering the adaptive capacity of coastal systems, settlements and societies. Indeed, this is an approach much more aligned with coastal management and planning approaches outlined in Chapters 3, 4 and 5.

How prepared society is to adapt to climate change is still unknown. In the face of such overwhelming consensus among scientists and policy makers, societies must begin to decide how to reduce the long-term impact of climate change and, in the short to medium-term, how to adapt to the impacts that are likely to arise. Policies and actions for adapting to climate change will need to be incorporated into coastal and planning activities at all levels if society is to continue to enjoy, and in some countries improve on, their current standards of living.

If adaptation is to be handled effectively, it is important that management plans made today contain the degree of flexibility necessary if they are to be adapted at a later date (Chapter 5). Options should be kept open where possible.

2.4 Administrative and legal issues

As this chapter has shown, there are many complex and overlapping problems along the world's coastlines. This complexity, linked with government administrative systems that are designed to addresses issues on a subject-by-subject basis, can create problems in the effective management of the coast. The implications of these administrative issues for the design of coastal planning and management programs are described in Chapter 3.

Governments throughout the world are dealing with an increase in the number of litigation cases in the courts. Often advice from the legal profession is sought for situation-specific cases and also to provide strategic advice to avoid and/or mitigate future incidents. This raises the question of liability, indemnity and compensation. Liability in the event of accidents or damage to property is a complex question. In many countries the agency which has vested control over the area is responsible for public safety and protection of property. Similarly, for major developments in the coast, it is often unclear who is responsible in the event of a natural disaster today or in the future due to climate change (Titus 1998).

2.5 Summary – coastal conflict

Today's coastal managers face a plethora of problems, challenges and demands, many of which were unheard of only a few decades ago. As coastal populations grow in both developed and developing countries, the scale and intensity of coastal issues is also likely to increase.

Two important conclusions can be drawn from this chapter. The first is that most, if not all, coastal management problems centre on the issue of conflict or trade-offs in a sea of uncertainty due to climate change. Obvious examples include the conflict between the conservation of mangrove areas and their conversion to shrimp ponds. A less obvious example is where land-based activities bring about a decline in water quality, creating a conflict with natural ecosystem values. Viewing the majority of coastal issues as conflicts is useful in that mechanisms for their management become in effect strategies for conflict resolution. Management can also be viewed as deciding on the trade-offs that society needs to make in meeting its sustainability goals. In the mangrove example, the trade-offs are ecosystem services such as biodiversity, pollution control and erosion control against short-term economic development, employment, shelter and food security.

The second, and perhaps the key conclusion, is simply that coastal issues are now recognised as problems for which solutions must be sought. Having crossed this threshold, the principal issue now is not what the problems are, but how they should be tackled. A holistic approach undertaken in an integrated and coordinated manner is fundamental to managing the coast. This approach enables managers to deal with issues or problems individually while being mindful of not exacerbating other issues. It also provides the framework to deal with current issues and to plan for sustainable coastal ecosystems well into the future. This orientation towards coordinated and integrated management action requires clear guidance, a well-organised government structure, and – most importantly – a well defined set of objectives and actions.

This conclusion forms a useful basis for describing and analysing the development of coastal planning and management practices, the subject of the next chapter.

Chapter 3

Concepts, terminology and organisation of coastal planning management

Planning and managing the coast only begins to be effective if those engaged in the processes have a common understanding of the concepts and terms used, as well as how the processes of planning and management are organised. As outlined in Chapter 1, the coast can be defined differently in order to enhance different coastal management outcomes. Similarly, concepts and terms such as coastal management, adaptive management and coastal planning may be presented and perceived in many different ways depending on a person's interests, perspectives and experiences. The same diversity in perceptions, goals, values and motivation of how coastal planning and management should be organised, implemented and evaluated also exists in most systems of management.

This chapter features the concepts that currently underpin the conscious planning and management of coastal areas, and attempts to examine them within the evolution of coastal management thought over recent years. In doing so, the chapter aims to provide insights into possible changes in the direction of coastal management concepts in the immediate future. It does so by examining and analysing coast-specific literature and case studies, and blending these with recent management and planning concepts applied globally to all environments, including specific concepts from fields as diverse as physiology to electronic engineering.

Many of the emerging philosophies that are considered to underpin current coastal management practice are discussed. It is recognised that while some of these elements appear to be increasingly accepted, they cannot be considered as a definitive theoretical paradigm without further discussion, analysis and most importantly on-the-ground validation. The chapter will approach this task through a dialogue between these conceptual elements, individually and then collectively.

The chapter has four main sections. First, key theoretical concepts are introduced and analysed with examples of how they have been interpreted and implemented by organisations with coastal management responsibilities. Second, the most important terms and guiding statements for coastal management and planning are outlined. Third, choices in the design of administrative arrangements to implement coastal zone management and planning programs are discussed. Finally, monitoring and evaluation of coastal programs are described and analysed.

3.1 Concepts of coastal planning and management

An analysis of underlying concepts of coastal management is an essential pre-requisite to move the practice of coastal management forward. We also need to develop confidence that what we are doing on a day-to-day basis is guided by some broader framework, some plan, perhaps some common theoretical principles. Does coastal management have a set of common underlying concepts based on philosophical and theoretical constructs? Is there one theoretical framework that underpins coastal management practice worldwide? Or is coastal management more a movement masquerading as a discipline?

While coastal management practitioners have fashioned a set of concepts to guide their actions, this cannot be construed to be a rigorous theoretical framework in the sense that, for example, physical laws govern a scientific discipline. The broadly accepted concepts used in coastal management described below are a combination of the general theory and practice of urban planning and resource management as applied to the coast, mixed with pragmatism. This mix provides a suite of coastal management concepts, which describe a set of practices that help to achieve desired management outcomes.

The broad concept of coastal management, a proactive approach that is distinct from simply managing activities at the coast, encompasses the management of everything and everyone on the coast within some form of a unified system or approach. So what makes the practice of coastal management distinct from other forms of resource management or urban planning?

The issue is not the extent of the coastal area involved, but that specific management initiatives are undertaken which focus on a defined region – the coast. This distinguishes coastal management initiatives from other organisational government programs, such as forestry and fisheries management, the provision of education and health care, for example, which are not targeted to the coast. However, it is important to stress that, especially in developing countries, it is expected that the improved management of coastal resources will improve the socio-economic and environmental conditions of coastal communities.

As previous chapters have shown, the coast has many unique attributes. The most important (and obvious) of these is the dynamic interaction of land and ocean. However, in terms of the overall concepts of coastal management, defining a geographic area – the coast – and then applying special coastal management tools is analogous to the management of other parts of the world which can also be separated geographically from one another. Examples include the management of mountain ranges, a catchment, or areas of significant groundwater resources. These can be mapped, and require sensitive and distinctive management arrangements. Perhaps the closest analogy to coastal management is river catchment management, with both concerned with the integrated management of land and water resources and the people and organisational systems that support this management perspective.

The point we want to emphasise here is that coastal management *per se* is not unique. There are management approaches and techniques for other environmental systems that attempt to explicitly manage human/environment interactions which bear close resemblance to the coastal planning and management tools and approaches described in this book. Hence, coastal management is concerned with the application of techniques that attempt to clearly focus the efforts of governments, industry and the broader community onto coastal areas. These techniques centre around ways to bring together disparate planning and management techniques on the coast, to form holistic and responsive coastal management systems.

Thus, it is the combination of developing adaptive and integrated, environmental, economic and social management systems focused on coastal areas, which are the core coastal management concepts.

What is called 'theory' today corresponds to what Aristotle called 'contemplative life' (Lobkowicz 1967). Thus, in the sense of Aristotle, examining theory is concerned with taking time to think about the world. Following this train of thought, coastal management needs theory if it is concerned not only with its day-to-day activities, but also with the question of why those activities are taking place at all.

In taking a considered view of coastal management, theory is also a useful way of helping the practice evolve. In the words of Eagleton:

> For much of the time, our intellectual and other activities bowl along fairly serenely, and in this situation no great expenditure of theoretical energy is usually necessary. But there comes a point where these taken-for-granted activities begin to ... run into trouble, and it is at these points that theory proves necessary.
>
> (Eagleton 1990: 26)

From the substantial list of issues outlined in the previous chapter it would appear that indeed the need for conceptual and theoretical analyses is justified.

However, it is sobering to consider the experience in the theoretical analysis of urban and regional planning. Despite the considerable amount of literature on the subject, for example Faludi (1973), Paris (1982), Alexander (1986) and Platt (1991), there is still no clearly defined or widely accepted set of planning theories. The reasons for this are clearly articulated by Campbell and Fainstein (1996, 2002), reproduced in Box 3.1.

Campbell and Fainstein (1996, 2002) add to their description of the difficulties, and maybe even the impossibilities, of delineating meaningful planning theory outlined in Box 3.1 by (ibid.: 2) describing planning theory as 'the assimilation of professional knowledge'. They also raise the question, does planning theory precede, lag or move in parallel with planning practice?

What then does this mean for coastal planning and management theory? Principally it must be recognised that there is no single unifying theory which guides coastal planning and management practice. Instead, there is a range of theories which have shaped coastal planning and management, and provide a 'menu' of theoretical approaches to choose from. These approaches can then be fashioned by coastal managers into approaches appropriate for particular cultural, economic, administrative and political circumstances – and of course, the issues being addressed by a coastal initiative.

Consequently, the coastal management planning approaches described in Chapter 5 tend to borrow from, and merge, a number of theories to provide the most appropriate planning and management approach for a particular stretch of coast. Indeed, this approach of introducing a menu of theoretical concepts, for the manager to piece them together into a meaningful framework, is undertaken here. The main elements of this menu are:

- rational, comprehensive planning;
- values-based planning;
- ecosystem-based management;

Box 3.1

The problems of defining planning theory (Campbell and Fainstein 1996, 2002)

Campbell and Fainstein (1996: 2) attribute the difficulty of defining planning theory to four principal reasons.

First, many of the fundamental questions concerning planning belong to a much broader inquiry concerning the role of the state in social and spatial transformations. Consequently, planning theory appears to overlap with theory in all the social science disciplines, and it becomes hard to limit its scope or to stake out a turf specific to planning.

Second, the boundary between planners and related professionals (such as real estate developers, architects, city council members) is not mutually exclusive; planners don't just plan, and non-planners also plan.

Third, the field of planning is divided into those who define it according to its object (land-use patterns of the built and natural environments) and those who do so by its method (the process of decision making).

Finally, many fields are defined by a specific set of methodologies. Yet planning commonly borrows the diverse methodologies from many different fields, and so its theoretical base cannot be easily drawn from its tools of analysis. Taken together, this considerable disagreement over the scope and function of planning and the problems of defining who is actually a planner obscure the delineation of an appropriate body of theory. Whereas most scholars can agree on what constitutes the economy and the polity – and thus what is economic or political theory – they differ as to the content of planning theory.

In providing more recent commentary to the above difficulties Campbell and Fainstein (2002: 12) state:

Much of planning theory is ... an attempt to bring our thinking of planning up to date and in line with other urban phenomena (cyberspace, globalization, etc.) or social theories from other fields (such as postmodernism or critical theory). In addition, the theory-practice time lag may run the other way round: the task of planning theory is often to catch up with planning practice itself, codifying and restating approaches to planning that practitioners have long since used (such as disjointed incrementalism or dispute mediation). Planning theory can therefore alternately be a running commentary, parallel and at arm's length to the profession; a prescriptive avant-garde; or instead a trailing, reflective echo of planning practice.

- adaptive/learning management and planning;
- systems theory and cybernetics;
- environmentalism;
- participation, consensus and conflict; and
- pragmatism.

These theoretical elements are outlined in the following sections.

3.1.1 Rational, comprehensive planning theory

Rationality has been the primary way western society has thought since the Renaissance era. This was the era of scientists, such as Galileo and Copernicus, who promoted a scientific approach to problem solving (Dawes 1988). In its simplest terms, 'rationality is a way of choosing the best means to attain a given end' (Alexander 1986).

When problems are relatively simple, one can quickly choose the best means to accomplish a given goal. This simple approach is termed 'instrument rationality'. Problems where this form of rationality is used generally have a determinate solution – a solution which is definite and can be defined or explained in tangible terms.

When rationality includes evaluating and choosing between goals as well as relating the goals to individual organisational or social values, it is termed 'substantive' or 'value' rationality. This form of rationality has a significant influence in planning, especially where there are conflicting and multiple objectives. Rational decision making assists planners to make choices within a framework which is consistent and logical; to validate assumptions about the problem and choices; to collect and analyse information, theories and concepts; and to provide a mechanism to explain the reasons for the choices made.

The rational decision model consists of a number of stages linking ideas to actions (Figure 3.1):

- identification of problems;
- defining goals and objectives;
- identifying opportunities and constraints;
- defining alternatives; and
- making a choice and implementing that choice.

Rational planning theory requires a very broad, comprehensive body of knowledge in order to make logical decisions when assessing all possible alternatives. Hence, the rational planning model is also called the 'comprehensive' model. Without 'perfect' knowledge, according to rational planning theory, there are inevitably value judgements made which reflect the biases and values of the decision maker. Generally, in coastal planning and management there is rarely complete information and understanding of all possible alternatives. In order to counteract these limitations of rational planning theory, some modifications have been adopted, including:

- considering the options one at a time with flexible goals and objectives which can be modified with the options considered – called 'satisficing'; and

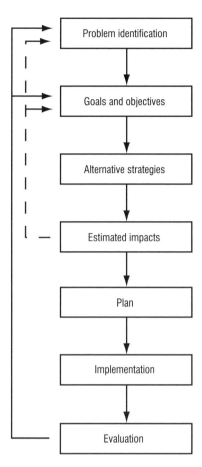

Figure 3.1 Rational (comprehensive) model of planning and decision making (Smith 1993)

- considering a few possible options which are formed and analysed based on their differences and the status quo – called 'disjointed incrementalism' (see below). This avoids information overload and also avoids suggesting radical solutions which may be socially or politically unacceptable.

Currently, the rational planning model generally applies only to the early stages of the coastal planning process – identifying problems, defining goals and objectives, defining opportunities and constraints and sometimes specifying alternatives. But selecting from different approaches and their subsequent implementation is often achieved with the assistance of other planning theories which explicitly recognise the influences of value judgements of the participants in the planning process.

(a) Incremental planning theory

Incremental planning is sometimes described as the 'science of muddling through' (Campbell and Fainstein 1996). It recognises the inevitability of decision-making strategies based on the limited cognitive capacities of decision makers and reduces the scope and cost of information collection and analysis. This method looks at alternatives, with limited deviation from the status quo. The main components of incremental planning theory are:

- choices are derived from policies or plans which differ incrementally from existing policies (i.e. the status quo);
- only a small number of alternatives is considered;
- only a small number of significant consequences is investigated;
- ends and means are adjusted to make the problem more manageable; and
- decisions are made through an iterative process of analysis and evaluation.

This model is considered by many as a better reflection of how planning decisions are actually made. However, a countering view is that incremental planning is focussed on managing present issues and not on the promotion of future goals or the ability to undertake large conceptual leaps. Nor does incremental planning theory consider fundamental re-evaluation of underlying planning principles and concepts. In this sense, incremental planning can be contrasted to adaptive/learning planning approaches that include continuing re-analysis and re-invention as one of their core principles.

3.1.2 Values-based planning

Values-based planning concepts can be thought of as the opposite of rational-comprehensive planning (Kay and Christie 2001; McKellar and Kay 2001). While rational planning assumes that all participants in the planning process operate objectively, values based planning assumes the complete opposite, or as Fekete (1988) cited in Guerrier *et al.* (1995: 3) states:

> Not to put too fine a point on it, we live, breathe, and excrete values. No aspect of human life is unrelated to values, valuations and validations.

Consideration of personal values provides a foil to the neutral, rational language of the rational planner, allowing consideration of social and individual aspirations, which in turn are determined largely by individual and community values shaped by spiritual and religious beliefs. Indeed, values-based planning concepts stress that planning is a social activity, in which values are used to set social agendas. Implicit in this process is that planning will determine the directions that reflect community values, thereby allowing activities to be implemented in a way and place that will not generate conflict. For example, in Australia Dutton *et al.* (1995):

> There is mounting evidence of public disenchantment and frustration with current approaches to coastal resource management in Australia and the processes employed and their outcomes. There are also lingering doubts about whether these processes adequately represent the diversity of views within the wider community.
>
> (Dutton *et al.* 1995: 246)

A useful definition of values was prepared during the development of the Australian coastal management program in the early 1990s. This divided values in coastal management used in the Australian context is between 'held' personal values and those values assigned to something of value – called 'assigned' values (Brown 1984; Resource Assessment Commission Coastal Zone Inquiry 1993).

Assigned values are those given to an object, depending both on the attributes of that object and what those assigning the value think and feel. For example, saying that a coastal landscape is 'beautiful' or 'remote' is an assigned value – 'beauty' and 'remoteness' have both subjective and objective attributes. Some analysts use the term Instrumental Values – the value of an object that can be used to attain some value. A beach is 'valuable' in this sense because it is used for walking, swimming and so on. Some analysts argue that coastal areas have 'intrinsic value'– there is value in the coast for itself rather than for what it is or could be used for. Further examination of the perspective that economists bring to values and valuation is given in Chapter 4.

'Held values' according to the Resource Assessment Commission Coastal Zone Inquiry (1993):

> ... serve to guide our choices and motivate behaviour. They arise out of our experience, education and socialisation and consist of both emotions (for example love, hate, fear, joy) and principles (for example 'individual freedom is paramount') which are used to appraise particular things or states of affairs. Consequently, to value something is to have a mixture of emotions, interests and beliefs which serve to accord importance to our lives.

Importantly, it is the very core of the human experience that becomes exposed when considering values in coastal planning and management (Hart 1978). Moreover, it is these held values that determine which assigned values are recognised, expressed and protected. Integrating held values with planning requires planners to address several critical issues, including the:

1   duty of care the planner or managers will hold for people who have exposed their values;
2   need to distinguish between the values of the planner or manager versus the values of the community and its constituent members;
3   need to ensure that all societal values are expressed and respected; and
4   need to apply conflict management processes suitable to resolve value conflict.

Importantly, people from different parts of the world, from different cultural and religious backgrounds and from different socio-economic circumstances will hold and express their values differently.

Cultural values, such as freedom of individual expression, are markedly different in many western countries from those of other cultures where values of respect to elders, saving face and clear social hierarchies are paramount. Understanding the subtleties of these important issues relies on the skills of the coastal planner and manager to interpret and merge with other conceptual elements outlined in this chapter to develop workable (and supported) coastal programs.

Kay (2000b) likens elicitation of personal values in the development of coastal programs to an iceberg. The metaphor was used because, like an iceberg, part of coastal planning remains invisible below the surface of a planning process (Figure 3.2).

Potential elements of values-based planning in coastal programs include (McKellar and Kay 2001):

- First, acknowledgement of the personal vulnerability implicit in value discussions, statement of the range of values and value holders, and exploration of any conflicts.
- Second, a clear link between values and vision, principles and strategic objectives and actions.
- Third, an assessment of the extent to which values would be and would not be served by the strategic objectives and actions.
- Finally, a discussion about how those values, which were not likely to be reflected in the plan, are to be honoured in other ways.

At present, values-based approaches in coastal planning and management concentrate on the development of statements that guide coastal programs. Examples of such statements are provided in Section 3.3.3. In addition, many coastal practitioners acknowledge the role that values and related 'under the surface' issues play throughout the various stages and components of coastal programs (Byron Shire Council 2000).

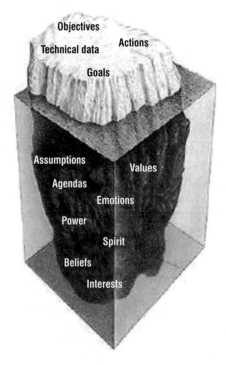

Figure 3.2 Indicative above and below the surface components of the coastal planning iceberg (Kay 2000) (Base iceberg graphic from: http://www.geocities.com/yosemite/rapids/4233/index.htm)

However, it is early in a coastal program, at the program definition phase, that concepts of values are perhaps the most important (Clark 2002).

3.1.3 Ecosystem-based management

The previous approaches of managing on a sector-by-sector basis, or on the basis of a single need or activity, have proven to be less than ideal in achieving sustainable development of the coast. Ecosystem-based management, which promotes a holistic approach to managing ecosystems, is seen as integral to achieving the goals and objectives of sustainable development for all ecosystems – broadly defined. There are various definitions of ecosystems, depending on the management context. However, Wang (2004) summarises the features common to the various definitions of an ecosystem. All are pertinent to coastal systems:

- An ecosystem exists in a space with boundaries that may or may not be explicitly delineated. Ecosystems are distinguishable from each other based on their bio-physical attributes, their locations and the spatial extent of their interactions.
- An ecosystem includes both living organisms and their abiotic environment, including pools of organic and inorganic materials.
- The organisms interact with each other, and interact with the physical environment through fluxes of energy, organic and inorganic materials amongst the pools. These fluxes are mediated and functionally controlled by species' behaviour and environmental forces.
- An ecosystem is dynamic. Its structure and function change with time.
- An ecosystem exhibits emergent properties that are characteristic of its type and that are invariant within the domain of existence.

Several international and national organisations promote the application of ecosystem-based management in marine and coastal management. The United National Food and Agriculture Organisation (FAO) Code of Conduct for Responsible Fisheries calls for countries to take an ecosystem approach to managing fisheries (FAO 2003). The Convention for Biological Diversity also recommends an ecosystem based approach as part of strategic planning at the national and regional levels (CBD Secretariat 2004).

There are various definitions that describe how such ecosystems can be managed. The Committee on the Scientific Basis for Ecosystem Management (CSBFEM) of the Ecological Society of America (Christensen *et al.* 1996) outlined the major elements of ecosystem management as:

1   intergenerational sustainability is a precondition;
2   goals are measurable for specific future ecosystems processes;
3   decision making relies on research performed at all levels of ecological organisation;
4   complexity and interconnectedness are integral to maintaining ecosystems;
5   ecosystems are dynamic;
6   context and scale are accounted for;
7   humans are a component of the ecosystem; and
8   approaches are adaptable and accountable.

These elements can be applied in the context of coastal management as shown in Box 3.2.

Importantly, an ecosystem-based management approach requires a number of conceptual building blocks. These include an adaptive management approach (Section 3.1.4), a holistic whole-of-system approach (Section 3.1.5), clear elicitation of environmental world views, including the role of humans in nature (Section 3.1.6) and pragmatism (learning-by-doing) (Section 3.1.8) (Grumbine 1994).

Incorporating ecosystem-based management in coastal planning initiatives will be demonstrated in subsequent chapters. Just as for the terms 'planning', 'management' and 'integration', discussed above, ecosystem-based management is also referred to as ecosystem management, integrated ecosystem management and total ecosystem management. In this book the term 'ecosystem-based management' will be used for consistency and clarity.

Box 3.2

Cooperative ecosystem management across the Canada–US border

An ecosystem approach is evolving to managing the Georgia Basin/Puget Sound area. The area encompasses the west-coast boundary waters between Canada and the United States and faces a number of issues such as population growth, water pollution, habitat and species losses. Prior to the 1990s transboundary, issues were managed on the basis of an individual species (e.g. salmon) or issue (e.g. water quality). Early in the 1990s the British Columbia (Canada) and the Washington State (USA) governments began to take an ecosystem-based management approach. The British Columbia government initiated the BC Roundtable on the Environment and the Economy in 1992 to manage the Georgia Basin as a complete unit and to include the United States in its shared management. The Washington State government initiated a similar program the Puget Sound Water Quality Management Plan, in 1996 to implement, the Puget Sound Water Quality Management Plan in managing and protecting the Puget Sound and for coordinating the various governments. The Plan was adopted in 2000 (Puget Sound Water Quality Action Team 2000). Other initiatives that encompass the Georgia Basin/Puget Sound followed these first initiatives on both sides of the border.

The approach was expanded in 2000 to include federal governments and First Nations' involvement with the signing of the Canada–United States Joint Statement of Cooperation on the Georgia Basin and Puget Sound Ecosystem. Federal agencies in the USA and Canada committed their support to transboundary arrangements and outlined common goals and objectives, recognised the special interests of the Coast Salish First Nations and Tribes, and established a formal Canada–United States mechanism at the ecosystem level to act on the challenges of sustainability.

The ecosystem approach through transboundary initiatives is just over ten years old and appears to be working. However, the governments and other stakeholders need to take a more comprehensive and fully integrated approach to managing the Georgia Basin/Puget Sound area.

3.1.4 Adaptive/learning management and planning

The concept of adaptive management was first popularised by Holling (1978). It is based on the concept of adaptive control process theory. This focusses on decision making founded on experience. In this sense, it borrows heavily from system theory (Section 3.1.5) and pragmatism (Section 3.1.8). In terms of coastal planning and management, adaptive management applies the concept of experimentation to the design and implementation of natural resource and environmental policies.

As new information is obtained, and current management processes are reviewed, new management methods are formulated. Adaptive approaches are based on the concept of learning from events of the past, and from experimentation, including recognising society's limited knowledge and uncertainties in predicting the consequence of using resources. The learning-by-doing aspects of adaptive management based on pragmatic concepts are outlined in Section 3.1.8. The three tenets of adaptive management were summarised by Torell (2000) as:

- adjustment of management processes and policy to the constraints of the situation as new information is obtained;
- management involves learning by doing and experimentation; and
- participatory process that actively engages significant stakeholders in management practice, collective enquiry and decision making.

As outlined above, a fundamental tenet of adaptive management is that all management is essentially an experiment (Lee 1993). In adaptive management policy, planning and subsequent management actions are viewed as being undertaken in a climate of incomplete knowledge of how systems function. As a result, management interventions will essentially be a best guess based on incomplete information. Adaptive management theorists strongly reject the notion of 'baseline studies' that imply complete understanding of the status and functions of systems.

An adaptive policy is designed, from the outset, to test clearly formulated hypotheses about the behaviour of a system; for example, an ecosystem (natural or urban) being changed by human use. In most cases these hypotheses are predictions about how one or more important species will respond to management actions. For example, commercial fishery regulation, monitored by a regulating authority, can readily be designed in an experimental fashion. If the policy has shortcomings, an adaptive design still permits learning, so that future policies and decisions can be based on a better understanding.

Adaptive planning is also an opportunistic form of planning. It is responsive to the prevailing management environment in which planning is taking place. It allows planners and managers to anticipate or take advantage of surprise, and views the outcomes of management activities as learning opportunities (McLain and Lee 1996). Some influential coastal management thinkers view the 'learning' aspect of adaptive management as fundamental to effective coastal management practice (Lowry 2003; Tobey and Volk 2003). As Lee (1993) states:

> Adaptive management is highly advantageous when policy-makers face uncertainty, as they almost always do in environmental planning. But the adaptive approach is

not free: the costs of information gathering and the political risks of having clearly identified shortcomings are two barriers to its use. Because the adaptive model of learning and management does not take into account the limitations of learning within human organisations, specific precautions should be built into the design of policies.

(Lee 1993: 54)

Other barriers to adaptive approaches include reluctance by managing agencies, and users of resources, to adopt experimental approaches to management. In addition, there may be suspicion of using non-scientific information, such as the perceptions and opinions of coastal users. Finally, adaptive planning requires a consensus on common values amongst diverse interests. This can contrast to the perception by some participants in the planning process, most often professional planners, that they 'know best'. In favour of adaptive management it is recognised that without experimentation reliable knowledge accumulates slowly, and without reliable knowledge there can be neither social learning nor sustainable development.

It is worth noting that while European Union principles for integrated coastal zone management (Box 3.4) include adaptive management among the principles, this is qualified by stating that adaptive management 'implies the need for a sound scientific basis concerning the evolution of the coastal zone'. While this is undoubtedly a core part of adaptive management, as outlined above, it could be interpreted as moderating the fundamentally experimental nature of 'pure' adaptive management.

Adaptive concepts have been used, for example, in the management of forest ecosystems and catchments in the Pacific north-west of the USA and British Columbia, since the mid-1980s (British Columbia Forest Service 2000). However, it was not until 1996 that adaptive management entered the lexicon of coastal managers with its inclusion in the GESAMP report (GESAMP 1996). GESAMP essentially adopted adaptive management principles through its recommendation of the 'policy cycle' (Figure 3.3).

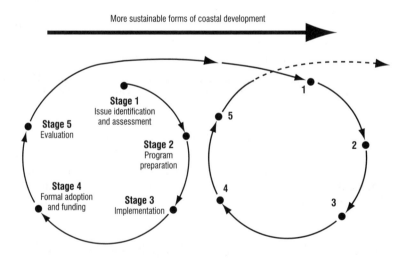

Figure 3.3 GESAMP policy cycle (GESAMP 1996)

The GESAMP implementation of adaptive management then flowed into a number of coastal management initiatives around the world. The recently adopted European Union ICZM process (Box 4.2) applied adaptive management as one of its principles. More recently, adaptive management concepts have been applied to the Australian national coastal management program (under the broader Natural Heritage Trust). As the agreement between federal and state governments states (Environment Australia 2003):

> The Parties agree to develop accountability arrangements, consistent with the principles of adaptive management, that do not preclude or impede any party from revising targets, strategies and timelines, at any level of delivery, as a result of new findings from evaluation, or new knowledge or data.

Adaptive management has also become a key part of the local-scale coastal planning process in the Philippines (Courtney and White 2000) (Box 5.3).

The advocates of adopting an adaptive approach in coastal planning and management state that the 'fundamental features of each step in the ICM policy cycle are widely accepted among ICM professionals' (Tobey and Volk 2003: 298). Indeed, the application of adaptive management to coastal contexts is difficult to argue against given both its intuitive and theoretical appeal. Nevertheless, as outlined above, there remain considerable practical difficulties with applying a 'pure' adaptive management model that views 'policy as an experiment' (Lee 1993). It appears that this view of adaptive management has itself been revised to blend rationalist and adaptive elements in order to be mainstreamed in coastal programs – especially in the developed world. Consequently, the widespread applicability of adaptive management is not yet clear. Given the rapid application of adaptive management principles, there is an urgent need to further test and consider adaptive management approaches within coastal programs operating at different scales, in different countries, and within different policy contexts.

The elements of Figure 3.3, together with the application of adaptive management into coastal programs, are discussed further in Chapters 4 and 5.

3.1.5 Systems theory and cybernetics

Systems-led analysis is seen as 'a framework for seeing interrelationships rather than things, for seeing patterns of change rather than static "snapshots"' (Stacey 1992: 68). Consequently, systems theory appears to have direct relevance to coastal management; yet to date it has played, at most, a marginal role in influencing coastal planning and management concepts (Kay *et al.* 2003).

Cybernetics is an approach to systems theory that focusses on information exchanges, as opposed to systems characterised primarily by exchanges of matter and energy (Bertalanffy 1969). The key feature of cybernetics is the issue of control, which is understood more in terms of influence or coordination than in the absolute determination of behaviour. Cybernetics emphasises those informational elements in a system that causes the system to respond appropriately to its environment in order to achieve its goals (Nauta 1972). In stressing goals, cybernetics is therefore most appropriate to those systems that have a goal (Bertalanffy 1969: 23), namely technological and social systems – including integrated coastal management.

Cybernetics addresses the internal components that mediate the relation of the system with the environment, and its capacity to change in response to it. Institutional arrangements in the cybernetic sense can be seen as control mechanisms for the systems they are intended to influence.

There are several key observations from cybernetics and broader systems theory that are of relevance to the concepts of coastal planning and management (Kay *et al.* 2003). Perhaps the critical lesson from cybernetics is that the complexity of modern coastal management systems must ultimately be reflected in the control mechanisms of those systems, if programs are to be effective. This is known in system theory as 'the rule of requisite variety' (Ashby 1969). As a result, institutional arrangements must be capable of accommodating a sufficiently broad range of changes external to the system to be able to maintain the system in a long-term, sustainable manner, within the parameters established by the program.

Cybernetic theory also requires that the mechanism controlling a system is part of the system, not external to it (Beer 1981). Thus coastal management systems, in this context, should be seen as part of the coastal ecosystem, as required in ecosystem-based management (Section 3.1.3). The essential processes in system control are the relationship between the control mechanisms and the elements that it controls through feedback loops. In this sense, system theory can be thought of as sharing the iterative-feedback requirement of adaptive management.

A critical differentiation between system theory and other management concepts is that cybernetic systems are self-organising. System theory holds that, while some components of a network will be more powerful than others, the decision-making process is one of mutual accommodation rather than central direction. Part of self-organising is decisions about system/context boundaries. This notion is in direct contrast to the 'mechanical' control and command approaches of modernist rational-comprehensive approaches in which the component parts of the system have a fixed relation to one another and the systems' primary function in the service of a human goal or need (Section 3.1.1).

System theory holds that a system can have highly disparate types of components – social, technological and environmental – so long as they form some kind of coherent whole. This avoids the limitations of approaches confined by discipline boundaries that can be reductionist, or confined by an inability to recognise phenomena beyond the scope of the discipline (Callon 1987). In addition, human systems exhibit reflexivity; that is, a system comprising another system and a person observing it. As a result, identifying the disparate components that form a coherent whole is largely a question of judgment (Vickers 1995). For coastal management a significant component of the task is to cooperatively reach a view of the respective whole and its components. A coastal management system, as viewed by a system theorist, is an appreciative system, meaning goal definition is at least as important as goal seeking.

Importantly, systems are considered to contain the ability to change their rules of operation, and their internal organisation, in response to changing circumstances. They do not change basic settings constrained by fixed rules or operation (Senge 1990). Consequently, system theory provides useful input into the conceptualisation of how changing circumstances, be they the results of monitoring programs or 'external' factors, impact on program evolution and adaptation.

An important emerging stream of system theory is 'soft systems methodology' (SSM). This deliberately brings systems thinking to bear on real-world problems (Checkland and Holwell 1998). The approach is particularly valuable in a coastal management context because it explicitly focusses on people and their place within managed organisations – termed 'human activity systems'. The common factor in all human activity systems, according to SSM, is that people are attempting to take some 'purposeful action'. Interestingly, in order for people to develop a set of actions they must clearly define their worldview (Section 3.1.6) that outlines their personal values, or a 'declared point of view'. SSM then approaches the development of actions through a set of consensual models that provide a coherent structure to discussion on accommodations between various options, and points of view, for moving forward (see Box 3.3).

Critically, SSM is a reflective action-learning system (Box 3.3). In addition, SSM takes important elements of the interpretive research approach in which the researcher is embedded within the social systems that are being studied. This approach has been termed 'action research' (Checkland and Holwell 1998).

3.1.6 Environmentalism

Environmentalism is the belief that humans are a part of nature and, as a result, they have a responsibility to ensure their existence is considered within the context of their environmental impact. Environmentalism is used here to introduce the stream of concepts developed in such diverse fields as environmental politics, environment and development, environmental ethics and environmental philosophy. These streams of thought have emerged as a powerful force since the 1960s (O'Riordan 1981) and now influence the implementation of many coastal management programs through environmentalism concepts being adopted by government and intergovernmental organisations, through direct citizen action or through the influence of non-governmental organisations and businesses (WBSC 2003).

Fundamental to the notion of environmentalism is the analysis of human interaction with, and conceptualisation of, nature – or environmental 'world views' (Redclift 1995). This conceptualisation of nature is rooted in our personal and spiritual beliefs, and shaped by our gender, age and our economic and social conditions. It is also deeply rooted in historical context, as clearly shown by the evolution of coastal management in tropical developing countries through the pre-colonial, colonial (centralised) and postcolonial eras (Christie and White 1997).

Environmental worldviews are the personal lens through which we see the world. As such, characterising them is important to an understanding of the perspectives brought by coastal management professionals to program development (Clark 2002). Moreover, consideration of the values, ideologies and philosophies held by stakeholders in a coastal program will underpin their relationship with a coastal program. These worldviews may be held by individual stakeholders or be expressed through the collective philosophies of stakeholder organisations, be they government, community-based or from the private sector. Consequently, individual and collective environmentalism will pervade all aspects of coastal program development: from its vision, through its guiding statements of principle, to its implementation (Section 3.3.3).

Box 3.3

The application of soft system methodology in the Malta coastal area management planning process

Soft systems methodology (SSM) was used as the basis for the development of a systematic approach to the assessment of sustainability (termed Systematic Sustainability Analysis (Bell and Morse 1999)) within the Malta Coastal Area Management Plan (CAMP) (Vella 2002). The CAMP was one of a series developed under the auspices of the Mediterranean Regional Seas Programme (see Box 5.17).

SSM was applied 'precisely because it is designed to deal with "messy" situations. It is based on the assumption that in many cases a well-defined problem is a rare luxury, and people involved often cannot agree on what is wrong or what should be done' (Bell and Morse 2003: 21). The application of SSM in the Malta coastal planning process used a structured group process shown in the figure below.

The application of SSM in the Malta CAMP process was found to be appealing because of its approach as a form of collaborative enquiry allowing all participants to address coastal sustainability issues in a structured learning environment (Bell and Morse 2003). A key outcome of its application was the development of a suite of simple national Sustainability Indicators (OECD 1998; Malta Observatory 2003) 'that extends the concept of sustainability down the line to specific stakeholders rather than retaining this concept at a theoretical level' (Vella 2002: 134).

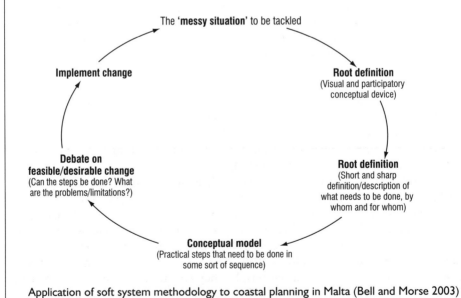

Application of soft system methodology to coastal planning in Malta (Bell and Morse 2003)

O'Riordan (1981) has provided an enduring view of how to classify environmental ideologies, with the fundamental contrast being between technocentricism and ecocentrism. A useful interpretation of O'Riordan's classification as provided by Pearce (1993) is shown in Table 3.1.

The discourse required to consider the role of worldviews in making sense of the relationship between people and coastal resources requires analysis of some fundamental philosophical concepts. This is required in order to make sense of how we see the world in general terms, and then through this approach make specific reference to human–environment interactions. These require consideration of:

* how reality can be known and the criteria for judging the truth of a statement about reality (epistemology);
* what it is possible to know (ontology); and
* the set of rules and procedures to guide enquiry (methodology).

Given the depth of philosophical enquiry that is needed to consider these issues, and the literally thousands of years over which philosophers have pondered such matters, it is easy to become mired in their deliberation. Nevertheless it is worth considering briefly one of the more recent philosophies adopted by environmentalists, critical realism (Archer 1998; Bhaskar 1998), in order to introduce an example of how environmental philosophies are described. Critical realism is an emerging philosophical perspective that can help 'provide insights into how social systems should evolve alongside biophysical systems' (Huckle and Martin 2001: 37) (see Table 3.2).

It is important to stress that debate on the role of environmental philosophies remains on the fringe of the coastal management mainstream (Kay 2000a). As such, there do not, to the authors' knowledge, appear to be case studies in the direct application of the environmentalist concepts outlined above in coastal management actions. At present their critical role in program development appears to be in the process of developing guiding statements (Section 3.3.3). As with values-based planning, consideration of environmentalism in this context provides critical guidance to the overall boundaries and intent for program development. Environmental philosophies also provide important tools for consideration of how conflict can develop between those individuals or organisations with differing worldviews. Tools to address this issue will be explored in more detail in Chapter 4 following its introduction in the next section.

3.1.7 Participation, consensus and conflict

> The emergence of consensus building as a method of deliberation has provided the opportunity to reformulate comprehensive planning.
>
> (Innes 1996: 461)

Consensual planning cannot be viewed as a separate planning theory, unlike those above, but it is perhaps only a matter of time until it is provided with a theoretical basis in the same way as other planning approaches. However, its widespread use in the coastal planning and management justifies a separate section here.

Table 3.1 Environmental world views (Cloke et al. 1999 adapted from O'Riordan 1995)

	Technocentric		Ecocentric	
	Cornucopian	Accommodation	Communalist	Deep ecologist
Green label	Resource exploitative	Resource conversationalist	Resource preservationist	Extreme preservationist
Type of economy	Anti-green; unfettered markets	Green: markets guided by marked instruments	Deep green: markets regulated by macro-standards	Very deep green: markets very heavily regulated to reduce 'resource take'
Management strategy	Maximise GNP: assumes human-environment resources are infinitely substitutable	Modified economic growth: infinitely substitutable resources rejected (some critical capital)	Zero economic growth: complete protection of 'critical natural' capital	Smaller national economy: localised production (bio regionalism)
Ethical position	Instrumental (man over nature)	Extension of moral considerability; inter-and intra-generational equity	Further extension of moral considerability to non-human entities (bio ethics)	Ethical equality (man in nature)
Sustainability level	Very weak sustainability	Weak sustainability	Strong sustainability	Very strong sustainability

Table 3.2 Components of critical realism

Component	Description	Coastal context
Epistemology	Knowledge is created by building models of how real processes shape events and experiences in the light of contingent circumstances	Numerous biophysical and socio-economic models that enlighten society of the importance of the coast and need to effectively manage it
Ontology	What exists are the related domains or levels of real processes, actual events and empirical evidence	Research and monitoring studies as well as traditional environmental knowledge and local environmental knowledge continually expand what is known and not known
Methodology	The building and testing of hypothetical models of how real processes shape events that we may or may not experience	Just now developing coastal ecosystem models that allow us to test and play out futures to see what may or may not be possible

Source: From Huckle and Martin (2001).

Consensual and participatory planning is now used in many coastal planning initiatives in developing and developed countries, including the United Kingdom, Australia, Indonesia, Sri Lanka, and the Philippines (Chapter 5). Its use has expanded rapidly in Europe and North America since the early 1990s. Consensus planning uses tools from dispute resolution, pragmatism and education which emphasise the importance of learning communities, empowerment and communicative rationality to effectively involve stakeholders (Innes 1996). It works best when stakeholders are fully informed, equally empowered and sincere about the plan. This represents the theoretical ideal for a consensus planning framework; however, rarely does this situation exist in real life. Consensual planning nevertheless draws on this theory's need for concerted deliberation between decision makers.

As the name suggests, consensual planning attempts to develop plans through the building of consensus between the various parties taking part in the planning process. This model is the nearest to a purely pragmatic planning model – that is, it deliberately approaches planning from the view that everyone taking part in the plan has an equally important role to play (Box 3.4). Through consensus building, the planning process strives to reach a win–win situation and to provide mutually beneficial outcomes (Susskind *et al.* 2000). This approach takes a deliberate 'learning' view of the planning process, by acknowledging explicitly that the participants will determine the final form of the plan. Because of this, any number of other planning models can be integrated into the consensual process, including rational, incremental and adaptive planning models.

3.1.8 Pragmatism

Pragmatism in coastal management can be summed up as: 'We will solve coastal management problems using whatever tools and techniques are found to work'.

Box 3.4

The consensus building process used in the Thames Estuary Management Plan (Kennedy 1996)

For the Thames Estuary Management Plan, information was gathered via the production of a series of ten topic papers, each paper drafted by a practitioner from an organisation with relevant expertise (e.g. Fisheries paper by the Environment Agency) under the guidance of a topic group. Topic papers were then integrated into a multi-use estuary management plan for the Thames.

One quite widely held concern about this process was that it would be difficult to integrate all of the papers fairly. The non-governmental organisations in particular felt that their views would not be heard when put up against the negotiating ability and financial weight of some of the other stakeholders. In order to allay fears and overcome this problem, the following steps were taken.

1   A small group was established. The group examined in detail a list of 'conflict habits' (Chapter 5) and between them tried to identify different scenarios under which project participants might adopt each of the different habits. From this exercise a list of Guiding Principles for Achieving Agreement (Chapter 5) was developed, each of which aimed at counteracting one or more of the more negative conflict habits.
2   The guiding principles were then presented to the project steering group. This generated a discussion on group dynamics (e.g. who is good at negotiating, how is the fact that conflict exists acknowledged, is compromise the best option?).

The steering group was signed up to respecting the guiding principles. This created a more level playing field and was also useful for the project manager to refer back to should any attempt be made to abuse the process.

3   In addition, the program for integrating topic papers had been carefully thought out with long periods of time set aside for debate, unoppressive venues selection, a proforma for rewording policies, etc.

Pragmatism is the theory that a proposition is true if holding it to be so is advantageous or practically successful (Maunter 1996). In a philosophical sense, a pragmatist's pursuit of truth is through the analysis of action. Pragmatist philosophers reject the uncritical use of dualisms in traditional philosophy, including the theory-practice, reality-appearance and mind-body dualisms. Pragmatism also rejects the 'spectator view' of knowledge that we are somehow observing knowledge as impersonal or objective facts (Magee 1998).

Indeed, a stated goal of many coastal research activities signifies pragmatic support of coastal management activities. For example, the US Cooperative Institute for Coastal and Estuarine Environmental Technology shows it is pragmatic by focussing on 'solving pressing issues facing today's coastal and estuarine resource managers' (CEET 2004).

Pragmatism provides a number of important elements for coastal management and planning. Perhaps the most important of these is a separation between theoretical analysis and practical action should not exist – that is, the theory–practice dualism must be avoided. In this sense a pragmatist would argue that theoretical coastal management is really an amalgamation of its activities: its concepts are defined by what it does. Consequently, as coastal management and planning practice evolves over time, as new tools and techniques are applied, so does its theory evolve.

Importantly, there is a direct operational chain from pragmatic principles to action learning and then to adaptive management. The action–learning cycle outlined in Section 3.1.4 is the direct result of applying pragmatic principles.

3.1.9 Section summary

This section has shown that coastal planning does not have a coherent set of theoretical concepts, but rather reflects a range of planning theories and practices. The overriding theme emerging amongst theorists is that planning theory and processes are inseparable from the culture, society and politics with which they are so closely tied. Therefore as a society changes, so will the approaches to coastal planning and management.

Indeed, a change over the past thirty years from rational planning theories to more participative approaches, including adaptive, collaborative and consensual planning, reflects the overall changes as to how societies relate to the environment, as outlined in Section 3.1.6 (van Gunstern 1975; Mintzberg 1994; King 1996) (Table 3.3).

Table 3.3 Changing coastal planning practices

Old planning practices	New or emerging planning practices
Mechanistic	Organic/cybernetic
Imposed control	Self-organising/adaptive
Compartmentalises	Interdisciplinary/holistic
Reductionist models	Complex/probabilistic
Closed systems	Open systems
Means-ends causality	(Sub) systems functions (multiple causation)
Elimination of uncertainty	Accept and learn from uncertainty
Planning creates order	Order is there already – work with it
Hierarchal order	Market type coordination
Avoid overlap	Semi-autonomous systems need to overlap
Ends given	Goals developed within process
Fixed course	Flexibility and learning
Exploitation of nature	Participation with nature – sustainable use
Programming the future	Flexible frameworks for a changing future
	Subjective judgements required
Consistent goals	Consensus building
Neutral to politics	Planning is politics
Power for others	Power with others
Institutional control	Self help with government
Government monolithic	Government of many departments, perspectives, agencies
Rational, linear	Intuitive and rational
Entrenched agencies	Experimentation encouraged
Either pragmatic or visionary	Pragmatic and visionary

Source: King (1996).

3.2 Coastal planning and management language

> One of the difficulties of writing about a process of management is that many of the words, which form the vocabulary of management are hopelessly overworked. Words of common usage have been taken and given a specific meaning by different authors: unfortunately they have not all been given the same interpretation. The result is a problem of semantics, which can act as a barrier to a common understanding.
>
> (Hussey 1991: 8)

A review of the words used by coastal managers and planners reveals that the same terms are frequently given different meanings. In most cases their usage makes it clear what is intended, but it nevertheless complicates any comparison of coastal programs from different parts of the world. Three areas of terminology used in coastal management and planning are discussed in turn below, and standardised terminologies are developed for use in later sections. These three groups of terms focus on the difference between coastal planning and coastal management; the meaning of integration; and statements, which provide guidance to coastal programs.

3.2.1 What is coastal planning, what is coastal management and what is the difference?

As with many widely used words, 'planning' and 'management' can have various meanings depending on the context in which they are used. Here we briefly discuss various interpretations and subsequently define the terms 'coastal planning' and 'coastal management' as they will be used in this book.

Everyone, everyday undertakes some form of planning. Looking at the choices of when to meet a colleague, or what time and where to go fishing, requires planning. So 'planning' is usually taken in everyday language to mean the process of charting future activities. To 'have a plan' is to be in possession of a way of proceeding. In this context planning has two core components. First, determining aims for what is to be achieved in the future; and second, clarifying the steps required to achieve those aims. These two components may be viewed as common to all plans and planning exercises. However, different types of plans and planning initiatives may interpret these two components in contrasting ways. An increasingly important third component of planning is the measures that must be put in place to monitor a plan's effectiveness in meeting its anticipated aims. Using the example above, the fisherperson may have decided to go fishing in three hours' time, but they would be checking the weather, tidal conditions and a host of other factors to determine exactly where to go.

There are perhaps as many types of plans as there are people attempting to classify them. Businesses produce business plans, operational plans, corporate strategies and so on. Some governments have a Department of Planning which, as the name suggests, has as one of its core activities the production and administration of formalised systems of planning – usually land-use planning and/or economic planning. However, despite the large number of plans, and different approaches to planning, the vast majority of plans and planning initiatives can be characterised as either strategic or operational.

Those that do not readily fall into either of these categories generally combine both strategic and operational components (Hussey 1991).

Strategic planning is the highest-order of planning; it attempts to provide a context within which more detailed plans are designed to set and achieve specific objectives. Strategic planning sets broad objectives and outlines the approaches required to achieve them; it does not attempt to give detailed objectives, nor to give a step by step description of all actions required to achieve the objectives.

There are two main types of strategic planning initiatives relevant to the management of the coast: namely, geographic focussed (integrated area plans) and sector based strategies (focussing on one subject area or the activities of on government agency). These types of strategic planning are described in Chapter 5.

In contrast to strategic planning, operational planning sets the directions and steps to achieve on-ground management actions. As the name suggests, operational planning dictates localised operations – such as the rehabilitation of a mangrove area, or the building of walkways through dunes. It has to detail exactly where, and how, operations will be carried out. The contents of typical operational plans include details such as site designs, costings and schedules of works, as shown in Chapter 5.

'Manage', like planning, also has a number of meanings. It can mean the ability to handle a situation (as in 'yes, I can manage'), or it can indicate control through the direct application of power, knowledge and skills, or through their delegation. Managers in business circles are people who are 'in control' of the organisation through the delegation of responsibility to employees and service providers.

Thus 'coastal management' could be interpreted to mean directing the day-to-day activities occurring on coastal lands and waters, or it could be used to mean the overall control of the organisations that oversee these day-to-day activities. Both of these interpretations are valid in different contexts. As is the case with planning, management can be divided into strategic and operational management, the former being the processes of determining an organisation/institution's long-term direction and performance with respect to the coast; the latter being the activities of controlling on-the-ground actions.

In this chapter the terms coastal planning and coastal management are taken to be inclusive of both strategic and operational components. This is partly for ease of use, and partly because the overall concepts of coastal planning and management described in Section 3.1 apply to both strategic and operational processes. Also, most of the literature describing the conceptual framework for coastal management and planning does not distinguish between operational and strategic planning or management, from which we may infer that the authors included both in their analyses. Where either operational or strategic planning and/or management is being explicitly described in this book, the relevant prefix is used; the implications of the use of the terms are also explained more fully in Chapter 5, where the division of both planning and management into strategic and operational components provides a very useful framework for the analysis of different types of coastal management plans.

Definitional blurring occurs when the overall approach to addressing coastal issues, or coastal programs, includes both planning and management components, but they are not defined as such. Under program-oriented terminology, individual coastal plans and/or coastal management activities are called projects. In this sense, a coastal planning program could be viewed as a 'meta' strategic plan in the sense that it is the strategic plan for strategic coastal plans that in turn cascade to operational plans and management actions. These issues are explored in depth in Chapter 5.

3.2.2 Placing an emphasis on 'integration'

Many governments and international organisations choose to include the word 'integrated' as a prefix to describe their efforts in bringing together the various elements of their coastal planning and management initiatives into a single unified system – hence the description of many coastal management initiatives as 'integrated coastal management'. Others choose to use 'coordinated' or similar words, while yet others are silent in this regard. Use of integrated in this way has been popular for many years, but has expanded greatly since its adoption in Agenda 21, where the introduction to the chapter on ocean and coastal management describes the need for new approaches to marine and coastal area management and development which 'are integrated in content' (UNCED 1992).

Interpretation of the word 'integrated' can have a bearing on whether governments choose to attach it to their program descriptions. For example, in much of the Pacific and south-east Asia the use of integrated has become widespread because many have found that it conveys an appropriate policy goal, is culturally and administratively appropriate and is widely understood. In contrast, Australian governments have historically chosen not to use integrated because of the inference that it could be interpreted to mean the amalgamation of different levels of government. The subtle differences between integration and coordination are defined by Kenchington and Crawford (1993) (see also Box 3.5) as:

> ... an integrated system is complete or unified although it will generally have subordinate components. A coordinated system involves independent, generally equivalent components working to a common purpose.
>
> (Kenchington and Crawford 1993: 112)

Interestingly, there are signs that the sensitivity to using 'integrated' in Australian coastal management has declined with the recent release of a discussion paper by state and federal governments entitled 'Framework for a Cooperative Approach to Integrated Coastal Zone Management' (DEH 2004).

Another way of looking at the use of integrated, coordinated and other descriptors of coastal management programs is outlined by Cicin-Sain (1993) who has set up a continuum of terminology describing the degree to which coastal programs bring together disparate elements (Box 3.6).

There are clear similarities between the various approaches adopted by Cicin-Sain (1993), Cicin-Sain and Knecht (1998), Kenchington and Crawford (1993), Scura (1994) and Olsen *et al.* (1997) to the use of integration and other words implying 'bringing together'. All approaches stress the amalgamation of disparate elements into a single coastal management system. Olsen *et al.* (1997) provide a variation on the theme by arguing that a transition from 'integrated coastal management' to 'integrated coastal zone management' takes place when the geographic scale of a coastal program expands from land and water adjacent to the coastal to a broader coastal zone (Box 3.6). The various words to describe this amalgamation concentrate on its degree and, to a certain

Box 3.5

The meaning of 'integration' in coastal management

Kenchington and Crawford (1993) cite the dictionary definition of 'integrate' and 'coordinate' to clarify their usage in coastal management:

Integrate – to combine to form a more complete, harmonious or coordinated entity;
Coordinate – to bring into a common action, regulate or combine in harmonious action.

Degrees of integration in coastal management (from Olsen (2003) and Olsen *et al.* (1997))

Continuum of policy integration in coastal management (Cicin-Sain 1993)

Each of the five locations on the continuum shown in the Figure above are described by Cicin-Sain as:

1 fragmented approach – presence of independent units with little communication between them;
2 communication – there is a forum for periodic communication/meeting among the independent units;
3 coordination – independent units take some actions to synchronise their work;
4 harmonisation – independent units take some actions to synchronise their work, guided by a set of explicit policy goals and directions, generally set at a higher level; and
5 integration – there are more formal mechanisms to synchronise the work of various units which lose at least part of their independence as they must respond to explicit policy goals and directions (this often involves institutional reorganisation).

Enhanced sectoral management

Focusses on the management of a single sector or topic but explicitly addresses impacts and interdependencies with other sectors and the ecosystems affected. Investments in coastal tourism and transportation infrastructure funded by development banks increasingly feature in this approach.

continued...

Box 3.5, continued

Coastal zone management

Multisectoral management focussed upon both development and conservation issues within narrow, geographically delineated stretches of coastline and nearshore waters.

Integrated coastal zone management

Expands the cross-sectoral feature of coastal zone management to consideration of the closely coupled ecosystem processes within coastal watersheds and oceans; it explicitly defines its goal in terms of progress towards more sustainable forms of development.

Box 3.6

Concepts of integration in coastal management

An interesting discussion and definition of 'integrated management' is provided by Scura (1994) in her work for the United Nations Development Program on integrated fisheries management. Her discussion has wide application to overall coastal management.

> The term integration is used differently by various disciplines. For example, at the micro production level, integration can focus on production technologies such as by-product recycling and improved space utilisation. Integrated farming also uses the term in a predominantly technical sense, where the focus is on the use of an output or by-product from one process as an input into another process. In a more macro sense, an integrated economy is one which is organised or structured so that constituent units function cooperatively. In a sociological or cultural sense, integration pertains to a group or society whose members interact on the basis of commonly held norms or values.
>
> A broad interdisciplinary definition of integration is adopted here, which incorporates several disciplinary and sectoral concepts. Integrated management refers to management of sectoral components as parts of a functional whole with explicit recognition that human behaviour, not physical stocks of natural resources such as fish, land or water, is typically the focus of management. The purpose of integrated management is to allow multisectoral development to progress with the least unintended setbacks.

extent, on the mechanisms by which it is achieved. Finding ways to achieve this amalgamation is a key theme of this book, and hence will be considered many times in the following chapters. However, the preceding discussion shows that the term integration has been used in such a variety of contexts that its strict meaning has become confused. So, to avert confusion, we deliberately avoid attaching any prefixes to the term coastal management unless quoting original sources. The terms 'coordinated coastal management' or 'integrated coastal management' will therefore only be used when referring to its use by other authors, or in Chapter 5 to described the integrated style of coastal management plans.

Cicin-Sain (1993), building on the work of Underdahl (1980), has undertaken a useful analysis of the meaning of integration as it applies to coastal management. Underdahl's work concentrates on 'integrated policy' in the sense that 'constituent elements are brought together and made subject to a single unifying conception' (Cicin-Sain 1993: 23).

According to Underdahl and Cicin-Sain, a coastal management approach qualifies as integrated when it satisfies three criteria: the attainment of comprehensiveness, aggregation and consistency (Table 3.4). For these criteria to be satisfied, 'integrated policy' (Underdahl 1980) must:

1   recognise its consequences as a premise for decision making;
2   aggregate the evaluation of policy alternatives from an overall evaluation perspective rather than the perspective of individual sectors or actors; and
3   penetrate all policy levels and all government agencies involved in its execution.

These three criteria are discussed later in this chapter, especially as they relate to the organisation of governments to assist in integrated decision making at the coast.

Table 3.4 Dimensions of policy integration

Stages in the policy process		
Inputs	Processing inputs	Consistency of outputs
Comprehensiveness	Aggregation	Consistency
Over time – Long-range perspective Space – Extent of geographic area for which consequences of policy are recognised as relevant Actors – Relevant interests incorporated Issues – Interconnected issues incorporated	Extent to which policy alternatives are evaluated from an overall perspective rather than from the perspective of each actor, sector, etc., i.e. basing decisions on some aggregate evaluation of policy	Consistent policy = different components accord with each other Vertical dimension – consistency among policy levels; specific implementary measures conform to more general guidelines and to policy goals Horizontal dimension – for any given issue and policy level, only one policy is being pursued at a time by all executive agencies involved

Sources: From Cicin-Sain (1993) following Underdahl (1908).

In the context of coastal management, Cicin-Sain (1993) interpreted Underdahl's dimensions of policy integration (Table 3.4), stressing that several groups of issues were important. These are listed below (Cicin-Sain 1993: 25).

1   Integration among sectors

  • among coastal/marine sectors (e.g. oil and gas development, fisheries, coastal tourism, marine mammal protection, port development);
  • between coastal/marine sectors and other land-based sectors such as agriculture.

2   Integration between the land and the water sides of the coastal zone.
3   Integration among levels of government (national, subnational, local).
4   Integration between nations.
5   Integration among disciplines (such as the natural sciences, social sciences and engineering).

A further concept in coastal management is the clear articulation of the overall philosophy of a coastal program. This philosophy, often called guiding principles, ethos or creed, underpins the entire basis of coastal programs. In the 1970s and 1980s the concept of 'balance' between multiple, competing uses of coastal resources was the dominant philosophy underpinning coastal management programs. Balance in coastal management programs attempts to weigh up, and reconcile, opposing or conflicting forces. Most often these opposing forces are those of conservation and development (Chapter 1). For example, although the United States' Coastal Zone Management Act (1972–1990) does not make specific reference to the concept of balance, this is widely seen as the CZM Act's intention (Keeley 1994). Indeed, the CZM Act was seen as striking the middle ground between earlier proposals for coastal management legislation in the United States which emphasised either conservation or development (Beatley *et al.* 2002).

In the late 1980s and early 1990s, balancing the opposing conservation and development forces in coastal management became viewed as being essentially fixed in time. The danger was that each 'balancing decision' was not seen in a long-term context of overall changes to the coast caused by incremental tipping of the balance in one direction. This was one of the many reasons why sustainable development became the principle underpinning most coastal management programs today. Sustainability is effectively the concept of balance extended to also include the notion of time dependency along with elements of social justice.

Since the Brundtland Report (World Commission on Environment and Development, 1987) and the Rio Earth Summit (UNCED 1992), and more recently the World Summit on Sustainable Development (UN 2002), sustainable development has been a central theme of numerous policy and planning initiatives throughout the world at all levels of government (see Section 1.4.1). The notion of balance was replaced during the 1990s as sustainability concepts entered the mainstream. The evolution from balance to sustainability can readily be seen in Box 3.12. The challenge facing those involved in planning for the coast is defining what the term sustainable development actually means in a planning context, and what are the practical steps required to 'achieve' sustainable development (Buckingham-Hatfield and Evans 1996). In practice, the 'translation' of sustainability into tangible actions is often focussed on the trade offs between various services provided by the coastal ecosystem. For example, converting coastal wetlands into agricultural land for increased food production may be traded off against long-term flood mitigation, nutrient cycling and biodiversity (Agardy and Alder, in press).

It is interesting to trace the development of these concepts over time using the history of the development of a coastal management in the Philippines as shown in Box 3.7.

Box 3.7

History of laws and policies on coastal resources in the Philippines (adapted from Milne et *al.* 2003)

Year	Event
1975	Presidential Decree (PD) 705 declares mangrove forests under Department of Environment and Natural Resources (DENR) jurisdiction but areas released for fish ponds under the Bureau of Fisheries and Aquatic Resources (BFAR).
1976	Water Code is enacted establishing easement/recreation zones in seashores and river banks; Pollution Control Law seeks to prevent and control water, air, and land pollution; DENR creates the National Mangrove Committee.
1977	Philippine Environment Policy and Philippine Environment Code are enacted; coral gathering is limited to educational and scientific purposes.
1978	PD 1586 establishing Environmental Impact System is enacted.
1979	22 government agencies form the Coastal Zone Management Committee.
1981	Province of Palawan and certain parcels of public domain are declared as mangrove swamp forest reserves.
1987	BFAR's administration, regulatory, and enforcement functions are abrogated and subsumed under Department of Agriculture (DA); the National Mangrove Research Program merges with the Forest Research Institute to the Ecosystems Research and Development Bureau.
1990	The Presidential Commission on Illegal Fishing and Marine Conservation is constituted to coordinate all government and non-government efforts in the planning and implementation of a national program for the conservation of marine and coastal resources.
1991	Republic Act (RA) 7160 (Local Government Code) is passed, devolving primary mandate for managing municipal waters to local government unit (LGU); RA 7161 bans cutting of all mangrove species.
1993	Coastal Environment Program established in DENR.
1995	Fisheries and Aquatic Resources Management Council created under Executive Order 241; community based forestry management institutionalised within DENR; coordination and funding are provided to implement Monitoring, Control, and Surveillance for the Conservation and Protection of Renewable Resources in the Philippines; the Philippines signs the Global Program of Action for the Protection of the Marine Environment from Land-based Activities.
1997	RA 8435 provides for integrated coastal management training.
1998	RA 8550, Fisheries Code, establishes coastal resource management as the approach for managing coastal and marine resources.
2000	DENR and DA sign Joint Memorandum Order on implementation of the Fisheries Code.

3.2.3 Guiding statements for coastal management and planning

Fundamental to the success of coastal planning and management programs is the use of statements which clearly enunciate the purpose, directions and expected outcomes of the program. Well planned coastal programs carefully consider such guiding statements so that stakeholders know exactly what ends they are working towards. Various terms are used to describe these direction setting statements, such as mission, vision, goals, principles, objectives, targets, and expected outcomes.

The choice of guiding statements depends on the particular coastal issues being considered, the political imperatives and the management scale. The choice will also be influenced by local languages and the cultural context. However, being clear about the purpose to which these phrases are to be put is more important than what they are to be called. As will be shown in Chapter 5, *the processes* by which these statements are derived is also important. A major exception to this is if guiding statements are to be used in legislation or other formal documents, where there may be specific legal requirements for the use of particular words to describe direction-setting statements, and reasons why others should not be used.

Despite differences around the world in the use of particular terms, there is general agreement that planning and management should use a hierarchy of direction-setting statements, following the traditional view of coastal planning and management as fundamentally a rational activity (see Section 3.1.1). A simplified version of such a hierarchy is shown in Figure 3.4.

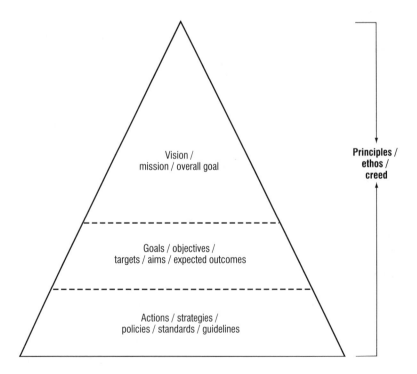

Figure 3.4 A simple hierarchy of direction setting statements for coastal planning and management

Overarching a hierarchy of direction setting statements are general expressions which describe the philosophy behind the direction of the coastal program. These are expressions of the philosophical background, which provides the basis to the implementation of a coastal program (Figure 3.4). In some cases these are statements of moral or ethical issues, which in the business planning world are often called statements of ethos or creed. However, for coastal programs they are most often called Statements of Principle. Statements of Principle often provide the philosophical climate for the development of a well-defined hierarchy of guiding statements.

At the top of the hierarchy is a statement which describes the overall direction, purpose, and which will guide all subsequent actions. Such a statement can be given various names, including Vision, Mission, or Overall Goal.

The choice of words will depend on the particular interpretations, interests and values attached to them by the program initiators. For example, the word 'vision' implies deliberate foresight, and some element of inspiration. An organisation or government may deliberately use Vision to imply that they have such attributes. The use of Overall Goal suggests that there is some overarching target which can be met. Likewise, a mission suggests that there is a well defined campaign ahead in order to develop and implement a coastal program.

The next, and probably the most important, set of guiding statements are those, which describe exactly what a coastal program is trying to achieve. Such statements are most commonly referred to as goals, objectives, targets, or expected outcomes. The critical issue in formulating these statements is the degree to which they are measurable, or specific as to time. For example, there is a distinct difference between describing an objective for the improvement of coastal marine water quality as 'safe for swimming', and defining specific targets such as 'ensuring the level in sea water is less than Faecal Coliforms <= 200/100 mL' (see Box 3.19). The latter objective is clearly something that can be measured, while the former would require additional performance standards to determine whether it has been met. The advantages and disadvantages of different types of goals/objectives/targets/expected outcomes are discussed further in Section 3.3.3.

At the lowest level of the hierarchy of coastal program statements are Action Statements. These translate the overall directions set higher in the planning hierarchy into tangible on-the-ground or on-the-water activities, and are designed to achieve the Goals/Objectives/Targets/Expected Outcomes that fulfil the Mission/Vision/Overall Goal. Where possible, action statements should be designed to meet specific Goals/Objectives/Targets/Expected Outcomes. This has the major advantage of clearly showing how the threads of a coastal program will be woven together by following, for example, the Mission Statement through to an Objective and then through to a set of actions designed to meet both the Objective, and subsequently the Mission. Examples of how these linkages are achieved in coastal programs are discussed in Section 3.3.2.

The above description of the hierarchy of guiding concepts for coastal management and planning is more easily achieved when there is a single organisational tier: a single recognisable organisational entity which can develop and implement a set of guiding statements for a coastal program. A single organisational tier is analogous to a self-contained business developing a business plan in which it can write various statements of Mission, Objectives, etc., and then implement these through its own business practices. However, this self-contained business environment is not usually the case for organisations managing the coast, where a single tier of government solely responsible for coast management is unusual. There may be constraints placed on, for example, local government by higher levels of government.

Coastal management goals and objectives may be written into national legislation, in which case local government has a limited ability to develop its own guiding statements. A national hierarchy of guiding statements may therefore include an interaction of guiding statements of different levels of government. Three such 'sub-hierarchies' may be required within a federal system of government (with national, state and local governments) in order to develop truly national guidance.

The concept of sub-hierarchies can also be applied within a single level of government, where various agencies may have their own guiding statements, such as performance criteria for fulfilling their specific coastal management and planning responsibilities.

Coastal programs around the world use different combinations of the guiding statements in each level of the hierarchy illustrated in Figure 3.4. There is no universal set of guiding statements; however, to simplify the use of language throughout this book the following standard set of terms will be used: Overall Goal, Objectives and Actions, guided by Statements of Principle.

How the above terminology is applied to actual coastal programs is described in Section 3.3.2.

3.3 Organising for coastal planning and management

The development of specific coastal planning and management initiatives is a common response to the many issues discussed in the previous chapter. Many coastal management issues will only be resolved effectively if managers develop a structured, rigorous and proactive approach. This section outlines methods for effective organisational design to facilitate delivery of coastal management and planning outcomes. The broad concept of coastal management, as distinct from simply managing activities at the coast, encompasses the management of everything and everyone on the coast, using some form of unified system or approach.

In the last ten years a number of governments and international organisations have developed guidelines on their perceptions of what constitutes appropriate concepts of coastal management. These internationally-focussed guidelines include those produced by different divisions of the United Nations (UN Department of International Economic and Social Affairs 1982; UNEP 1995; IWICM 1996; Henocque and Denis 2001). They also include those from other international organisations such as the Organisation for Economic Cooperation and Development (OECD) (OECD 1993), the International Union for the Conservation of Nature (IUCN) (Pernetta and Elder 1993) and the European Union (2002). There are also other guidelines produced by donor organisations such as the United States Agency for International Development (1996), and recently the Convention on Biological Diversity (CBD 2004). These guidelines generally begin with the philosophy underlying the coastal program, followed by a list of guiding statements, issues to be addressed, and steps to be taken to tackle these issues. The guidance documents of the World Bank and the European Union provide a useful example of the current thinking on the concepts guiding coastal management (Box 3.8).

Box 3.8

Two examples of the concepts of coastal management (adapted from World Bank 1993; European Union 2002)

Currently accepted principles and characteristics associated with the Integrated Coastal Zone Management (ICZM) Concepts are that ICZM:

- focusses on three operational objectives:
    - strengthening sectoral management, for instance through training, legislation, staffing;
    - preserving and protecting the productivity and biological diversity of coastal ecosystems, mainly through prevention of habitat destruction, pollution and overexploitation; and
    - promoting rational development and sustainable utilisation of coastal resources.
- moves beyond traditional approaches which tend to be sectorally oriented and fragmented in character, and seeks to manage the coastal zone as a whole using an ecosystem approach where possible;
- is an analytical process which advises governments on priorities, trade-offs, problems and solutions;
- is a dynamic and continuous process of administering the use, development and protection of the coastal zone and its resources towards democratically agreed objectives;
- employs a holistic, systems perspective which recognises the interconnections between coastal systems and uses;
- maintains a balance between protection of valuable ecosystems and development of coast-dependent economies;
- operates within established geographic limits, as defined by governing bodies, that usually include all coastal resources;
- is an evolutionary process, often requiring iterative solutions to complex economic, social, environmental, legal, and regulatory issues;
- provides a mechanism to reduce or resolve conflicts which may occur at various levels of the government, involving resource allocation or use of specific sites, and in the approval of permits and licences;
- promotes awareness at all levels of government and community about the concepts of sustainable development and the significance of environmental protection;
- also embraces certain general principles in the course of developing the program by a given nation, including:
    - the precautionary principle;
    - the polluter pays principle;
    - use of proper resource accounting;
    - the principle of trans-boundary responsibility; and
    - the principle of intergenerational equity.

continued...

Box 3.8, continued

European Union (European Union 2002)

Principles in the recommendation of the European Parliament and of the Council concerning the implementation of integrated coastal zone management in Europe outlined that member states should take a strategic approach to the management of their coastal zones, based on:

- protection of the coastal environment, based on an ecosystem approach, preserving its integrity and functioning, and sustainable management of the natural resources of both the marine and terrestrial components of the coastal zone;
- recognition of the threat to coastal zones posed by climate change and of the dangers entailed by the rise in sea level and the increasing frequency and violence of storms;
- appropriate and ecologically responsible coastal protection measures, including protection of coastal settlements and their cultural heritage; and
- sustainable economic opportunities and employment options.

In formulating national strategies and measures based on these strategies, member states should follow the principles to ensure good coastal zone management:

- a broad overall perspective (thematic and geographic) which will take into account the interdependence and disparity of natural systems and human activities with an impact on coastal areas;
- a long-term perspective which will take into account the precautionary principle and the needs of present and future generations;
- adaptive management during a gradual process which will facilitate adjustment as problems and knowledge development. This implies the need for a sound scientific basis concerning the evolution of the coastal zone;
- local specificity and the great diversity of European coastal zones, which will make it possible to respond to their practical needs with specific solutions and flexible measures;
- working with natural processes and respecting the carrying capacity of ecosystems, which will make human activities more environmentally friendly, socially responsible and economically sound in the long run;
- involving all the parties concerned (economic and social partners, the organisations representing coastal zone residents, non-governmental organisations and the business sector) in the management process, for example by means of agreements and based on shared responsibility;
- support and involvement of relevant administrative bodies at national, regional and local level between which appropriate links should be established or maintained with the aim of improved coordination of the various existing policies. Partnership with and between regional and local authorities should apply when appropriate; and
- use of a combination of instruments designed to facilitate coherence between sectoral policy objectives and coherence between planning and management.

The international-level guidance has been complemented by national and/or provincial/state level initiatives in a number of jurisdictions. The number of such documents is growing rapidly. Recent examples include those in Western Australia and the Philippines. Crucially, the development of guidelines to guide coastal management programs generally evolves iteratively, and not in a simple global to national to sub-national cascade (and vice versa).

The concept of coastal management has at least three dimensions – spatial, sectoral and organisational. Coastal areas are not completely closed systems and therefore factors outside of the immediate coastal management area will be important. The flows of water, air, sediments, people, pollutants and goods between coastal and non-coastal areas need to be considered. Thus, coastal planning and management considers an open system – not one that is totally focussed on the coast.

Any system of management only survives in the long term when a great deal of attention is paid to its organisation. This is especially true of coastal management, where the range and complexity of issues involves many players, operating over long time scales. These players include those charged with legal responsibilities for managing the coast, such as government agencies with land or waters under their direct control (such as marine parks, or public beaches), and coastal industries which may be required by law to restrict pollution into coastal waters. People who live on the coast or use coastal resources for recreation, either as individuals or as community-based organisations, have also become important in the design of coastal programs. Specific interest groups, be they conservationists or sector-based advocates, also have vital roles to play. All participants in coastal management programs and initiatives are commonly termed 'stakeholders', to stress that they have a stake in the future of the coast. Examples of coastal stakeholders include coastal residents, those earning a living from the exploitation of coastal resources, or it is their job to administer rules and regulations controlling coastal use. Stakeholders also include vicarious users who may never use or access the coast but still value it, and those who may not reside on the coast but recreate there.

There are currently estimated to be around 700 coastal zone management programs operating at international, national or sub-national levels (Sorensen 2002). This figure represents a 2–3 fold increase in activity during the last ten years, with the exact level of growth open to interpretation (Box 3.9). These programs are at various stages of development and implementation (Olsen 2003), meaning that there is a large pool of information to draw on in order to analyse the performance of the different institutional arrangements used to develop and implement coastal management programs. Indeed, there is now a relatively large body of literature reporting such analyses, as outlined in Section 3.4.

This section first analyses the various ways to organise those institutions with government responsibilities to deliver coastal programs, then discusses mechanisms for linking coastal users and residents with those institutions. However, before proceeding it is worth reiterating the factors, which are distinctive to coastal management programs and their administration. These have been summarised by Sorensen and McCreary (1990) as:

1   Initiated by government in response to very evident resource degradation and multiple-use conflicts.
2   Distinct from a one-time project: it has continuity and is usually a response to a legislative or executive mandate.

Box 3.9

The growth in coastal programs worldwide (Sorensen 1993, 1997, 2002; Hildebrand and Sorensen 2001)

In 1993 it was estimated that there were 142 coastal management initiatives outside the USA and twenty international initiatives, based on an international questionnaire using letters and faxes (Sorensen 1993). These estimates were later corrected to include fifty-five initiatives in the USA, making an estimated total of 197 initiatives globally in 1993 (Sorensen 1997).

In the lead up to the 2000 Coastal Zone Canada Conference, Jens Sorensen was asked to update his 1993 survey through the Baseline 2000 (B2K) study. B2K used the Internet extensively, a tool not available for the previous survey. When presented at the Coastal Zone Canada conference, B2K found a total of 447 initiatives globally, including forty-one at the global level (Hildebrand and Sorenson 2001). This dramatic increase in activity was attributed both to new initiatives that had started since 1993 and to the improved ability to find coastal management initiatives though the use of the Internet.

Sorensen continued his research and expanded his criteria to coastal management activities to include those managing large inland seas. The latest of his surveys, published in February 2002, estimates there are a total of 698 coastal management initiatives (see Figure) operating in 145 nations or semi-sovereign states, including seventy-six at the international level (Sorensen 2002).

As Sorensen stresses in all his survey work, these are estimates based on the best available information at the time and based on specific criteria. As shown in Chapter 5 there is ongoing debate on classification systems for coastal programs; as such there is likely to be continued re-examination in coming years of the global status of coastal management initiatives, and historical re-analysis of the growth of coastal initiatives over the past ten years and how this growth is monitored.

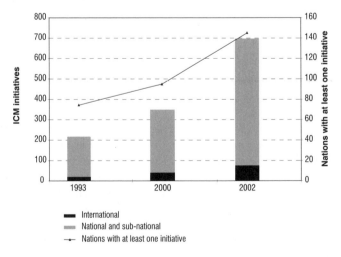

Estimated worldwide growth of coastal management initiatives (adapted from Sorensen 2002)

3   Geographical jurisdiction is specified. It has an inland and an ocean boundary.
4   A set of specified objectives or issues to be addressed or resolved by the program.
5   Having an institutional identity – it is identifiable as either an independent organisation or a coordinated network of organisations linked together by functions and management strategies.
6   Characterised by the integration of two or more sectors, based on the recognition of the natural and public service systems that interconnect coastal uses and environments.

The background to points 1–4 (above) was provided earlier in this chapter, along with an introduction to points 5 and 6. The latter is the current focus of attention. The two key issues drawn from these points are that coastal management programs should be clearly identifiable. This distinct identity could be a description of a coordinated network of organisations, or a separative organisational unit within an institution's within an administrative system. These two issues form the basis for discussion in the next section.

Importantly, it is now commonly accepted that there is no unique combination of elements that will lead to an optimal or 'best' (Jones and Westmacott 1993) suite of coastal management institutional arrangements. Instead, the contemporary focus on institutional arrangements for coastal management is outcome oriented, in that 'the goodness' of an institutional arrangement can best be judged by the effectiveness and efficiency with which coastal use conflicts are resolved' (Jones and Westmacott 1993: 130). These issues are explored first through the organisation of governments for the delivery of coastal management and planning programs.

3.3.1 Institutional and governmental arrangements

> Coastal nations should be in a position to develop Integrated Coastal Zone Management structures uniquely suited to that nation – to the nature of its coastal areas, to its institutional and governmental arrangements, and to its traditions and cultures and economic conditions.
>
> (World Bank 1993)

Many coastal nations have developed, or are in the process of developing, their own approaches to coastal management. This necessarily involves a key role for government. This section describes and analyses the common threads from these approaches.

The role of government is doubly important because of the dominance of common property in at least the oceanic component of the coast, especially in developed countries (Boelaert-Suominen and Cullinan 1994). A central question for the administration of coastal management and planning programs is, then, how government is organised to deliver its programs and how these programs interact with the private sector and the wider community.

The core issue with organising government to contribute efficiently and effectively to the delivery of a coastal program is focussing the activities of many different government sectors in a manner that views government as a whole, and not a set of uncoordinated separate elements. This is not an easy task. The majority of governments are established along sectoral divisions, assigning the responsibility for delivering services and functions to different government agencies. This concept of 'differentiation' is one of the central elements of how most public-sectors around the world are organised (Heady 1996) and is based firmly on modernist, rationalist management principles (Kraus and Curtis 1986). The notion of differentiation is then applied within government agencies to further segment operational responsibilities (see Figure 3.5). Analysis of the differentiation between and within government agencies is an important first step in understanding how different agencies contribute to land and water management and planning. Often such institutional analysis is accompanied by an understanding of the legislative mandates under which line agencies function, and how budgets are apportioned. Also, because differentiation is such a well-understood concept its analysis often forms a neutral entry point into governmental institutional analysis. This is aided by the ubiquity of its visualisation tool: the organisational chart. Differentiation analysis can also be very instructive in providing information on the forces that work to push agencies apart, into their respective management 'silos'. This is often useful when considering mechanisms aimed at drawing agencies together to develop more holistic coastal management approaches.

Differentiation analysis of coastal programs often divides management responsibilities into both 'horizontal' and 'vertical' components (Figure 3.5). Levels of government are shown as the vertical component, while the different sectors comprising a single level of government form the horizontal component. In the example shown in Figure 3.5, there are three levels of government, as is common in many large and/or populous countries. In many countries the division of power between levels of government is not purely linear, in the sense that higher levels of government exert power over lower levels of government, as inferred in Figure 3.5. Thus, in many federal systems of government the different governments (federal, state/provincial and local) are termed 'spheres' in order to reflect their non-hierarchical nature.

Horizontal components of government are separated according to function, which in turn is reflected in division of government into various agencies and departments. This has been termed the 'Management Accountability Hierarchy' (Jaques 1996). For example, a government may choose to create separate departments, such as environment, transport, energy and primary industry. This horizontal differentiation, while important for line management and accountability, can lead to gaps and overlaps between the various government departments with responsibilities for coastal management.

Roles and responsibilities for coastal management are usually divided both horizontally and vertically as a result of the way in which governmental systems apportion power and responsibility. That is, some activities will be carried out by one level of government, and not another (vertical division), while others are carried out by one particular sector of government (horizontal division). In reality, this horizontal and vertical differentiation is very complex. This complexity is heightened by trends of outsourcing government functions to the private sector and also by the increased role of non-governmental organisations, both in an advocacy role and also as a surrogate for the provision of government functions. Indeed, this complexity often provides one of the prime motivations for developing a coastal management system in the first place, in response to the need for cooperation and coordination.

The horizontal/vertical differentiation of responsibility is a useful starting point given that this is perhaps the most visible expression of how governments organise.

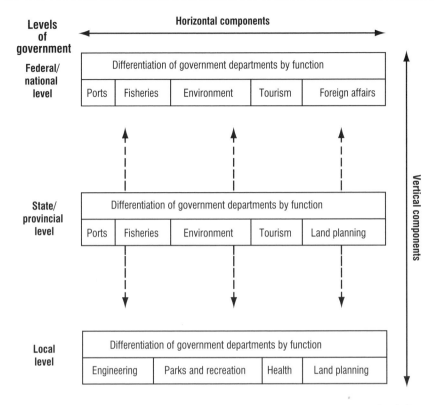

Figure 3.5 Example of national system of coastal management governance divided into vertical and horizontal components

However, there is an increasing realisation that this 'mechanical' concept of organisation does not fully reflect the inter-relationships of people, management functions and power between and within organisations. The resulting concepts of how organisations are designed are thought to be more organic, network-oriented and adaptive (Hurst 1995; Daft 2000).

In practice, differentiation often occurs in governmental approaches to coastal management through one agency being provided a mandate to coordinate other agencies. For example, the Indonesian Government has mandated the Department of Marine Affairs and Fisheries to lead coastal management but implements programs using a combination of line agencies, coordinating agencies and non-governmental organisations. In Indonesia, line agencies have legislated responsibilities for management of various coastal resources or sectors. Coordinating agencies, despite having no legislated powers, have the government mandate to bring various line agencies together with other relevant parties and formulate coastal management initiatives. In addition, NGOs provide cost effective debate and support to a range of coastal initiatives especially at the local level (Box 3.10).

As the Indonesian experience has highlighted, it is important that the design of institutional arrangements within a level of government include fostering cooperation and/or coordination between government agencies with responsibilities on the coast

Box 3.10

Example government organisational structure – Indonesia (updated from Sloan and Sugandhy 1994)

Line agencies	Responsibilities relevant to coastal management
• Department of Marine Affairs and Fisheries/All Directorate Generals, and the Agency for Marine and Fisheries Research	• Fish and aquaculture management and research, marine protected areas, empowerment of coastal communities, development of marine and fisheries technology, management of marine and fisheries resources in relationship with regional autonomy
• Department of Forestry/Directorate General of Forest Protection and Nature Conservation	• Marine conservation, mangrove management
• Department of Communications/ Directorate General of Sea Communication	• Ports, shipping, navigation, safety including emergency responses
• Department of Mining and Energy/ Directorate General for Oil and Gas	• Gas and oil exploration and production
• Department of National Education and Universities	• Education and research
• Navy/Hydrographic and Oceanographic Service	• Territorial water security, hydrography
• Department of Industry and Trade	• Development and waste management
• Department of Resettlement and Regional Infrastructure/Directorate General of Regional Plan	• Engineering and erosion control
• State Ministry for Culture and Tourism	• Tourism and cultural heritage

Coordinating agencies	
• Ministry of State for Environment	• National coordination, Environmental Impact Assessment
• National Development Planning Board	• National development plans
• Department of Home Affairs/ Directorate General of Regional Development	• Regional development
• Ministry of State for Science and Technology/Technology Assessment and Application	• Natural resource inventory
• National Coordinating Agency for Surveys and Maps	• Coastal mapping
• Indonesian Institute of Science/Research and Development Centre for Oceanology	• Marine science and research
• Coordinating Committee for National Sea Bed Jurisdiction	• National boundaries and Law of the Sea
• Coordinating Board for Marine Security	• Security in national waters
• Indonesian Maritime Board	• Marine development

continued…

Box 3.10, continued

Non-governmental organisations

• Indonesian Forum for Environment	• National coordination of Non-Governmental Organisations
• World Wide Fund for Nature	• Marine conservation, public education
• Asian Wetland Bureau	• Coastal wetland management, aquaculture development, EIA
• United Nations Educational, Scientific and Cultural Organisation	• Marine pollution, education
• International Union for the Conservation of Natural Resources	• Marine protected areas
• Clean Coastlines and Beaches	• Public awareness, coastal litter and pollution
• Mangrove Foundation	• Mangrove conservation and sustainable use
• Indonesia Green Club	• Sea turtle conservation; public awareness

---

(Rogers and Whetten 1982). This can be perceived as a threat to the power and autonomy of individual agencies; however, if well managed and/or the coordinating agency has limited legislated power, then the power struggle problems can be reduced. The critical issue for the line agencies is being clear on the benefits to them in working within a coordinated approach. As outlined above, the forces working against coordination must be understood in order to design approaches that will act to bring line agencies closer together.

Nevertheless, the relative power of line agencies versus coordinating agencies or coordinating bodies is generally skewed markedly towards the line agencies. The power of horizontally-oriented line agencies has been likened by Tasque Consultants (1994) as being 'rods of iron' versus the 'threads of gossamer' which act to pull them together. There are a number of mechanisms used to balance these relative powers, as described later in this chapter.

In organising governments to develop and implement coastal management the issue is not how institutions are arranged, but rather what is achieved through those institutional arrangements. This focus on outcomes is the reason for Jones and Westmacott (1993) concluding that there is no 'best' way to organise governments in order to manage the coast. In practice the diversity of cultural, social, political and administrative factors around the world confirms that there is indeed no single best way. Instead, the designers of the administrative arrangements for new coastal management programs tailor administrative structures to take advantage of the particular cultural, social and political factors within their jurisdiction as they interact with issues. For example, what may be the best system of coastal zone management program governance for a European coastal nation may be disastrous for a country in the Pacific, and vice versa. This section therefore concentrates on addressing the factors, which are usually considered in the design of the administrative arrangements for coastal management programs.

Detailed analysis of institutional arrangements for coastal programs was initiated by Sorensen *et al.* (1984), and subsequently updated by Sorensen and McCreary (1990). These two texts remain the standard works on institutional arrangements for coastal management and are drawn on in the following discussion. In addition Sorensen (2002) has developed additional conclusions on institutional arrangements following two global surveys of coastal program activity undertaken in 1992/3 and repeated in 1999/2002 (see Box 3.9).

Sorensen *et al.* (1984: 1) describe institutional arrangements as 'the composite of laws, customs, organisations and management strategies established by society to allocate scarce resources and competing values for a social purpose, such as to manage a nation's coastal resources and environments'.

Mitchell (1982) developed a classification method of the governance arrangements for coastal management. Mitchell analysed national coastal management systems according to three criteria:

Coastal focus	either coastal specific programs or coastal issues are addressed as part of overall agency responsibilities.
Strength of national control	strong or weak national government control.
Policy orientation	programs oriented towards economic development or environmental/amenity considerations.

Mitchell used these three criteria to develop an eight-fold classification. For example, Sri Lanka's coastal program at the time (Box 5.12) was classified as 'coast specific, with a strong national structure and environmental orientation' (Mitchell 1982).

Sorensen *et al.* (1984) adapted Mitchell's classification system to include five 'types' of governance arrangements. These effectively combine the degree of integration between government sectors with the degree of the program's coastal focus.

Sorensen (1997, 2000, 2002) has updated his previous approaches to institutional analysis to consider coastal management governance arrangements that can be classified according to two key factors:

- regulatory and planning boundaries;
- reliance on planning, or on regulation, or a combination of planning and regulation.

Sorensen's first factor of boundaries is used to make a distinction between integrated coastal management and integrated coastal zone management. The former applies only if the land and waters immediately adjacent to the coast are the focus of management efforts. This distinction is also made by Olsen *et al.* (1997). Integrated coastal zone management, according to this train of thought, applies when areas further inland, such as catchments, and offshore waters are also considered.

The second of Sorensen's factors is the degree to which a coastal management program includes regulation and/or planning measures. Sorensen's distinction between direct and indirect regulation is made on the basis of whether the part of government with the mandate for coordinating coastal management efforts issues coastal development permits directly or indirectly to other agencies. In this sense there is an indirect delegation of coastal management plans or policy implementation. Sorensen makes four such distinctions:

1   Integrated direct regulation only (e.g. Spain, Turkey).
2   Integrated planning and direct regulation (e.g. Great Barrier Reef Marine Park Authority (Box 4.3) and many of the US state programs under the US Coastal Zone Management Act – CZMA – Box 3.12).
3   Integrated planning and indirect regulation (e.g. CZMA state programs in Connecticut, Florida, and Massachusetts).
4   Integrated planning only (e.g. Venice Lagoon and watershed and Regional Sea Action Plans – Box 5.8).

The above factors have been applied by Sorensen to produce nine differentiators of program organisational design.

Another way of assessing coastal zone management governance is to focus on how various coastal management activities are controlled (Born and Miller 1988). This approach has been used to classify American state coastal programs developed under the Coastal Zone Management Act (Knecht *et al*. 1996). Using this method, two main types of governance are identified:

Networked   Existing government sectors and institutions remain. No new specific coastal management legislation is enacted. Sector coordination is improved though 'networking' of existing legislation and policies.

Legislative   New specific coastal management legislation is enacted. This legislation can have a variety of purposes. New institutes, or the enabling of existing ones, enacted.

Networked coastal management programs are those which bind together a range of pre-existing coastal management initiatives into a well defined coastal program (Taussik and Gubbay 1997). The networked approach was originally developed in the USA (Box 3.12) and has been adopted by other coastal nations around the world, for example Australia (Kay *et al*. 1997). Born and Miller (1988) distinguish four attributes of the networked coastal programs in the USA:

1   The program emphasises making pre-existing authorities work better and in a more coordinated manner.
2   The designated 'lead agency' has broad policy formulation and coordination responsibilities and a horizontal (cross-cutting) orientation.
3   The 'lead agency' tends to be an executive department and not an operational agency.
4   The lead agency relies significantly on other agencies and/or different levels of government (dispersed program management) especially regarding regulatory powers.

An important addition to the above four points is the role of coordinating committees, or councils, which help pull together the various threads in the network (Box 3.12). These coordinating groups play a vital role in the success of networked coastal management systems, especially if the membership and/or mandate of the groups is powerful enough to ensure the cooperation of its member government agencies, such as land use planning, land management, environmental protection, transport, infrastructure development, primary industry or mining. For example, some Australian state governments rely on Coastal Councils to coordinate their coastal programs. Membership of those committees is contrasted with the membership of the Californian Coastal Commission in Table 3.5.

Perhaps the central concern in any networked system of coastal management is the critical role of the people involved in tying the network together. By its very nature, a network is a system, which requires the commitment of the people staffing the various groups and agencies within the network. Without this commitment, at a personal and agency level, the functioning of the network becomes vulnerable to failures since no one participates and, if they do, they do so ineffectively. Conversely, this reliance on goodwill can also be the network's greatest strength, but only if network members recognise the importance of their participation

Coastal management systems relying on the networked approach may appear to be less efficient than those relying on a legislative-based approach, especially when there is not an apparent force of law to ensure contributions from the key players. However, comparative studies of the various approaches taken in the USA have indicated that networked systems may be more (Born and Miller 1988) (or at least as) efficient (Knecht *et al.* 1996; Hershman *et al.* 1999) than fixed, legislated programs. The advantages and disadvantages of networked systems require further evaluation, however, before final conclusions can be reached on their efficiency relative to legislative approaches.

Separate coastal management legislation, or coast-specific sections of broader legislation, can be enacted to assist programs in a number of ways. The most common legislative mechanism is to pass enabling legislation, which defines the form of a coastal program. This is often through specifying the vision, goal and/or objectives of a coastal program and outlining mechanisms for delegating power and/or money to a lead agency or coordinating body. In many cases, sufficient regulatory and enforcement mechanisms are already present within a jurisdiction; for example, planning and environmental impact assessment requirements, and where the purpose of legislation is to define mechanisms for enhanced coordination. Where existing regulatory and enforcement mechanisms are not present, as in many developing countries, coastal management legislation can be enacted to establish various new forms of regulatory instruments such as (Jones and Westmacott 1993):

- licenses and permits, e.g. for construction and concessions;
- physical planning regulations for the establishment of developments, water supply of conservation zones, and setbacks;
- standards for a range or parameters including water quality (related to environmental and/or human health, construction, the provision of amenities);
- quotas, such as on fish catches.

An example from Sri Lanka of a legislated basis for coastal management is shown in Box 3.11. The use of these various legislative tools, including those used in Sri Lanka, is described in more detail in Chapter 4.

The United States' coastal management system allows flexibility as to whether each state bases its coastal management efforts on legislative or networked approaches (Box 3.12). The two cases of California and Florida show that both state programs conform

Table 3.5 Current membership of two Australian state government coastal councils compared with the California Coastal Commission

California Coastal Commission	Western Australia Coastal Zone Council	Victoria Coastal and Bay Management Council
Twelve voting members appointed equally (four each) by the Governor (2-year term), Senate Rules Committee (4-year term) and the Speaker of the State Assembly (4-year term)	Independent Chair (appointed by minister)	Chair (appointed by minister)
Six public members from the general public	Local Government (Perth Metropolitan Region)	Department of Planning and Development
Six locally elected officials (county supervisors or city council members) from six coastal regions specified in the California Coastal Act.	Local Government (Country Region)	
		Department of Transport
Four non-voting members are secretaries of:	Western Australian Tourism Commission	Department of Conservation and Natural Resources
Resources Agency	Department of Planning and Infrastructure	Municipal Association
Business, Transportation and Housing Agency	Department of Conservation and Land Management	
Trade and Commerce Agency	Department of Environmental Protection	Six representatives of the community with experience in conservation, tourism, recreation, commerce, issues relating to indigenous peoples, community affairs or coastal engineering
Chair or the State Lands Commission	Department of Industry and Resources	
	Fisheries Department of WA	
	Department of the Environment	
	Three community/industry representatives from coastal community groups, coastal industry or environmental conservation	

Source: Updated from Kay et al. (1997).

Box 3.11

Governance arrangements for the Sri Lankan Coastal Management Program

Sri Lanka has made significant advances in coastal resource management in comparison with most other countries in the Asia-Pacific region. There is a need to consolidate the gains made through further refinement and sophistication of strategies adopted in coastal and marine resources management.

(IUCN 2002)

Unlike many of its neighbours, Sri Lanka has developed a national coastal zone management program in response to its numerous coastal management problems (Box 2.6). The Sri Lankan approach to coastal management was developed in partnership with international aid agencies and coastal specialists, most notably those from the USA, Germany, Denmark and Holland (Lowry and Wickramaratne 1987; Kahawita 1993; Lowry 2003).

Sri Lankan coastal management has been undertaken through various government initiatives since the early 1970s. These initiatives during the 1970s concentrated on the management of critical coastal erosion problems, and hence focussed on planning coastal engineering works and attempting to place controls on the construction of new building at the coast. There was a change in emphasis from 'coast protection' to 'coastal zone management' in 1981 with the passing of the Coast Conservation Act (Kahawita, 1993 repeated in Clark 1996: 580–6). The Coast Conservation Department was mandated under the Act with responsibilities for the administration of a permit system for development control, the formulation and implementation of works and research within a 'coastal zone', the boundaries of which are shown in Figure 1.2 (Kahawita 1993).

The Act also enabled a range of coastal planning and management tools to be used, which are listed in Table 4.1. These tools were found to address certain issues well, such as ensuring well planned major coastal development (e.g. tourist hotels). However, a broader more encompassing approach was required to ensure that issues external to the permitting system and outside the legal 'coastal zone' could be addressed. The 'second generation' program was founded on a hierarchy of national, provincial, district and local coastal management plans, linked with enhanced institutional capacity, public education and community involvement (Olsen *et al.* 1992). This strategy, called 'Coastal 2000', was endorsed in 1994 and contained a range of strategies described in Box 5.12 (Coast Conservation Department 1996).

Sri Lanka is currently within its third major cycle of coastal planning and policy reform in order to consolidate the progress made to date, to learn from experience and to adapt to the changing socio-political realities of the country and evolution in international practice (IUCN 2002). The rationale for the new law, to be called the Coast Conservation and Coastal Resource Management Act, is (Asian Development Bank 1999):

continued...

Box 3.11, continued

to provide a stronger legal basis for the implementation of the policies and strategies for coastal environment and resource management. The present Act has been able to implement parts of the Coastal Zone Management Plan, but needs to be strengthened for better coordination and enforcement. Further, it is necessary to address the problems arising from overlapping or contradictory statutes, which give similar powers over the coastal zone to various ministries or authorities, and which have, in effect, caused the powers, functions, and duties of Coastal Conservation Division to be eroded over the years.

The proposed new law has recently been drafted and is awaiting finalisation.

Box 3.12

Legislative and networked coastal management program structure in the United States (updated from Fisk 1996)

At the heart of coastal zone management in the United States is an agreement between federal and state governments. The federal government offers financial and technical assistance to states that voluntarily choose to develop and implement coastal management programs, but they must adhere to a set of minimum federal standards. These standards are made up of four core issues:

*   protection of coastal resources;
*   ensuring public access to the coast;
*   managing development along the coast; and
*   managing coastal hazards.

As an added incentive to the states to develop coastal programs, the federal government assumes the legal responsibility that all of its activities including permitting activities are consistent with approved state programs, subject to certain appeals and reviews in case of disagreements between the states and the federal agencies (Archer and Knecht 1987; Beatley *et al.* 2002).

Coastal management in the United States is both process and policy oriented. States can tailor their coastal programs to fit particular state needs and can adopt a wide range of policy implementation tools. At the time the federal Coastal Zone Management Act was passed in the early 1970s, several states had already enacted state-wide comprehensive laws and policies for management of their coastal zones that conformed well with national guidelines. However, the majority of states asserted control over coastal activities through sector-specific statutes and laws. For these states, the federal government conceived the networking approach. Networking gave states the option of incorporating existing state laws and policies into a network of management controls for the coastal zone, assuming that a lead state agency was named to oversee operation of the program.

continued...

Box 3.12, continued

The networking concept was very popular, making up the legal basis of more than half of the current participating state programs. Only a handful of the participating states base their programs on comprehensive pieces of coastal legislation or policy. Other states share elements of both approaches: specialised coastal laws and policies networked with other state legislation. To illustrate the difference between these two approaches to program structure in the United States, two states are discussed. The first case, the California Coastal Management Program, is a good example of a program with a legislative approach. At the opposite end of the spectrum, the Florida Coastal Management Program epitomises the networked approach with more than twenty-three single-purpose statutes bundled together to establish its legal basis.

Case I: California

The California Coastal Management Program (CMP) is divided into two segments. The first segment, the San Francisco Bay and surrounding area, is managed under the San Francisco Bay Conservation and Development Commission (BCDC). The remainder of the coastal area (extending seaward from the coast three nautical miles and landward 1,000 yards so as to include coastal estuaries and recreation areas) is managed by the second segment, the California Coastal Commission (CCC). The CCC is the lead regulatory body in the state that operates the statewide permit system. A third coastal body, the California Coastal Conservancy, is a non-regulatory body involved with land acquisition, ensuring public access and critical area restoration (NOAA 1996: 58–60).

The CCC administers the California Coastal Act of 1976, which, as amended, requires all coastal cities and counties to develop their own Local Coastal Program (LCP) through a land use plan or zoning ordinance. The CCC then determines if these plans and ordinances conform to state standards set out in the 1976 Act. Once a county or local plan has been approved, municipalities can issue their own permits for coastal building. As a result, the bulk of land use planning and day-to-day coastal management activities occurs at the local level.

Following implementation of the 1976 Act, federal approval of the California Coastal Program followed shortly thereafter in 1978. Since that time the program has made great strides in ensuring public access to the coast, protecting valuable wetlands, and preserving coastal areas of archaeological significance. By means of the consistency provision, offshore petroleum leasing and development by the federal government have been halted or occur with the necessary provisions to protect the environment. The CCC conducts several major education and awareness initiatives related to coastal protection, beach clean-up, and marine conservation (NOAA 59). Recently the CCC has completed a report 'Desalination and the California Coastal Act' to guide decision making for desalination plants along the state's coast.

continued...

Box 3.12, continued

Case II: Florida

The Florida coastal zone is defined to include the entire state and its coastal waters (three nautical miles seaward on the Atlantic coast; nine nautical miles seaward on the Gulf of Mexico coast), making it the second largest in the USA.

The Florida Coastal Management Program (FCMP) is based on twenty-three existing statutes. To obtain federal approval of their coastal program, these laws were networked together, and a new Florida Coastal Management Act of 1978 was enacted. The 1978 Act established the lead coastal agency, the Department of Environmental Protection (DEP), to coordinate and review current plans and to provide a clearing house and other services requested by the other eight state agencies, five water management districts, and local governments that administer laws and regulations relevant to the coastal zone. By a joint resolution of the Governor and state cabinet, an interagency management committee was formed including the heads of all FCMP agencies and a non-government chairperson from the Florida Citizens Advisory Committee on Coastal Resource Management. The Florida State Clearinghouse works to coordinate policies and resolve disputes among the various agencies and user groups. Based on these program improvements, the FCMP was approved to enter the national coastal zone management program in 1981.

The DEP is the lead agency for the FCMP and hosts the Florida State Clearinghouse. The DEP and the Department of Community Affairs, which is the state's lead agency in land planning, closely coordinate their respective activities through a memorandum of understanding. The FCMP has a number of assistance programs, federal and state funded, for local governments to develop and implement various coastal management initiatives (Florida DEP 2004).

With the help of federal funds, the FCMP has helped to prepare coastal counties against the threat of hurricanes and other coastal hazards with hurricane evacuation plans. Critical estuaries and aquatic preserves have been protected, as have wetland resources through the implementation of the comprehensive Henderson Wetlands Protection Act (1988). The FCMP remains a major player in both the National Estuarine Research Reserve Program (operating four sites) and administration of the Florida Keys National Marine Sanctuary.

---

to the minimum standards set out in the United States federal Coastal Zone Management Act (1972–1990) and have had success, despite major differences in program structure (legislative versus networked) and level of implementation (local versus state).

A further method for classifying the institutional arrangements of coastal programs is based on the primary level of implementation. State programs in the USA were classified by Knecht *et al.* (1996) according to whether they were implemented primarily at state level, or at both state and local levels. Initial findings of programs which were implemented at these two levels suggest that there is no measurable difference in performance between the two. However, there has been no systematic analysis of how the various types of coastal planning initiatives analysed in Chapter 5 influence these outcomes.

The culture and social structures of a coastal nation are often the hidden determinant of its organisational approach to coastal management everywhere in the world. This influence is most clearly seen by those examining the organisation of coastal management programs in countries other than their own; particularly when there is a strong cultural contrast, for example a westerner with a strong rationalist, technocentric worldview visiting a Pacific island nation. It would be tempting for this visitor to interpret the coastal management efforts in the Pacific as 'culturally driven', while classifying those of the western nation in which they live according to the various schema described above. It is intriguing to consider what a Pacific islander visiting a European country, for example, would make of how the cultural setting influences the organisation of coastal management there. It could be argued that most, if not all, coastal programs are influenced by cultural beliefs. Nevertheless, given the importance of customary beliefs in many developing countries, and the difficulty of classifying them as either networked or legislative approaches by using essentially a western cultural context, coastal management programs largely controlled by customary beliefs may provide a different category of coastal program organisation. However, this area of research has not been comprehensively analysed within coastal management institutional design, and hence no firm conclusions can be drawn at present.

The importance of tailoring a coastal management and planning program to reflect cultural and social conditions can be illustrated by considering the development of a coastal program in Western Samoa, where there are two styles of government: a Westminster style parliamentary system of national government, superimposed upon the traditional village-based decision-making structure of the Fa'a Samoa (Box 3.13).

Partnerships between organisations, a form of institutional arrangement, have grown rapidly in recent years. These can take the form of agreements for agencies to work together on coastal management issues of shared concern through agreements or memoranda of understanding, such as the Coastal America Partnership (Coastal America Partnership 2001). They can also be agreements for agencies to form partnerships through the creation of separate organisational entities (McGlashan and Barker, in press). This is the model used in the United Kingdom for estuary management, as shown in the example from the Thames (Boxes 2.1, 5.23, 5.30).

Finally, an alternative method for analysing coastal program design has been developed by considering various program 'orientations' (Figure 3.6) (Scura 1993, adapted by White 1995). Some of these orientations reflect previous analyses based on program focus, but also inject a component of coastal planning, especially the balance between planning and the implementation of plans. Thus, a useful link between the development of coastal management programs and their planning components is formed. This is explored in later chapters.

Box 3.13

Coastal zone management decision-making framework in Samoa

The customary system of decision making in Samoa predominates over most of the island's coastal land, and is effectively semi-autonomous from national decision making (Cornforth 1992) (see Figure below). Only in the capital city, Apia, is the customary decision making system of reduced importance.

Coastal resource management decisions are therefore expressly made by villages. National-level decisions are made by the elected members of parliament.

At the village level, decisions made by the Village Council of Chiefs and Orators are usually expressed by the formulation of rules. Rules can be either long standing, forming an integral part of village culture, or short term in response to immediate village concerns. Rules are enforced and policed through the village Council of Chiefs and Orators (*fono*) and heads of families, and non compliance results in punishment depending on severity. Punishment can be though various forms of shaming, and in extreme cases banishment from the village. Rule breaking is not usually referred to the police, and is instead settled according to custom.

During the early 1990s Samoa developed a national legislative approach to sustainable management. This paralleled many such initiatives around the world stimulated by the UNCED process. However, during the early 1990s environmental management legislation in Western Samoa found that very few 'laws are complied with, and even fewer enforced' (Cornforth 1992). Since this time there have been considerable improvements in Samoan environmental governance, with a considerable increase in the capacity of government and a clarification of ministerial responsibility and accountability. There is currently a widespread belief that harmonising the two systems is gradually being achieved in Samoa, paralleled by an increasing focus on environmental governance throughout government (Cornforth 2004).

Indeed, this progress was recently built on through two donor-funded coastal-oriented initiatives in Samoa.

The first was the creation of two multi-use, community-based marine protected areas at Aleipata and Safata. These are IUCN/World Bank-sponsored sites as part of the The World Bank/International Union for the Conservation of Nature report to develop a Global Representative System of Marine Protected Areas (MPA). Twenty villages adjacent to the marine protected areas developed a partnership approach with the project through district committees. The five-year project has been operational since January 2000. To date is has established the MPAs and governance arrangements. The local management committees have made a number of important decisions, including banning scuba fishing and sand mining within the MPAs (IUCN 2002).

A second project is the ongoing World Bank funded Coastal Infrastructure Management Strategy initiated in 2000. The strategy relied heavily on the use of the village based community structure to develop Coastal Infrastructure Management Plans (Taylor and Roberts 2004). The plans focus on the development

continued...

Box 3.13, continued

of transport, utility and coastal engineering infrastructure within the context of the vulnerability of coastal communities to hazards, including greenhouse-induced climate change and sea-level rise. The plans have provided valuable input into the developed of the Samoan National Adaptation Programme of Action (NAPA) (Crawley 2003). NAPA is a UN development programme/global environment fund funded initiative to develop strategies for developing the long-term capacity of developing countries, such as Samoa, to adapt to climate change (UNITAR 2004).

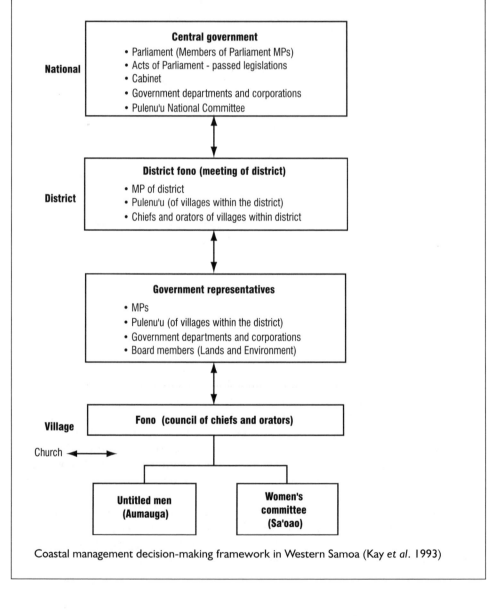

Coastal management decision-making framework in Western Samoa (Kay *et al*. 1993)

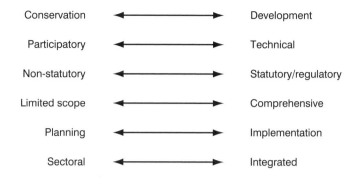

Conservation	←——————→	Development
Participatory	←——————→	Technical
Non-statutory	←——————→	Statutory/regulatory
Limited scope	←——————→	Comprehensive
Planning	←——————→	Implementation
Sectoral	←——————→	Integrated

Figure 3.6   Range of orientation of coastal management programs (Source: Scura 1993, adapted by White 1995)

(a) Integration and coordination between levels of government for coastal management

The previous section described and discussed the overall approaches to the administration of coastal management programs. In this section discussion focusses on the coastal management roles and responsibilities of different levels of government. General principles, and case studies of how different powers for coastal management are shared between levels of government, will both be examined.

Effectively dividing vertical responsibilities for coastal management activities between levels of government is often much more difficult than resolving problems within a level of government. Political, administrative and budgetary clashes between levels of government often lead to confusion in allocating responsibilities. Such vertical imbalances of power, money, and differences in political affiliation often dictate the overall shape of the coastal management governance of a nation, both horizontal and vertical. This is because horizontal differentiation will, to a large extent, be controlled by the relative degree of vertical power held by a particular level of government. Thus a lower, poorer level of government will be unable to create a complex horizontal differentiation of its sectors, in contrast to a larger and more influential higher level of government.

A central issue to the vertical distribution of management authority is the degree of centralisation in decision making – a fundamental management question not restricted to the coast. The advantages and disadvantages of centralism and localism are summarised in Table 3.6.

The European Union uses the term 'subsidiarity' to describe the concept of the locally-focussed nature of decision making, but with accountability to higher decision-making levels (European Union, unknown).

One common compromise is to attempt to delegate decision-making powers to the lowest level of decision making, consistent with the scope of the problem, but to constrain those decisions within a framework articulated by the next higher level. Some coastal management programs attempt to achieve a 'controlled devolution' of powers through planning. In these cases integrated or sector plans are formulated jointly by different levels of government. Once a planning framework is established,

Table 3.6 Advantages and disadvantages of centralism and localism in coastal management

Advantages of centralism	Advantages of localism
Increased general perspective	Intimate knowledge of the problems
More abstraction	More localised outlook
More experts available	Greater likelihood of living with the effects of a decision, creating an incentive to be successful
Increased funds	
Greater political will	A stronger need for integration as one approaches more concrete environmental and social situations
Greater specialisation	Less fragmentation of occupational specialisations

Source: Adapted from Ketchum (1972).

ongoing management activities directed by local-level decision makers can take place, provided they are consistent with the plan. The development of such coastal planning frameworks is described in more detail in Chapter 5.

3.3.2 Linking government, the private sector and the community

By the late 1990s the relative roles of government, industry and the community had changed in most developed countries due to the increased environmental awareness of community and industry, privatisation of government services, and the desire of both to be more closely involved in decision making. Governments, private industry and the wider community now tend to work more closely together in developing and implementing coastal programs – albeit with a guiding, and often firm, hand from government. Of course, this does not negate the realities of power, and the influences that powerful government and private sector interests can have in coastal programs. Nevertheless, the growth of community advocacy groups, and the increasingly rapid access to information by community members, has changed the relative balance of power in coastal program development and implementation in many coastal nations.

A similar changed relationship between government, industry and the wider community has also occurred in many developing countries as a result of the re-emergence of interest in indigenous cultures. Consequently, the relative degree of power between the formalised systems of government, often established by colonial powers, and customary land tenure and resource-use practices, has altered – and continues to alter. Community participation has become crucial in the development of such plans. Box 3.14 shows an example of community participation that is woven through a plan production process.

The degree to which the general public and/or indigenous people take part in planning and decision-making processes has been described according to a spectrum of citizen participation (Arnstein 1969). Arnstein drew the analogy of a ladder of citizen participation, as shown in Figure 3.7.

Box 3.14

The special area management framework used in Sri Lanka

The national Sri Lankan coastal planning system (Box 5.10) moved in the early 1990s from relying solely on national-level strategies to multi-level planning approaches. Local-level plans are called Special Area Management Plans (SAM)[1] (Lowry and Wickramaratne 1997). SAMs are community-oriented integrated management plans focussing on resolving conflicts within defined areas using a total ecosystem approach (including humans). SAMs concentrate on local areas, usually in the order less than 10 km of coast (White 1989). SAMs were required in Sri Lanka because of the 'inability to mobilize the support and commitment of the local communities for the implementation of national policies' (White and Samarakoon 1994: 20). Funding for the initiation and development of the SAMs is provided by the United States international aid agency (USAID).

The general processes used in each SAM, and the relationship between these and the national coastal zone management program, are shown in the table below.

Special area management framework for Sri Lanka (White and Samarakoon 1994, with assistance from L. Hale)

Policy steps	Outputs	Process/methods
Consensus on use of SAM by national agencies	Concept paper approved	Discussion and workshops
Site selection and criteria of choice	Two sites approved based on issues and practicability	Discussion and workshops
Issue identification and analysis for each site	List of issues and causes	Local workshops, interviews, training and education
	Environmental profile on immediate and surrounding area of management	Secondary information, key informants, rapid area assessment, local government and non-governmental participation
	Boundaries for area of work, planning needs and research identified	Planner analysis and interagency discussion
Goals and objectives for resource management	Clearly stated objectives and indicators for completion	Techniques to gain consensus through community planning and dialogue with government
Policy selection for resource management	Appropriate policies and their implications	Legal advice and planner analysis, consultation with community and local government
Management strategies and actions	Draft management plan	Workshops, and inter-agency coordination, local participation

continued...

Box 3.14, continued

Special area management framework for Sri Lanka, continued

Policy steps	Outputs	Process/methods
Implementation	Field project for education, training, research, people, organisation, small public works, resource management	Public involvement, political support, professional assistance as needed
Evaluation	Monitoring of key indicators and trends, information to revive management plan	Participatory monitoring with professional guidance
Readjustments to plan and implementation	Revised plan and procedures	Planners and local participants

Two SAM sites on the south Coast of Sri Lanka were chosen to initiate the SAM process. One of these sites, at Hikkaduwa, is described in Box 5.18. More recently the Coastal and Marine Resources Management and Poverty Reduction in South Asia initiative has been extended the SAM process to a further five areas considered to be of high conservation, social and economic significance, namely Pasekudah Cluster, Puttalam Lagoon, Gulf of Mannar, Chilaw Lagoon and Bentota Estuary (IUCN 2003).

Note
1 'SAM' is used by the Coastal Resources Centre (CRC) of the University of Rhode Island (USA) and its consultants to describe integrated local area plans. SAMs have been developed in a number of countries in which the CRC operates.

Figure 3.7 Arnstein's ladder of citizen participation in decision making (Arnstein 1969)

The top two rungs of the ladder are thought of as 'rubber stamp committees'. Here the community's opportunity to participate is only allowed if there is agreement with those in power. The degree of interaction between stakeholders and decision makers increases down the next three rungs of the ladder. 'Informing' identifies citizen's rights and options, while 'consultation' allows for citizens to express their concerns. 'Placation' allows for citizens to advise on management decisions, but decision makers do not necessarily act on these concerns. These three levels are characterised by people being tolerated by those in power. At the 'partnership' level, citizens participate actively in decision making through negotiations or 'trade-offs' with managers. On the next rung, 'delegated power', citizens are given management power for selected parts of a program. On the last rung of the ladder citizens have total control of the decision-making process.

Private industries are also increasingly playing a key role in supporting the development and implementation of coastal initiatives. Clearly private companies, which depend on coastal resources, have a keen interest in how the coast is managed and planned. The traditional adversarial role between government and industry is beginning to be broken down in some parts of the world, with government and industry forming partnerships for coastal management initiatives. A good example of this comes from the Thames Estuary, as described in Box 3.15 (Kennedy 1996).

The partnership approach shown in the case of the Thames Estuary (Box 3.15) has been extended in some parts of the world to full community and industry participation in coastal management decision making. Also, management arrangements for jointly managing resources, called collaborative management, are becoming increasingly successful in coastal management. These initiatives are discussed in more detail in Chapter 4.

Box 3.15

Coastal management and the commercial sector – the case of the Thames Estuary (Kennedy 1996; Stratford 2004)

An important challenge that British coastal managers have faced is ensuring the private sector becomes involved in projects, alongside public and voluntary organi-sations. Usually an element of persuasion is required to get companies fully on board with the coastal planning process, especially as they are also frequently asked to contribute financially towards projects. The following list, based upon experience on the Thames, highlights some of the commercial reasons for becoming involved in coastal management initiatives:

- greater coordination of planning and management activities will lead to a reduction in time and resources required by individual organisations when planning their own activities and consulting on their proposals;
- more certainty for private sector interests looking to develop, alter or change uses in particular areas of the estuary;
- more efficient and responsive action by management agencies to proposed actions;

continued...

Box 3.15, continued

- greater clarity on the subjects that will need to be explored as part of environmental assessments on specific proposals;
- once established, coastal plans will provide an information source that may be used by the commercial sector, thus cutting consultancy costs;
- a strong partnership will allow information to be disseminated between participating organisations that might otherwise need purchasing and collating, again often by consultants;
- the strength of the environmental movement is such that it has a large influence over competitive ability to the extent that it is no longer economically viable to work against it;
- can see how maintenance of a healthy environment is critical to overall regional economy (e.g. revenue generated by tourist industry, water quality and fisheries);
- can assist in the identification/protection of development land for industrial/port uses that require a coastal location; help to guard against being priced out of the market by other commercial uses with higher land values (e.g. housing development such as waterfront gentrification);
- increased professionalism within the environmental movement – more of a force to be reckoned – with increased recognition of this professionalism; and
- assistance with ensuring that access to the waterfront is managed more strategically, without *ad hoc* requests that potentially can disturb and/or have safety implications for port operations.

3.3.3 Guiding statements for coastal programs

Many organisations chose to clarify their administrative arrangements for coastal management by articulating what a coastal program is attempting to achieve. As described above, there are a variety of ways to formalise this process; for example, legislation and the production of various types of documents to guide networked approaches. These documents generally begin with the philosophy underlying the coastal program, followed by a list of guiding statements, issues to be addressed, and steps to be taken to tackle those issues.

Statements which guide coastal programs are usually separated into a hierarchy. Section 3.2.3 covered a number of terms used to describe the various statements in this hierarchy, and developed the following standardised terminology: Overall Goal, Objectives and Actions, guided by Statements of Principle.

How various governments and international organisations have worded these guiding statements is examined in the following sections.

There can be significant advantages to producing formal written statements of program philosophy and guidance. As will be demonstrated later in this section, in most cases the advantages outweigh any disadvantages. Nevertheless, it is important to weigh up the pros and cons of formalising a program's goals, principles, objectives and actions, both for the organisations involved in the program and for the stakeholders (who are often the key to a program's ultimate success or failure) charged with its implementation (Steers *et al*. 1985). While these issues are discussed throughout the following sections, it is worth summarising these pros and cons at this point (Table 3.7), if only to focus attention on the realities of operationalising the various coastal zone management concepts.

Table 3.7 Advantages and disadvantages of formalising organisational objectives

Policy steps*	Outputs	Process/methods
Consensus on use of SAM by national agencies	Concept paper approved	Discussion and workshops
Site selection and criteria of choice	Two sites approved based on issues and practicability	Discussion and workshops
Issue identification and analysis for each site	List of issues and causes	Local workshops, interviews, training and education
	Environmental profile on immediate and surrounding area of management	Secondary information, key informants, rapid area assessment, local government and non-governmental participation
	Boundaries for area of work, planning needs and research identified	Planner analysis and inter-agency discussion
Goals and objectives for resource management	Clearly stated objectives and indicators for completion	Techniques to gain consensus through community planning and dialogue with government
Policy selection for resource management	Appropriate policies and their implications	Legal advice and planner analysis, consultation with community and local government
Management strategies and actions	Draft management plan	Workshops, and inter-agency coordination, local participation
Implementation	Field project for education, training, research, people, organisation, small public works, resource management	Public involvement, political support, professional assistance as needed
Evaluation	Monitoring of key indicators and trends, information to revive management plan	Participatory monitoring with professional guidance
Readjustments to the plan	Revised plan and procedures implementation	Planners and local participants

Source: Adapted from Steers *et al*. (1985).

Note
* Lynne Zeitlin-Hale, formerly of the Coastal Resources Centre of the University of Rhode Island, and now the Nature Conservancy, contributed to this framework.

(a) Coastal program principles

Statements of principle in a coastal program describe the program's overall philosophy (Section 3.2.3; Figure 3.4). The most pervasive coastal management principles in use today are sustainability and ecosystem management.

Box 3.16

Example guiding principles for coastal management programs

Vision and principles of coastal and ocean management in the Pacific

The following vision and guiding principles were presented at the First Pacific Island Regional Ocean Forum in February 2004 (First Pacific Island Regional Ocean Forum 2004). The vision and principles were developed to assist in developing a framework for national- and regional-level action to implement the Pacific Islands Regional Ocean Policy that was adopted by the thirty-third Pacific Islands Forum held in Fiji in August 2002.

Vision

Healthy coastal ecosystems sustaining Pacific Island communities

Guiding principles:
*   Inspired and innovative *leadership* supporting integrated and coordinated legislation, policy, planning, and implementation.
*   Empowered communities working in collaboration with other stakeholders to ensure *equitable benefits* of sustainable coastal development for present and future generations.
*   Applied and locally appropriate systems of *good governance* demonstrating participation, transparent and accountable decision making and linkages between local, national and international policy priorities.
*   Timely and *appropriate information* generated and effectively *communicated* to support decision making and the fostering of partnerships.
*   Valued and applied *traditional knowledge* and customary practices.
*   Manage using the principles of *adaptive management* based on continuous learning and the *ecosystem* approach.
*   *Building capacity* strategically in an ongoing process and at all levels.

Guiding principles for regional marine plans under Australia's ocean policy (Commonwealth of Australia 1998; Alder and Ward 2001)
*   Maintenance of healthy and productive marine ecosystems.
*   The distribution of benefits from ocean use and management responsibility should be shared by all Australians.
*   Internationally competitive and ecologically sustainable marine industries are needed for economic development.
*   Economic, environmental, social, and cultural aspirations should be accommodated through multiple-use planning and management.

continued...

Box 3.16, continued

Thames Estuary Partnership aim and guiding principles for management action
(Thames Estuary Partnership 1999)

Aim:
- To consider the estuary as a valuable resource in terms of its biodiversity, natural and built heritage, environmental quality and a focus for economic growth.

Guiding principles:
- The rationale of sustainable development should underpin all management decisions along the estuary.
- Treat the estuary as a single unit across interests and organisational boundaries, through a coordinated management approach.
- Support the use of the estuary and rail to transport materials, goods and passengers within and through the estuary, wherever possible. Assess any local disadvantages against the wider benefits provided by these sustainable transport modes.
- Encourage efforts to ensure that all development proposals balance the economic and environmental needs of the estuary.
- Maintain and enhance opportunities for recreation and leisure on the estuary to provide suitable conflict-free access for all.
- Ensure that the existing network of designated sites and other areas of important habitats are safeguarded and managed appropriately, and that opportunities for habitat re-establishment are taken to retain and enhance the estuary's diversity of wildlife.
- Ensure the protection, promotion and understanding of the historical and cultural resource, including safeguarding and augmenting the existing network of designated sites and promoting careful evaluation of any proposals for development that may affect them.
- Develop and establish comprehensive knowledge about the natural physical processes of the estuary.
- Consider flood defence on an estuary-wide basis and explore the full range of options.
- Support and promote measures to reduce pollution throughout the estuary.

(b) Overall goal in coastal management programs

Achieving a balance between development pressures and conservation needs has been used by many coastal nations as the centrepiece of their coastal zone management efforts. Sustainable development principles in coastal programs (Section 3.2) essentially subsumed the earlier notions of balance, but the balance concept remains in use by some coastal nations, as shown by the examples in Box 3.17.

The vision statement chosen for the Sulawesi Selatan Province (Indonesia) Coastal Strategy forms part of a hierarchy of actions (Figure 3.8) (Bangda 1996). It is guided

---

Box 3.17

Examples of overall coastal management goals

Draft Western Australian state government vision and goal for coastal management (Western Australian Planning Commission 2001)

The Government's vision for coastal zone management in Western Australia is that:

- the principle of ecological sustainability and a commitment to maintaining healthy functioning ecosystems will underpin all planning, policy and management decisions about the coastal zone in Western Australia;
- coastal zone management will be coordinated across all of government leading to sensible balanced decisions and the most efficient use of resources, and will be undertaken in partnership with the community; and
- the management of Western Australia's coastal zone will engender international respect and admiration.

    Achieving the vision – a goal for coastal zone management

- The Government's goal for coastal zone management in Western Australia is to manage the coast sustainably for the long-term benefit of the community, by protecting environmental quality, biodiversity and features of cultural significance, and providing for social and economic needs.

Purpose of the coastal resources management program in Ecuador (Robadue 1995)

- The preservation and development of the coastal resources of the coastal resources in the provinces of Esmeraldas, Manabí, Guyas, El Oro and Galápagos.

Vision for the Sulawesi Selatan Province (Indonesia) Coastal Strategy

Coastal resources of Sulawesi Selatan will be managed on an integrated basis so as to ensure optimisation of the economic, social and ecological benefits they provide.

---

by national strategies and policies and in turn guides regional and local coastal initiatives. Similarly, the Central Coast Regional strategy in Western Australia developed a set of Founding Principles guided by an overall statement of purpose.

The Indonesian and Western Australian case studies represent the geographical, cultural and administrative spectrums. At one end, Indonesia represents a tropical archipelagic nation of over 200 million people, with a diversity of peoples who intensively use their coast, and a highly centralised government. The Central Coast is a

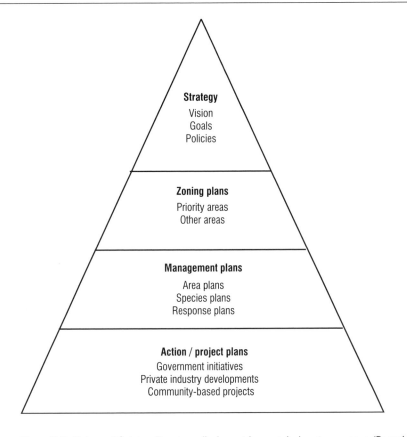

Figure 3.8 Sulawesi Selatan Province (Indonesia) coastal planning system (Bangda, 1996)

length of sub-tropical coast in a sparsely populated but developed country (Australia) where the major issues are linked to growth management and conservation. Nevertheless, all the case studies contain similar planning goals, based on concepts of sustainable development. Both planning systems are described at greater length in Chapter 5.

(c) Coastal program objectives

The objectives of a coastal zone program are, in many ways, the program's cornerstone. Badly framed objectives can fundamentally weaken the status of a program, and seriously hamper its implementation. However, defining program objectives that strike the right balance between clarity of purpose on one hand, and becoming overly rigid on the other, is often not straightforward. The issues which influence the balance are described in this section.

The objectives of coastal programs are important for a number of interrelated reasons: program development, implementation and evaluation, and the role of program stakeholders. The most important issues are discussed below:

- whether or not to develop quantifiable objectives;

Box 3.18

Central Coast (Western Australia) Regional Strategy purpose and founding principles (Western Australian Planning Commission 1996)

The background to the Central Coast Strategy is given in Box 2.4 and its planning context and the approach to its development shown in Box 5.12.

Primary purpose:

- to provide a link between state and local planning which is based on balance of economic, social and environmental considerations.

Broad founding principles relate to:

- ecologically sustainable development;
- regional identity;
- managing natural resources;
- facilitating the development of community facilities and social services;
- fostering economic development and promoting diversification; and
- coordinating and integrating regional planning and development.

- that coastal programs generally aim to achieve multiple objectives;
- given the multiple-objective nature of most coastal programs, that there can be conflicting objectives; and
- that there are different methods of setting objectives to address coastal issues (including implied versus stated objectives).

Choosing whether or not objectives should be quantifiable depends on their intended purpose. Coastal programs focussed on achieving outcomes generally develop quantifiable objectives, which can help measure if the program has been successful. Other programs may decide to develop a small hierarchy of objectives statements, called operative and operational objectives. These are defined by Steers *et al.* (1985) as:

Operative objectives: Represent the real intentions of an organisation. They reflect what an organisation is actually trying to do, regardless of what it claims to be doing.

Operational objectives: Have built-in standards that can be used to determine if objectives are being met.

Whether the objectives of a coastal initiative are operative or operational largely depends on its geographic coverage and focus. Water quality management plans may, for example, have very specific operational objectives relating to the attainment of certain minima for pollutant loads in the water column, or the extent of marine ecosystem damage from changed water quality. Most other types of coastal plans and programs generally employ operative objectives, because of their wider-reaching aims. Examples of both types of statements are shown in Box 3.19.

In general, the greater the geographic coverage of a coastal initiative the less the likelihood that its objectives will be operational in nature. The broad coverage of national coastal programs or regional plans or policies means that specific targets are difficult to write into objective statements. Specific targets are often left to a lower level in the hierarchy of guiding statements. For example, the objectives statements in the Thames Estuary Management Plan stated a set of general objectives (aims), and left the development of targets consistent with those objectives to specific actions listed within an action plan (Figure 3.9). These actions specify timelines for their achievement, and list the 'partners and players' required to implement each action. As such, the actions specified by the Thames Estuary Management Plan contain the built in standards (time frame and responsible agencies) required for operational objectives.

Box 3.19

Example of operative and operational objectives for coastal management programs

Goals (objectives) for the Sulawesi Selatan (SulSel) province (Indonesia) Coastal Strategy

- To provide for the present and future interests and aspirations of SulSel residents with respect to coastal resource use, access and enjoyment.
- To protect, restore and enhance coastal ecosystems of SulSel and the ecological processes which sustain them.
- To encourage optimal, efficient and sustainable use of coastal resources.
- To ensure that coastal management and planning activities are undertaken on an integrated basis and that management resources are used efficiently and effectively.

Operational objectives, Vancouver Port

The Burrard Inlet, which is part of the Vancouver Port, is important for a number of uses. The water quality management objectives are focussed on maintaining water quality for recreation, aquatic life and wildlife. The objectives are achieved through setting of water quality objectives for a number of parameters that include (BIEAP 2001):

Parameter	Objective
Microbiological	Faecal coliforms <= 200/100 mL
Suspended solids	10 mg/L maximum increase
Dissolved oxygen	6.5 mg/L minimum
Total copper	<= 2 micrograms/L mean and 3 micrograms/L maximum
Total arsenic	20 micrograms/g dry weight maximum
Total mercury (in sediment)	0.15 micrograms/g dry weight maximum
Total mercury (fish flesh)	0.5 micrograms/g weight

| **Our vision** |
| The sustainable development of the Thames |

| Introductory text | Example: Recreation |

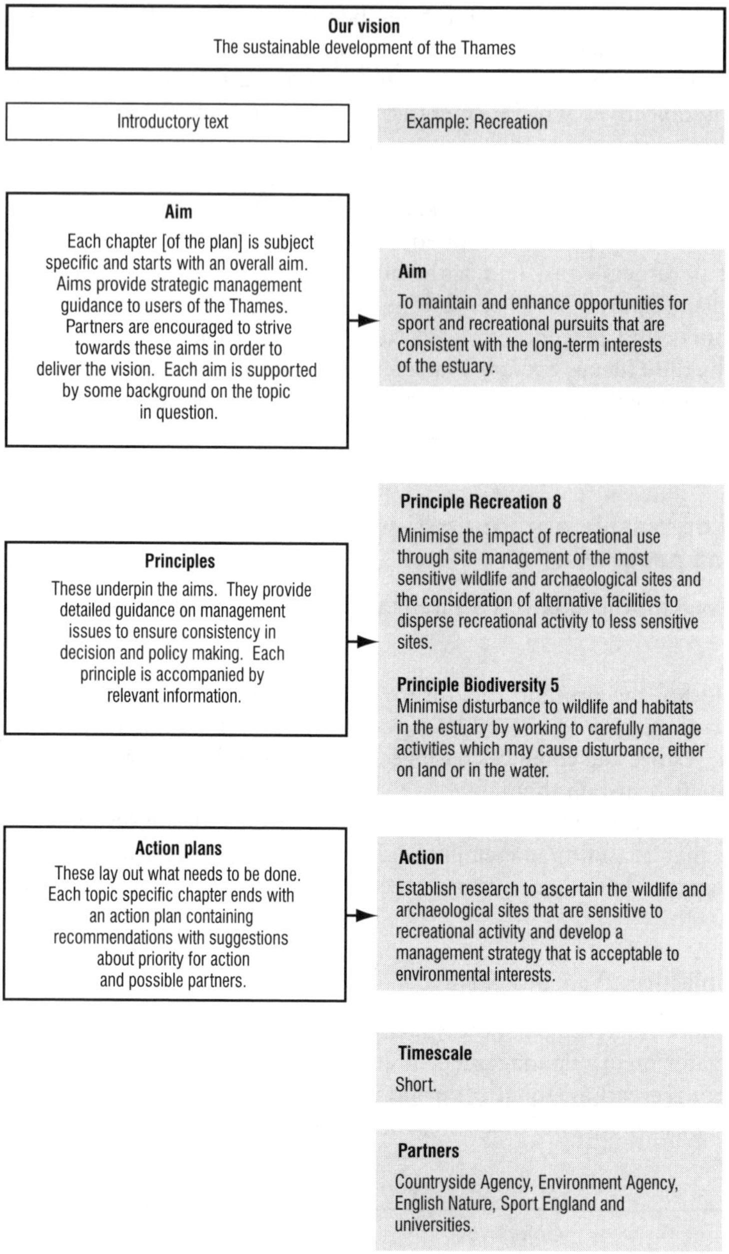

Aim

Each chapter [of the plan] is subject specific and starts with an overall aim. Aims provide strategic management guidance to users of the Thames. Partners are encouraged to strive towards these aims in order to deliver the vision. Each aim is supported by some background on the topic in question.

Aim

To maintain and enhance opportunities for sport and recreational pursuits that are consistent with the long-term interests of the estuary.

Principle Recreation 8

Minimise the impact of recreational use through site management of the most sensitive wildlife and archaeological sites and the consideration of alternative facilities to disperse recreational activity to less sensitive sites.

Principles

These underpin the aims. They provide detailed guidance on management issues to ensure consistency in decision and policy making. Each principle is accompanied by relevant information.

Principle Biodiversity 5
Minimise disturbance to wildlife and habitats in the estuary by working to carefully manage activities which may cause disturbance, either on land or in the water.

Action plans

These lay out what needs to be done. Each topic specific chapter ends with an action plan containing recommendations with suggestions about priority for action and possible partners.

Action

Establish research to ascertain the wildlife and archaeological sites that are sensitive to recreational activity and develop a management strategy that is acceptable to environmental interests.

Timescale

Short.

Partners

Countryside Agency, Environment Agency, English Nature, Sport England and universities.

Figure 3.9 Thames estuary management plan explanation of guiding statements and an example (Thames Estuary Partnership 2003)

In general, local coastal initiatives are more likely than those at national or regional level to contain specific operational objectives. However, if coastal plans or programs at any scale are wide-ranging, and aimed at assisting in integrated coastal management, their objectives are likely to be too broad to be directly implementable. In these cases too, the standards required for ensuring the effective implementation of a particular objective are written at the action statement level (Figure 3.4).

More often than not, coastal management programs have multiple objectives, as shown earlier in this chapter. One reason for this is the complex nature of the exercise, as neatly summarised by Owens (1992):

> from the outset the purpose of coastal zone management in the United States has not been simple or straightforward. It includes multiple goals, both preserving and developing coastal resources, some of which are invariably conflicting.
>
> (Owens 1992: 144)

The objectives of coastal programs generally fall into four groups:

* environmental;
* economic;
* social and/or cultural; and
* administrative.

The groupings, and the objective statements within them, will depend on the overall goal of the particular program. The wording of the overall goal will also affect whether subsequent objectives conflict with each other. In cases where either balance or sustainability is a key aim, multiple and conflicting objectives will be inevitable. This is because balance and sustainability goals promote both conservation and development, and objective statements will contain objectives that encourage both the development of coastal resources and their conservation. Resolving this inherent complexity is commonly one of the most important roles of coastal planning and management activities. Addressing the issue of conflicting objectives is often central to overall program design, and to the various program components, especially in coastal management plans. It is consequently a topic which is discussed throughout this book.

(d) Coastal program action statements

Action statements are closer to the 'doing' part of coastal programs. They specify what tasks and activities will occur, when and by whom in order to meet the program's objectives, and ultimately its overall goal, which is being steered by its principles.

The form of action statements will depend to a large extent on the geographic coverage of the particular coastal program. For example, if a coastal program covers an entire nation, action statements are most likely to be oriented towards the development of new plans, or the implementation of broad management initiatives, such as the development of a dune rehabilitation program. In contrast, a more localised coastal program may include action statements defining specific on-the-ground/on-the-water coastal management actions, such as the provision of access ways in certain locations, particular dune management works or the installation of public boat moorings. To be effective, action statements should define operational matters such as those responsible for carrying out the action, the time within which an action should be completed, the required budget, and so on. Indeed, these are the essential elements of well-developed action statements: they are clear and unambiguous, providing all the required elements to ensure that they are actually carried out.

(e) Ownership of guiding statements in coastal programs

It is tempting to think that after reading this chapter, or leafing through the various publications of national governments and international organisations, that writing a set of guiding statements for a coastal management program is relatively easy. Indeed, it would not take much effort to put together a set of statements by drawing on the many examples in current programs, including those used here as case studies. In most cases, however, this would at best be likely to result in a short-term improvement in the management of the coast, and at worst a rejection by many stakeholders and a moratorium on coastal planning and management. Poorly defined objectives may contribute to the actions of those involved in a coastal program being unproductive, or even counter-productive. One reason for this is to do with ownership.

Because most coastal management programs attempt to tackle a multitude of issues which occur over different and often overlapping spatial and temporal scales and affect multiple sectors, simply developing a set of guiding statements – and then telling stakeholders what they are – can do more harm than good. This is often referred to as the 'top down' management approach; that is, government agencies impose their ideas on those affected by the decisions of government. There can also be disadvantages in developing guiding statements purely from the 'bottom up', due to local biases and problems in how non-expert opinions are formed.

Successful coastal management programs use a systematically inclusive process to develop guiding statements. These attempt to integrate the values and expectations of both the top and bottom decision-making levels with those of additional key stakeholders. There are various ways to achieve this through consultative processes, including key stakeholder workshops, seminars or structured formal enquiries. Importantly, the choice of particular technique(s) aims to actively facilitate the meaningful involvement of all stakeholders. The decision about process is often made within the context of reviewing a coastal management program, during the development of a coastal planning strategy, or during its evaluation, as outlined in the next section.

3.4 Evaluating and monitoring coastal management programs

Imagine a recently elected government minister responsible for coastal planning and management asking government officials, 'before I give you more funding, tell me if the coastal management program is successful or not.' Assume that these officials have included monitoring and evaluation criteria and processes as part of the coastal program's design and have undertaken regular performance reviews. The minister's question could be answered through the presentation of monitoring and evaluation results and how management objectives are being met throughout the lifetime of the program (King *et al.* 1987).

This hypothetical example highlights both the practical and political importance of program evaluation and monitoring. The emphasis on 'proving' that coastal programs have achieved desired outcomes is becoming increasingly important as a means to maintain program funding, and to secure the continuing involvement and commitment of all program stakeholders.

Monitoring and evaluation are not simple tasks, and a systematic ongoing approach to them throughout its lifetime is a vital part of any successful program (Herman *et al.* 1987; Rossi and Freeman 1993).

An important role for program evaluation is the completion of the coastal management cycle, shown in Figure 3.10. This does not mean the planning and management cycle ends, but rather a new round of the cycle begins. Figure 3.10 shows two interpretations of the cyclic nature of planning initiatives. The top row shows three separate planning cycles. In this example, Plan 2 occurs a number of years after Plan 1 is completed, and so on. This separated plan cycling is demonstrated in the Thames Estuary Management Plan, for example. In contrast, the second row in Figure 3.10 shows that Plan 2 is initiated soon after the completion of Plan 1, effectively linking them. This close linkage between plans may be required where issues are particularly complex, or where evaluation requirements are stringent. However, as will be demonstrated in this section, evaluation should not be considered only at the end of the policy cycle, but rather during all its stages. This is particularly important within the context of programs using adaptive management principles that require ongoing evaluation as a prerequisite for program learning and continuous improvement.

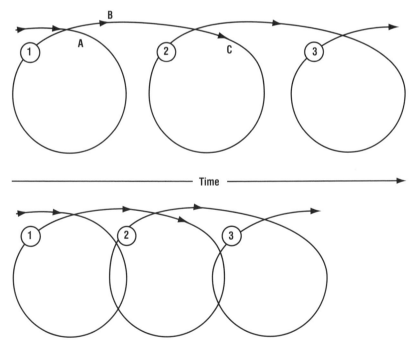

Figure 3.10 Plan production cycles (adapted from Hale 1996; Olsen *et al.* 1996)

Monitoring and evaluation are processes which assist in answering the question 'is the program working?' and, if not working, what future actions are needed to improve performance of the program. If a coastal program has included measurable objectives, and criteria to evaluate the plan's ability to meet those objectives, ongoing monitoring may provide the information required to evaluate the effectiveness of the program. Monitoring and evaluation also contribute to decision making, as noted by the National Research Council (1995):

> monitoring can narrow the uncertainty associated with decision making, but it can not eliminate it, and monitoring contributes to understanding change and ascribing causes to these changes. Monitoring results are also useful in weighting the societal benefits of management alternatives.

The terms 'program monitoring' and 'evaluation' need to be defined in a management planning framework. Monitoring is a process where repetitive measurements are taken in order to indicate natural variability, and changes in environmental, social and economic parameters. Measuring these changes contributes to the information base managers need to evaluate a program's effectiveness. Evaluation is analysing information, some of it gained through monitoring, and comparing the results of the analysis against predetermined criteria. Optimal program evaluation therefore depends on using quality information, with much of this information resulting from monitoring. Monitoring tools and approaches to support coastal management program evaluation are outlined in Chapter 4, including reference to the innovative approach to monitoring and evaluation adopted in the Philippines.

Program evaluation is well developed in the field of business and organisational management (e.g. Mukhi et al. 1988), but in the area of coastal planning evaluating the effectiveness of plans relative to a set of performance indicators (Table 3.8 and see Chapter 5) is a relatively recent development that has seen significant progress only in recent years. The evaluation of coastal plans and programs is a result of progress in several fields of coastal management, increased community participation, and the accountability requirements of funding organisations.

There has been a considerable amount of emphasis in international development on assessing, quite simply, 'does aid work?' (general references World Bank 1998; Marsden and Oakley 1990 and specific coastal evaluations Lowry et al. 1999; Christie, in press). These initiatives have been mirrored by the requirements for increased accountability within governments for program expenditures, including those aimed at managing the coast; for example in the United States, as shown in Box 3.20.

Advances in other fundamental aspects of management such as administrative arrangements, institutional development, planning approaches and monitoring methods, have provided managers with the scope to address the question of program evaluation in the coast. As community involvement in decision making has increased, so has their questioning of the consequences of decision making. Accountability in government and funding organisations has also increased, with the general trend of evaluating programs and strategies now being the policy of many agencies. These factors have all contributed to the increased interest in coastal program evaluation. In addition, there has been a significant increase in the application of program evaluation techniques to program comparisons, in order to draw conclusions on best-practice approaches to coastal program design (Olsen et al. 2002; Lowry 2003).

Box 3.20

Timeline in the development of an evaluation framework for activities under the US Coastal Zone Management Act (CZMA) (adapted from Shea 2003)

- 1997 The Department of Commerce Inspector General (DOC/IG) releases a report noting 'only anecdotal evidence' of the accomplishments of the CZM program and calls on the US government to develop a strategy to measure the effectiveness of the CZM program.
- 1999 Responding to the DOC/IG report, the US government commissions the University of Washington and colleagues to undertake a comprehensive study of the effectiveness of US coastal zone management. One key finding calls for the development of a common set of indicators that would link state management activities to national CZMA objectives (Hershman *et al.* 1999).
- 2001 (July) NOAA initiates development of a national performance measurement system for coastal zone management by commissioning a study through the Heinz Center.
- 2001 (December) Coastal Resources Conservation Act of 2001 introduces to amend and reauthorise the CZMA that directs 'the Secretary of Commerce … (to) submit a common set of measurable outcome indicators to evaluate the effectiveness of state coastal zone management programs in the achievement of the national policy…' and '… establish a national coastal zone management outcome monitoring and performance evaluation system using the common set of indicators prepared …' (Gilchrest 2001). The bill is referred to a House committee, where a subcommittee hearing is held. No committee action is taken (Heinz Center 2003).
- 2003 (May) Heinz Center publishes The Coastal Zone Management Act: Developing A Framework For Identifying Performance Indicators. The report recommends an outcomes-based performance evaluation framework including common indicators for the CZMA (Heinz Center 2003).
- 2004 (April) US Ocean Commission releases its preliminary report calling for a major overall of the CZMA in order to improve the ability to monitor and assess its performance in meeting new measurable goals (US Commission on Ocean Policy 2004).

The recent increased interest in program evaluation for coastal initiatives has now provided a useable set of tools for managers to approach the problem. However, given the general paucity of monitoring data available to support longitudinal program evaluation, even in developed countries (Hershman *et al.* 1999), it will be some years before these evaluation tools can themselves be evaluated.

Anyone contemplating program evaluation might usefully start by asking the set of questions framed by Owen and Rogers (1999) (a characterising key word for each question is bracketed):

Table 3.8 Advantages and disadvantages of setting performance indicators

	For organisations	For individuals
Advantages	• Focusses attention to common goals • Rationale for organising and prioritising programs • Provides a set of standards for assessing program effectiveness • Source of legitimisation • Recruitment of staff through identification of program priorities	• Focus attention • Rationale for working with an organisation • Vehicle for personal goal attainment • Personal job security • Self identification and status with an organisation
Disadvantages	• Means to an end can become the real goals • Measurement stressed quantitative goals at expense of qualitative • Goal specificity problem (ambiguous goals fail to provide direction; highly specific goals may constrain action and creativity)	• Rewards may not be tied to goal attainment • Difficulty in determining relevant performance evaluation criteria • Inability of individuals to identify with abstract goals • Organisational goals may be incongruent with personal goals

Source: Adapted from Owen and Rogers (1999).

- What is the ultimate reason for undertaking an evaluation (orientation)?
- What is the state of implementation of the program (state)?
- What aspect(s) of the program should the evaluation be concentrated on (focus)?
- What is the temporal relationship between the evaluation and program development and delivery (timing)?
- What is the appropriate underlying evaluation approach, and what are acceptable methods of collecting and analysing relevant information consistent with this approach (approach)?

There are two main foci for program evaluation – objectives-based and needs-based (Owen and Rogers 1999). As the name suggests, objectives-based evaluation examines how a program has performed in relation to its stated objectives. This is the most common type of program evaluation (Cronbach 1992). In an alternative approach, assessment is used to determine if a program meets identifiable needs; for example, whether the needs of key stakeholders in the program are being met. The critical evaluation issue is the attribution of 'inputs' into the coastal program (e.g. money, staff, management tools, location) through the 'chain of results' (namely activities, outputs and outcomes) to the eventual impact of the program (Smutylo 2001).

The right type of objectives-based evaluation will depend on what exactly the evaluation is trying to achieve, the status of the program, and how the results of the evaluation will be used. For example, the statements of program objectives may have been written with evaluation in mind (termed 'operational objectives' – Section 3.3.3). Such a program is 'objectives led'. In this case, an objectives-based evaluation would be the most appropriate measure of that program's performance. Owen and Rogers (1999) divide the many different types of evaluation into five 'evaluation forms'. The important features of each type of evaluation form, in relation to the major components of any evaluation program, are shown in Table 3.9 and discussed below.

Table 3.9 Forms of program evaluation

Evaluation Forms

	Impact	Monitoring	Interactive	Clarificative	Proactive
Orientation	Justification/fine tuning	Justification/fine tuning	Improvement	Clarification	Synthesis
State (of program)	Settled	Settled	Development	Development	None
Focus	Outcomes/delivery	Outcomes/delivery	Delivery	All elements	Context
Timing (*vis-à-vis* program delivery)	After	During	During	During	Before
Key approaches	• Objectives-based • Process-outcomes studies • Needs-based • Goal free • Performance audit	• Component evaluation • Devolved performance assessment • Systems analysis	• Responsive • Quality review • Action research • Developmental • Empowerment	• Evaluability assessment • Program logic/theory development • Accreditation	• Needs assessment • Review of best practice • Research synthesis

Source: Adapted from Owen and Rogers 1999: 54.

A common perception is that evaluation can only take place as the last step in the policy cycle, i.e. after a program has been completed. As shown in Table 3.9, this is not the case. Only one form of evaluation, impact evaluation, usually occurs after program completion. The other four forms of evaluation are undertaken either before the program has started or during its operation. Thus, the evaluation of coastal programs should be viewed as a process which occurs before the program is initiated, continues throughout its life, and is completed after the program is finished. A wide range of techniques can be used to undertake the different forms of evaluation depending on the orientation, focus and timing of the evaluation. Detailed information on various approaches to evaluation can be obtained from specific texts on evaluation techniques (e.g. Owen and Rodgers 1999; Cronbach 1992).

In evaluating environmental objectives such as maintaining water quality, international, regional and national standards or criteria can provide some guidance. But for less tangible objectives – such as maintaining biodiversity – deciding what to measure and the criteria to use is more difficult; when social and economic parameters are included, the problem becomes even more challenging. Specific tools for developing meaningful indictors to support coastal program evaluation have recently been summarised by the Intergovernmental Oceanographic Commission (Belfiore *et al.* 2003).

The problem of evaluating coastal programs is highlighted in the United States, where coastal management is undertaken primarily by state and local governments, with federal financial and other assistance (Box 3.12). Major problems have been experienced in evaluating the state coastal management programs since the inception of the US Coastal Zone Management Act in 1972 (Box 3.12). In order to overcome these evaluation problems researchers have used surrogate measures of coastal program performances, including opinions and perceptions of those knowledgeable about a program (Born and Miller 1988; Center for Urban and Regional Studies 1991; Knecht *et al.* 1996; Hershman *et al.* 1999). For example, the results from the study of Knecht *et al.* (1996) using surrogate evaluation measures found considerable variation between sample groups (academics, coastal program managers and coastal interest groups), but no systematic difference between the performances of different program structures. The Coastal Zone Management Effectiveness Study by Hershman *et al.* (1999) similarly developed a set of surrogate indicators of performance noting that the study 'emphasised that outcome monitoring is the missing piece in the 20 years of assessing performance of the CZM system' (Hershman *et al.* 1999: 133). Indeed, the lack of effective performance evaluation data in the US coastal program was a key finding of the recent Ocean Commission preliminary report. Importantly, the Ocean Commission recommended that the US Coastal Zone Management Act be significantly updated (re-authorised) to include a range of evaluation and performance assessment measures, as shown in Box 3.21.

The recent recommendations of the US Ocean Commission (Box 3.21) highlight the importance of considering evaluation at different stages of the program, including during the program design stage, during its implementation, and after its completion (Box 3.22 and Table 3.9). These evaluation themes will be revisited throughout the book, especially in relation to the using of planning approaches in coastal programs (Chapter 5). Unfortunately, program evaluation is often the last component to be thought about in the design and implementation of a coastal initiative, rather than being part of the design phase. Indeed, as will be shown in Chapter 5, 'monitor and evaluate' is often stated as the last step in the planning process.

Box 3.21

Extract of preliminary US Ocean Commission Report (US Commission on Ocean Policy 2004)

Congress should re-authorize the Coastal Zone Management Act (CZMA) to strengthen the planning and coordination capabilities of coastal states and enable them to incorporate a coastal watershed focus and more effectively manage growth. Amendments should include requirements for ... the development of measurable goals and performance measures, improved program evaluations, ... , incentives for good performance and disincentives for inaction ...

Specifically, CZMA amendments should address:

- *goals* – state coastal management programs should develop measurable goals based on coastal resource assessments that are consistent with national and regional goals. State coastal programs should work with local governments, watershed groups, non-governmental organisations, and other regional entities, including regional ocean councils, to develop these goals.
- *performance measures* – state coastal management programs should develop performance measures to monitor their progress toward achieving national, regional, and state goals.
- *evaluations* – state coastal management programs should continue to undergo periodic performance evaluations by the National Oceanic and Atmospheric Administration. In addition to the existing evaluation criteria, the performance measures developed by state programs should also be reviewed. The public, representatives of watershed groups, and applicable federal program representatives should participate in these program evaluations.
- *incentives* – existing incentives for state participation – federal funding and federal consistency authority – should remain, but a substantial portion of the federal funding received by each state should be based on performance. Incentives should be offered to reward exceptional accomplishments, and disincentives should be applied to state coastal management programs that are not making satisfactory progress in achieving program goals.

Perhaps the most pragmatic reason for program evaluation is to justify future funding by governments and international donor agencies that want to know their money is being well spent, and that the program is meeting the expectations of coastal communities. There are also some political issues related to monitoring and evaluation which should be borne in mind by potential evaluators. Coastal management programs are essentially run by governments, with varying degrees of involvement of private industry and the community (Section 3.3.2). Thus any evaluation of the successes and

---

Box 3.22

Program evaluation stages (Owen and Rogers 1999)

- Consideration of evaluation procedures should occur before the program starts (*proactive*). This pre-assessment should include an emphasis on the wording of statements of objectives and actions.
- Early in the implementation of the program a preliminary evaluation should be undertaken in order to check on the overall design of the program, and how easy it will be to evaluate its performance once it is in full operation (*clarificative*). Modifications to the program can be made to ensure the program performs as expected.
- During the implementation of the program, periodic evaluations should be undertaken both for purposes of accountability and in order to highlight possible areas for improvement (*interactive and monitoring*).
- Finally, once the program has matured or has been completed, the justification of the program should be tested (*impact evaluation*). This evaluation could focus on either objectives- or needs-based evaluation. This will depend on how the program was designed, and how previous evaluations have been carried out.

---

failures of a coastal program can be viewed as essentially an evaluation of the performance of government itself. This can be particularly so in nations with a strong coastal focus, where much of the nations' infrastructure is located.

Thus, coastal program evaluations can rapidly take on a high level of political importance, attracting the attention of politicians and senior government officials. As a result, there may be an impression formed that it is better not to do an evaluation, simply because the results may not reflect well on the government of the day. This impression may be enhanced by key interest groups who may be benefiting from current coastal management arrangements, and who may feel they would lose out from any actions which flow from the evaluation's results (Table 3.8). Some government officials working on a coastal program may share this perception of program evaluation. A negative evaluation, they may feel, could reflect poorly on their performance as professionals. In the recent evaluation of the US coastal program, this issue was managed through the process of the Federal Government itself establishing an independent panel of experts to undertake the program evaluation (see Box 3.21). This approach has the dual benefit of government endorsement while retaining professional independence and credibility.

There are also forces working in favour of coastal program evaluations, most notably within governments and international donor organisations that borrow much of their philosophy and practice from the business world. With such key players there is an increased awareness of setting and attaining targets, and of performance measurement and accountability. Although this encourages evaluations to be carried out, there can be disadvantages in focussing on outcomes which are easy to measure. This has the potential to downgrade the importance of other important issues, such as coastal environmental quality or scenic beauty. These may require more concerted efforts to yield meaningful evaluation criteria. This issue is explored in more detail in Chapter 4.

3.5 Chapter summary

The concepts underpinning the practice of coastal planning and management have undergone something of a revolution in the past ten years or so. What was, thirty years ago, a relatively simple blend of rational comprehensive planning, with the concept of balance between conservation and development, is now a blend of diverse theoretical and philosophical concepts. This is perhaps to be expected, given the diversity of challenges facing coastal managers, and the sheer physical, biological and socio-economic diversity of the coastal zone. At present this blend of theories appears to include rational comprehensive planning, values-based planning, ecosystem-based management, adaptive/learning management and planning, systems theory, public participation, consensus, and conflict management. These concepts overlie constant core principles of clear, transparent, accountable, focussed and enforceable coastal management and planning approaches.

This chapter has attempted to summarise these major conceptual building blocks of the development of effective coastal management and planning programs.

While the theoretical basis for coastal planning and management has evolved over the last thirty years, there has been little debate as to which theory or set of theories is most appropriate to take coastal planning and management into the next decade. It must be borne in mind that, given the nascency of theoretical dialogue in coastal planning and management, there are likely to be new and emerging conceptual elements that are bound to require consideration.

The pressing needs of the coast and its inhabitants call for management now (if not yesterday). Consequently, there is a requirement for managers and planners to use the best available tools, techniques, social and political analysis underpinned by well articulated and understood theories and philosophies. Indeed, coastal management is perhaps ultimately a pragmatic discipline, using whatever approaches, whatever theories and whatever tools are most effective in the achievement of given management objectives. In this sense, coastal planning reflects the broader concerns of urban planning theorists; namely that coastal planning and management theory is likely to mirror coastal planning practice. It is this dynamic management climate – the tension – between management and planning, between concept and practice (if indeed there are differences between concept and practice) that requires considerable and concerted attention.

Finally, it is worth reflecting on the possible impacts of considering the conceptual and theoretical concepts in coastal planning and management. Sound conceptual analysis cannot take place without awakening us to what we do, and why we think we do it (Kay 2000a) – and this 'may always raise the possibility that we should do something else for a change' (Eagleton 1990: 27).

In conclusion, we believe that much more time should be given to reflecting on the below-the-surface issues of why we are doing coastal management in the way we do. If this is done, conceptual and theoretical thinking will play an important role in assisting coastal managers face day-to-day challenges. Without such self-reflection there is a risk of the practice of coastal planning and management becoming empirical – devoid of any theory and as a result dependent entirely on trial and error and personal experience. But it is also worthwhile considering that taking a reflective view of coastal management may lead to beneficial unexpected consequences. We may become enamoured with new ideas, new ways of managing the coast; or we may strengthen and deepen our relationship with existing conceptual principles, renewing our collective justification for the way we do things. There is a risk, albeit a minor one, of thinking too deeply about coastal management theories to the detriment of management actions. However, the urgent imperative of better management of the coast should prevent this from happening.

An important building block in the conceptual framework outlined in this chapter is pragmatism. While pragmatism can be criticised as the 'do whatever it takes' school of management, this is in fact what is faced by coastal managers on the ground charged with addressing enormous daily challenges. Indeed, while the conceptual analysis outlined in this chapter is critical for thinking about what managers are doing, and how they are structuring their approach to management challenges, pragmatism is often their defining 'theory': enabling theories and techniques to be chosen that suit local circumstances. It is how these tangible actions can be structured and delivered through organised coastal programs that is critical, and what tools are used to deliver the required management outcomes. A variety of such coastal management tools is introduced in the following chapter.

Chapter 4

Major coastal management and planning techniques

A wide range of techniques is commonly used in coastal management and planning. They can be used individually to address specific problems, combined to address more complex issues, or used as part of a coastal management planning process. The number is enormous, and effectively covers all the techniques available for the management of the natural environment, urban centres, and systems of government, and for managing shared action between multiple groups and stakeholders.

In order to narrow down the range of choice we have selected the coastal planning and management techniques which are the most common and/or important to assist in the sustainable development of coastal areas. They include those used today – the mainstays of coastal planning and management – such as policy, environmental impact assessment, economic assessment and public participation. Also included are those techniques that have emerged in recent years including customary (traditional and indigenous) management practices, visual analysis techniques and the use of information technology.

Though we have chosen to focus on the most important techniques, the number is still relatively large, meaning that the description of each will be necessarily broad. Nevertheless, each section describing a technique is structured to allow an introduction to the main factors important in its application to coastal planning and management, and is illustrated through the use of case studies. Sources of further reference are given throughout to enable additional detail on each technique to be readily obtained.

The major techniques are grouped into administrative, social and technical. This grouping is undertaken to highlight the similarity between some techniques, while showing the differences between others. This grouping is useful if at times somewhat artificial in that there are techniques which contain elements of more than one group. For example, environmental impact assessment is a government process, a technical procedure, and also involves social components. The chapter concludes with a section on techniques of monitoring and evaluation, which constitute part of all coastal management and planning techniques.

As in previous chapters, case studies are used to demonstrate the application of each technique to actual coastal management problems and issues.

4.1 Administrative coastal management and planning techniques

Governments can assist in improving the management of coastal areas in a variety of ways: by encouragement, through force or through the use of research and information.

Approaches include using policies or general guidelines, or much more targeted means such as enforcing regulations or issuing permits and licences. Increasingly, a softer, less authoritarian approach than emphasising coastal management problems is being taken through education and training programs.

4.1.1 Policy and legislation

'Policy' and 'legislation' are two words easily recognised by the public. When managers or politicians announce the passing of new policy or a new piece of legislation it is a visible sign that the coast has a high priority for decision-makers. And depending on their implementation and enforcement powers, policy and legislation can be powerful tools for managing the coast.

Policy and legislation as described in this section are used by most coastal nations, but in different combinations and to varying degrees. To a large extent this reflects economic, cultural and political circumstances and also the length of time coastal programs have been active. In some cases it reflects the maturity of a nation's coastal planning initiatives. As will be shown through case studies, coastal programs, especially in developed countries, have tended to evolve through early controlling stages founded on policy or legislative control (government dominated) into communicative and participatory stages where education and other techniques dominate. Indeed, such evolution in coastal programs in many cases cannot take place without first establishing a clear set of operating parameters, often established through policy and/or legislation.

(a) Policy

Politicians, administrators and managers often cite 'policy' as a basis for decision making. But what exactly is policy? Clark (2002) uses the definition of Lasswell and McDougal (1992) that policy is a social process of authoritative decision making by which members of a community clarify and secure their common interests.

Policy is about guiding decisions (Figure 4.1), specifically about decisions regarding choices between alternative courses of action (Colebatch 1993). Policy therefore is deeply rooted in decision-making processes and hence is interwoven within the

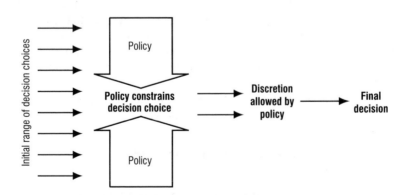

Figure 4.1 Policy and discretion in guiding decision making (adapted from Mukhi *et al.* 1988)

mechanics of organisational behaviour, that are in turn embedded within social systems – public and private, large and small. Consequently, there is a risk that analysis of policies in coastal planning and management becomes no more than sweeping generalisations for looking at the way decision-making processes operate. As described by Davis *et al.* (1993) in the Australian governmental context:

> The idea of 'public policy' works on a range of levels. It can simply mean a written document expressing intent on a particular issue, or imply a whole process in which values, interests and resources compete through institutions to influence government action.
>
> (Davis *et al.* 1993: 7)

Nevertheless, the importance of policy to the effective management of the coast is so high that such an analysis must be undertaken here. In this section issues of policy will be linked wherever possible to other chapters where government processes are discussed, most notably Chapter 3.

Policies important in the management of the coast can broadly be divided into public policy, that is the policies of government agencies and their staff, and non-public policy. The latter refers to the polices of all organisations not part of the public sector, and their staff – including private businesses, non-governmental organisations and community groups. In practice, there is little or no difference between the concepts of policy development and implementation between the public and the non-public, but the distinction allows the extensive literature on public policy, most notably from the USA (Parsons 1995), to be divided from that on policies in the private sector (Christensen 1982; e.g. House and Shull 1988; Considine 1994).

The broad notion of policy described above shares common elements with the general definition of planning adopted in Chapter 3, the most important being that both planning and policy assist in setting some conscious course of action. As outlined by Clark (2002, citing Parsons 1995):

> Like philosophy, policy wrestles with fundamental problems about people, how we live, how we find meaning and how we go about making important decisions.
>
> (Clark 2002: 5)

There is no distinct boundary between planning and policy formulation; indeed, in some cases coastal plans may be considered as spatially-oriented policies. Policies attempt to steer a course of action by deliberately affecting decision making; planning attempts to do the same. Both attempt to produce structured, deliberate and consistent decisions by first clearly stating goals, principles and objectives, then actions in order to achieve those objectives.

In practice, the similarities between policy and planning increase as the geographic coverage of each increases. At the national and international level especially, coastal management plans and policies provide guidance as to how decisions are made – generally there is discretion to allow decisions to be made at regional and/or local level. At this level of planning the difference between planning and policy can become merely semantic, and does not necessarily reflect true differences in approach.

A useful way of describing policy in coastal management is through the use the terms 'expressed' and 'implied' policy used in business management (Mukhi *et al.* 1988):

> *Expressed policies* are written or oral statements that provide decision makers with information that helps them choose among alternatives.
> *Implied polices* are not directly voiced or written. They lie within the established pattern of decisions.

The use of expressed policies in coastal management is widespread. Coastal programs, for example, may choose to specify a set of general statements of policy. Such policies may operate at a range of geographic scales, from international to local. They can have a broad range of applications, and degrees of prescriptiveness. Examples or policies developed for the Sri Lankan coastal management program (Table 4.1) demonstrate one possible range of application. A further example of expressed policies is taken from the New Zealand Coastal Policy Statement which lists the policies developed for the management of coastal hazards (Box 4.1).

The vast majority of expressed policies allow a degree of discretion in decision making. Allowing the professional staff of organisations to make decisions within the broad confines of expressed policies is one of the underlying principles of many

Table 4.1 Management techniques used in the Sri Lankan coastal management strategy

Policy	Management technique*
Erosion control	
Regulate development suitability at specific sites	Education, permit
Insure proper location in relation to the shoreline	Setback, education
Regulate amount, location and timing of sand mining	Permit, devolution
Build coast protection structures at appropriate locations	Master plan for coast erosion management
Regulate private construction of groynes, revetments	Permit
Limit construction in erosion prone areas	No-build zones
Habitat protection	
Regulate location/use of development activities relative to valued habitats	Education, permit Special area management
Regulate discharges from development which may affect habitats	Permit, education
Reduce resource use conflicts	Special area management
Coastal pollution	
Regulate effluent discharge of new development activities	Permit
Archaeological, historical, cultural and scenic sites	
Regulate development activities in relation to valued sites	Permit, education

Source: White *et al.* (1994); Coast Conservation Department (1996).

Note
* More than one management technique is normally used to implement a given policy; only primary techniques are listed.

Box 4.1

Expressed policies for coastal hazard management in New Zealand

The New Zealand Coastal Policy Statement (Department of Conservation 1994) lists a number of specific requirements of New Zealand governments. For example, for the management of the impacts of coastal hazards and potential sea-level rise the New Zealand policy contains six specific policies:

Policy 3.4.1   Local authority policy statements and plans should identify areas in the coastal environment where natural hazards exist.

Policy 3.4.2   Policy statements and plans should recognise the possibility of a rise in sea level, and should identify areas which would as a consequence be subject to erosion or inundation. Natural systems which are a natural defence to erosion and/or inundation should be identified and their integrity protected.

Policy 3.4.3   The ability of natural features such as beaches, sand dunes, mangroves, wetlands and barrier islands, to protect subdivision, use, or development should be recognised and maintained and, where appropriate, steps should be required to enhance that ability.

Policy 3.4.4   In relation to future subdivision, use and development, policy statements and plans should recognise that some natural features may migrate inland as the result of dynamic coastal processes (including sea-level rise).

Policy 3.4.5   New subdivision, use and development should be so located and designed that the need for hazard protection works is avoided.

Policy 3.4.6   Where existing subdivision, use or development is threatened by a coastal hazard, coastal protection works should be permitted only where they are the best practicable option for the future. The abandonment or relocation of existing structures should be considered among the options. Where coastal protection works are the best practicable option, they should be located and designed so as to avoid adverse environment effects to the extent practicable.

The New Zealand Coastal Policy Statement is currently under review (see Box 5.10). It will be interesting to track how the New Zealand response to the potential coastal impacts of sea-level rise evolves in response to greater scientific understanding combined with lessons learned in implementing the above policies.

organisations. Within governments discretion has been described as an 'inevitable, inescapable characteristic' (Bryner 1987: 3). One way of visualising the role of policy and discretion in decision making is shown in Figure 4.1 which highlights the role of policies constraining the range of possible decision-making choices. Figure 4.1 shows a policy acting to reduce the range of possible decisions. In this visualisation the degree of discretion narrows as the width of the gap constrained by policy reduces.

In many cases the link between expressed and implied policy is blurred with the discretionary powers of an organisation's staff intertwined with that organisation's culture or unwritten rules. The result can be a substantial grey area between expressed and implied policies. The grey area often occurs in cases where decision-making authorities are required to make individual decisions in the absence of expressed policy. Such situations can occur where formal expressions of policy have not yet occurred in newly established authorities, where decision-making powers have extended beyond the boundaries of existing policies, or where day-to-day decisions have been made with the assumption that expressed policies existed because 'that is always how things were done'.

For example, a permitting authority is developing 'policy on the run', because once a decision is made to allow a particular activity at a particular location policy has been set to allow others to undertake the same activity. However, this is not an expressed policy, unless there is a process to formally document that decision as a precedent that will be applied uniformly to all subsequent permit decisions.

There are significant advantages and disadvantages of implied policies (Table 4.2). Their major disadvantages include being hidden from public scrutiny, and hence the communication of them to stakeholders involved in decision-making processes possibly being poor. Implied policies can also lead to *ad hoc* and sometimes inconsistent decisions. This can be exacerbated if informal policy formulation is undertaken by a few individuals without consideration of the flow-on effects.

In conclusion, policy-making is one of the central components of many coastal programs around the world. The expression of formal policies can act as a guide to decision makers by helping to choose between actions. In addition, many coastal initiatives contain unwritten (implied) policies which can be a critical part of how programs operate in practice. The interaction between these different types of policy with legislation for coastal management is described in the next section.

(b) Legislation

In western-style democracies, legislation is the government of the time's response to community demands for government action or management of particular issues, areas

Table 4.2 Advantages and disadvantages of implied policy-making in coastal management

Advantages	Disadvantages
Provides flexible decision making	Open to decision-maker bias
Can be quick	Decision making may be inconsistent
Can be used to test future expressed policy	Lacks documentation and transparency
	Not discussed with the broader community

or activities. Legislation or law is defined through a parliamentary or legislative process and the outcome is often expressed as an Act or law and associated regulations. Before the assenting/passing of an Act or law considerable debate in parliament and the community generally takes place. The government and community view legislation as a long-term approach to management of issues, areas or activities irrespective of the ruling political party. Because the formulation, passing and amending of legislation consumes considerable staff and financial resources, changing the law is often subsequently avoided. Indeed, this is perhaps the greatest strength, and the greatest weakness, of legislated approaches to coastal management. Once legislation is enacted it brings with it its own inertia, sheltering it from the winds of political change. For example, since its enaction in 1972 the US Coastal Zone Management Act has survived a number of attempts to strike it down (Beatley *et al.* 2002). Yet, the ongoing problems with evaluating the effectiveness of the legislation have also proved difficult to enact, leading to recent calls for its fundamental overhaul in order to develop clear performance and accountability criteria (see Chapter 3) (US Commission on Ocean Policy 2004).

Legislation has a number of functions in coastal planning and management, especially in translating concepts, as discussed in Chapter 3, to plans and management actions. Most importantly it sets out the broad purpose for managing the coast and the guiding principles for planning and management. It enables governments to incorporate sustainable development and ecosystem management principles, including the precautionary principle and intergeneration equity, into a formal and authorised management framework, thereby establishing a basis for sustainable use of the coast while meeting international and national obligations. Also, in some countries legislation is used to spatially define the coast (see Chapter 1) for the purposes of legislative action.

Legislation can define or clarify institutional arrangements; or, if a new coastal management agency is required, it can specify how that agency will be formed, resourced and operated. If a new agency is not formed, legislation can specify the linkages and interactions of the various institutions. Kenchington (1990) suggests using existing institutions where possible and using interagency agreements to effect these institutional arrangements. Specific legislation to achieve this purpose is considered only if other coordination mechanisms have failed. Legislation can also specify the basis, scope and nature of planning and management. It can detail the steps undertaken to declare an area for specific planning attention and planning processes, including the requirements for public involvement, and performance monitoring specifications for stakeholder participation. It can include the type of plans that can be produced, such as zoning plans, and make provisions so that plans also have the force of law, greatly aiding implementation and enforcement.

An Act or law can make provisions for the basis for management; it can also facilitate the use of specific mechanisms for management such as permits, licences, enforcement, education, monitoring and evaluation; and it can specify how the Act or law will be enforced and who will enforce it. Similarly, legislation can facilitate the formulation of regulations so that provisions in the Act or Law can be implemented and that day-to-day management activities in the coast can be undertaken as highlighted in Chapter 5. Finally, legislation can specify the resourcing of planning and management activities.

4.1.2 Guidelines

The term 'guidelines' is used here to describe a group of documents which are less prescriptive and/or forceful than formal legislation, policies or regulations, but nevertheless guide the actions of decision-makers. Clearly, there are many ways to 'guide' decisions, such as using advertising campaigns. This section does not focus on these, but rather examines the informal, yet structured, approaches used by governments to the production of guidance documents.

A useful way to consider the range of ways decisions may be guided was developed by Kay *et al.* (1996b) for examining the variety of approaches available to guide the examination by governments of potential future coastal vulnerability to climate change and sea-level rise.

The concept in Figure 4.2 is a spectrum of guidance which varies according to levels of prescriptiveness, direct applicability, flexibility and extent of required local knowledge. The practical outcome from the consideration of such a spectrum is that the form of guidance could range from guidelines, through broadly structured frameworks and manuals, to methodologies (Figure 4.2).

At one end of this guidance spectrum are very broad, flexible and non-prescriptive guidelines. For example, sea-level rise vulnerability assessment guidelines could describe the range of possible assessment techniques and approaches for different biophysical, governmental, social, economic and cultural settings. Such guidelines would have to be interpreted according to need. Although the degree of flexibility is high, the level of direct applicability is low (Figure 4.2). At the other end of the guidance spectrum are highly prescriptive methodologies which aim to be directly applicable, but by their very nature are inflexible and require little local knowledge for their implementation.

Midway in the vulnerability assessment guidance spectrum are documents which allow some degree of flexibility while maintaining some direct applicability. Such documents include 'frameworks' and 'manuals'.

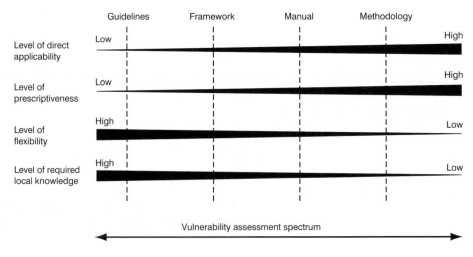

Figure 4.2 Schematic coastal vulnerability assessment guidance spectrum (from Kay *et al.* 1996b)

A good example of an effective manual is the Western Australian Coastal Planning and Management Manual (WAPC 2003). The manual clearly describes the range of approaches available to coastal managers, and discusses their strengths and weaknesses within the context of the State's coastal planning system. Another manual from Australia, with an engineering focus, is the older New South Wales Coastal Management Manual, (New South Wales Government Department of Public Works 1990) currently in the process of being updated.

The choice of guidance document types will be determined in part by the advantages and disadvantages shown in Figure 4.2, and in part by the way they are intended to fit within the broader coastal management system. In some cases the use of a manual will simply be explaining a range of techniques which may be available to implement a particular policy, legislative requirement or coastal management plan; in which case the manual is being used as an implementation tool that may supplement, or replace, the need for more detailed site-level planning. In other circumstances an education program may require additional material which explains things such as the approach of governments in its coastal management efforts.

The European Union has recently taken a 'guidance' approach to the development of a Europe-wide approach to coastal management among its member states. The outcomes of 35 'demonstration projects' from around the European coastline and six thematic studies fed into the Communication from the Commission to the Council and the European Parliament on 'Integrated Coastal Zone Management: A Strategy for Europe (European Parliament and Council 2000)' and the subsequent proposal for a European Parliament and Council Recommendation concerning the implementation of Integrated Coastal Zone Management in Europe (European Parliament and Council 2000). This recommendation was adopted by European Council and Parliament in May 2002. The adoption signalled a major impetus for coastal management in the member states of the European Union.

Within the European Union a recommendation does not carry with it statutory force; it 'invites' member states to implement the principles of good coastal zone management and recommends general steps, including the development of national strategies for ICZM (Belfiore 2001; Burbridge and Humphrey 2003). The principles of the program are shown in Box 3.8 while its operational elements are outlined in Box 4.2. An important element of the recommendation is that while there is considerable discretion in how member states choose to interpret its requirements, there are time frames specified for reporting back to the European Commission for evaluation and further action if required (Box 4.2).

An interesting dimension to the proactive use of guidelines is shown from the Philippines (Box 5.3) where guidelines have been developed within a standards framework.

4.1.3 Zoning

Zoning is one of the simplest and most commonly used tools in coastal planning and management. It is also one of the most powerful. Zoning, which is based on the concept of spatially separating and controlling incompatible uses, is a tool which can be applied in a range of situations and which can be modified to suit varying social, economic and political environments.

Box 4.2

European Union ICZM recommendation

The EU has taken a two-stage approach to the implementation of the ICZM recommendation (European Union 2002). The first stage is to undertake a national 'Stocktake'. The requirements for the Stocktake are defined in the EU recommendation including (Gubbay 2002; Atkins 2003):

- consider a broad range of sectors (including but not be limited to: fisheries and aquaculture, transport, energy, resource management, species and habitat protection, cultural heritage, employment, regional development in both rural and urban areas, tourism and recreation, industry and mining, waste management, agriculture and education);
- cover all administrative levels;
- analyse the interest, roles and concerns of citizens, NGOs and the business sector;
- identify relevant inter-regional organisations and cooperation structures;
- take stock of the applicable policy and legislative measures.

The Stocktake process has been started in some EU Member States, including the United Kingdom and Spain. The UK stocktake was the first to be completed in April 2004 (Atkins 2004a, 2004b). The UK stocktake found (Atkins 2004b):

> ... a mixed picture of how the principles of ICZM (as set out by the European Commission) are being implemented in the UK. There are examples of good practice but the current framework reflects the sectoral approach to managing coastal issues in the UK and, like many other European nations, this framework is not representative of true ICZM principles.
>
> The findings suggest that even without an integrated national framework, local ICZM initiatives have developed around the UK to address specific issues. This local commitment has been driven by the need to find a practical way of resolving conflicts in the coastal zone. However, this progress has been uncoordinated and many initiatives have been constrained by the lack of long-term resources and commitment by some stakeholders.
>
> This need for a more secure funding arrangement to support ICZM, coupled with stronger leadership at all levels (national, regional and local) is a key message of the stocktake. A second message is that more coastal stakeholders at all levels need to engage in ICZM activity.

The second stage in the new EU approach to coastal management is the development of a national strategy or, where appropriate, several strategies, to implement the principles for integrated management of the coastal zone (Gubbay 2002). These strategies should:

continued...

Box 4.2, continued

- identify the roles of the different administrative actors;
- identify the appropriate mix of instruments for implementation of the principles;
- develop or maintain national and other policies and programmes which address both marine and terrestrial areas of coastal zones together;
- identify measures to promote bottom-up initiatives and public participation in integrated management of the coastal zone and its resources;
- identify sources of durable financing;
- identify mechanisms to ensure full and coordinated implementation and application of community legislation and policies that have an impact on coastal areas;
- include adequate systems for monitoring and disseminating information to the public about their coastal zone; and
- determine how appropriate national training and education programs can support implementation of integrated management principles in the coastal zone.

Member states are also encouraged to discuss and implement existing conventions with neighbouring countries to establish mechanisms for better coordination of responses to cross-border issues. They should also work actively with the community institutions and other coastal stakeholders to facilitate progress towards a common approach to integrated coastal zone management, examining the need for a European coastal stakeholders' forum.

Member states must report to the commission on experience in implementation of this recommendation forty-five months after adoption (February 2006). The reports should be available to the public and the commission should review this recommendation within fifty-five months (December 2006) and submit a report to the European Parliament and the Council for evaluation accompanied, if appropriate, by a proposal for further action.

Given that this process of developing national ICZM strategies and stocktakes is currently underway, it is too early to determine their overall effectiveness in dealing with coastal management challenges in Europe. Nevertheless, the initiative is an important new element in the global coastal management landscape and it will be of great interest to track its development in the coming years.

Zoning grew from the 'nuisance' crisis in urban management in newly industrialised cities in Europe and North America, especially in relation to health, sanitation and transportation problems. These problems were exacerbated early in the twentieth century by the advent of the new technologies of the motor car, electricity, telephones, elevators; and the new construction methods, most notably steel-framed modular construction, which allowed high-rise buildings for the first time (Leung 1989; Campbell and Fainstein 2002). Zoning was promoted in the USA as a form of 'scientific management' for urban areas (Cullingworth and Caves 2003). The result was that zoning became one of the founding principles of land-use planning systems in Europe and the United States. For the latter country, Haar (1977) cited in Cullingworth (1993) described zoning as the workhorse of the planning movement. According to Hall *et al.* (1993):

> In Britain, as elsewhere, town planning had grown up as a local system of zoning control designed to avoid bad neighbour problems and to hold down municipal costs.
>
> (Hall *et al.* 1993: 19)

The use of zoning in land-use planning in the United States is summarised by Cullingworth (1993) as:

> the division of an area into zones within which uses are permitted as set out in the zoning ordinance. The ordinance also details the restrictions and conditions which apply in each zone.
>
> (Cullingworth 1993: 34)

Thus zoning provides a simple mechanism for urban planners to integrate complex and often competing demands and land uses onto a single plan or map; and zoning plans provide an effective tool for communicating implicit and often complicated management objectives to the community in an easily understood form.

The widespread use of zoning schemes in urban planning has spread into larger scales of regional planning, where broad-scale land-use zones can be identified. Use of zoning has broadened considerably from urban planning through its use in ecological conservation, especially in protected area management where the 'biosphere' model of core, buffer and utilisation zones is used to manage and protect biodiversity (Gubbay 1995). Zoning is also used extensively in the management of ocean space under international maritime regulations, which ensure the spatial separation of marine traffic in order to avoid collisions at sea. The use of zoning in urban planning, described above, has expanded greatly past restriction through the issuing of permits to being the primary land-use control mechanism. Zoning in many coastal management schemes now involves the three categories of 'allowed', 'permitted' and 'restricted' use.

(a) The mechanics of zoning

Zoning manages an area (land or marine) using management prescriptions which apply to spatially defined zones. Activities within a zone are managed by either specifying which activities are:

- allowed, or allowed with permission; and if an activity is not specified it is assumed not allowed unless permission is given; or
- prohibited, or allowed with permission, and if an activity is not specified it is assumed to be allowed.

It is worth noting these two approaches since they will influence how activities will be managed. In the first, and more common approach, new activities can be managed since a permit will only be issued if that activity meets management objectives. In addition, the permit may contain conditions which minimise the impacts of the new activities. Under the second approach new activities are allowed unless management can demonstrate that they are inconsistent with management objectives or have significant adverse environmental impacts. This approach is not used very often since it is costly and time consuming for managers to demonstrate the inconsistencies associated with each new activity.

Zoning as a concept can be applied at varying planning scales. Zoning plans can be formulated for broad geographical areas spanning political boundaries, or for a small area of only a few hundred square metres. The types of zones, the management objectives within the zone and the types of activities managed within these zones will, however, vary with scale. Zones such as 'tourism', 'agricultural' and 'industrial' are effective for broad management of a region or district, but are ineffective in managing conflicting recreational uses along a narrow beach.

There are a number of discrete steps in developing a zoning scheme in coastal management. The application of these steps depends on the existence of legislation to give effect to the zoning plan. In some cases, such as that governing the management of the Great Barrier Reef Marine Park (see Box 4.3), the legislation specifies the types of zones and the purposes for which they can be used. Such legislative prescriptions are more common for the land component of coastal areas, enforced through land-use planning legislation. Where land-use zoning legislation applies there may be very detailed zoning requirements in place which prescribe details of permitted and/or excluded activities.

The scale of management and the objectives for each zone underpin the formulation of a zoning plan. Again these objectives may be pre-determined by legislation, policy, or policies. In cases where the objectives are not predetermined, there is scope for clearly stating why a particular zone is being developed (see Chapter 3 for details on objective setting in coastal management). Where management is at the broad regional scale, zones will be defined to manage a range of uses, will have broad management objectives, and will cover broad areas. As the scale decreases, the range of uses is likely to decrease, and management objectives usually become more specific and operate at a fine scale in order to render the zoning provisions more readily understood by the community.

Existing environmental, social and economic information combined with community input on the current and future use of the area forms the base information for the establishment of a zoning plan. The complexity of the information required varies according to the intensity of use of an area and complexity of the zoning plan.

Finally, zones generally define the appropriate uses within a given area. Where possible, issues, activities or uses which can be differentiated into separate spatial areas should be allocated to appropriate zones. For example, if the risk of an accident for water skiers and windsurfers is an issue for a particular area, motorised and non-motorised water sports zones may be an option for managing the area. The non-motorised vessel area may also protect areas of higher conservation value because the damage caused by propellers is reduced.

When zoning is used to manage an area, the zoning scheme should be as simple as possible and the number of zones should be kept to a minimum. As more complex zoning with more numerous zones is introduced, the difficulty in implementing the plan increases and the community's understanding and support for the plan may decrease. An example of a zoning scheme applied to the management of the Australian Great Barrier Reef is shown in Box 4.3.

The Great Barrier Reef approach to zoning integrates the Biosphere Model with the zoning of protected areas. In this model a core zone is used to give a high degree of protection to a specific area. The core zone is then surrounded by a buffer zone which allows limited use of the area while providing some protection This buffer zone is surrounded by a utilisation zone where there is limited or no protection (Figure 4.3).

The Biosphere Model is one of the simplest zoning plans and because of its simplicity is used by many agencies for protected area management (e.g. Indonesia where it forms the basis of all protected area zoning plans including MPAs). The definitions of core zones (which consist of a network of research and reference sites), buffer zones (which manage human impacts for sustainable use and ecological function), and a utilisation zone to manage conflicting uses by spatial separation, are simple. These broad definitions enable planners to either use the broad objectives without modification, to redefine the objectives in light of local needs, and to use these zones as a basis for a more detailed zoning scheme. The details of developing a zoning scheme, oriented towards the management of marine protected areas, are provided by Kenchington (1990) and Gubbay (1995).

An area is 'zoned' using criteria which the planning team have developed in consultation with the community. The criteria are based on a range of ecological, social and economic values including: conservation and the presence of threatened or endangered species; access, recreation, traditional use and proximity to urban centres; and existing and potential commercial and industry development in such areas as tourism, fishing, mining, port development, or mariculture/aquaculture.

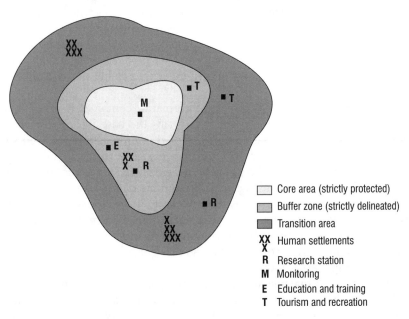

Core area (strictly protected)
Buffer zone (strictly delineated)
Transition area
XX
X   Human settlements
R   Research station
M   Monitoring
E   Education and training
T   Tourism and recreation

Figure 4.3 The 'biosphere' model of zoning marine protected areas (Gubbay 1995)

Zoning boundaries should be clear and consistent. Setting the boundaries of the zones must also be considered, especially where zones extend into the marine environment. Zone boundaries can be precisely defined using geodetic reference points, but this may be of limited use to users who do not have the equipment or skills to locate these points. Geophysical features may be used, such as depth, high/low water mark, streets, depth/elevation, vegetation line, etc. The disadvantage of many of these features is that they are subject to change. Often the two are combined, with geophysical features the preferred method and reference points used when features are not available. This approach is used in the establishment of the zones on the Great Barrier Reef (Box 4.3 and Box 4.4).

Box 4.3

The broadscale zoning scheme of the Australian Great Barrier Reef

Zoning at varying scales is used in managing the Great Barrier Reef Marine Park (GBRMP). The GBRMP is large (348,700 km²) and to undertake operational management on a park-wide scale is difficult. To overcome the problem of size, the park is divided into Sections, which have the capacity to manage or regulate impacts, and to buffer the more highly protected areas from impacts originating outside the Marine Park (Kenchington 1990). Within a Section of the park, a zoning system is used (see table below).

Original and modified zones within the Cairns section of the Great Barrier Reef Marine Park

Original zones	Date defined	Current zones	Date defined
General use 'A'	1981	General use	1992
General use 'B'	1981	Habitat protection	1992
		Conservation park	1992
Marine national park 'A'	1981	Buffer	1983
Marine national park buffer	1982	Scientific research	2003
Marine national park 'B'	1981	National park	1992
Scientific	1981	Preservation	1992
Preservation	1981	Commonwealth island	2003

The initial zoning scheme was based primarily on extractive uses and minimising these uses while providing for reasonable use (Kenchington 1992) (see table above). As issues, uses and community expectations and perceptions of the reef's management have changed, zoning has changed accordingly. The table shows how the names of the current zones have less focus on use, but greater emphasis on using other zones for habitat and resource protection to ensure general use zones are sustainable. In turn this reflects the evolution of management objectives. Since 1981, the purposes of zones have changed. For example, a scientific research zone was originally restricted to non-extractive research activities; in the early 1990s it was rarely used and in 2003 the new Great Barrier Reef Marine Park

continued...

Box 4.3, continued

Zoning Plan (Great Barrier Reef Marine Park Authority 2003c) allowed limited impact research (extractive) in addition to limited impact research (non extractive) as well as low impact recreation that is non-extractive.

Zoning which manages uses over a broad area may not suitable for managing activities at a specific site. For example, tourism is allowed by permit in a number of zones, and the zoning plan does not specify the nature and intensity of tourism throughout the park or within a specific zone. As a consequence, zoning alone cannot manage the tourism at a specific site; it has the potential to allow nearly every site to be intensively developed for tourism in an *ad hoc* manner. Permits issued to tourist operators give some degree of flexibility in managing the impacts of these activities, but do not provide much scope for managing at the site level. Plans of management can be used which provide more detailed management of the use of a particular area of the park or the conservation of species or ecological communities within the park (GBRMPA 2003a).

How the broad-scale zoning provisions outlined above relate to the zoning plan for Green Island in the Great Barrier Reef region is discussed in Box 4.4.

Box 4.4

Reef activities zoning plan of Green Island, Great Barrier Reef (Zigterman and De Campo 1993)

Within the Cairns/Cooktown Management Area of the Great Barrier Reef Marine Park, a management plan for Green Island and Reef, a popular tourist destination, is used to intensively manage tourism at the site (Mau 2003). The site is zoned national park and the overall purpose of this zone is to provide for the protection of areas in a natural state while allowing for public appreciation of natural features which are relatively undisturbed; and to provide for traditional fishing, hunting and gathering (GBRMPA Zoning Plan 2003a). Within this zone tourism is an acceptable use, but the zoning system does not make any provisions for determining the level, form and intensity of tourism.

In the site plan a number of strategies are used to manage tourism: restriction of the amount and types of use through limiting the number of day visitors to the site to a daily maximum of 2,240 up from 2,015 in the original Plan (Zigterman and De Campo 1993); limiting the number of permitted operators at the site, and a form of tourism facility zoning; reduction of the impacts of uses which are allowed; hardening of the site; and monitoring. The management plan for the site includes the use of precincts (zones) to separate conflicting uses. Three precincts are used: conservation, recreation and infrastructure (see figure below). These precincts complement or reflect the purpose and use of the national park zoning (see below). Implementation of the Green Island Plan commenced in 1993 and the use of zoning appears to have addressed many of the issues associated with conflicting use (Mau 2003).

continued...

Box 4.4, continued

Green Island zoning plan

Precincts	Purpose
Infrastructure	To provide for the development of permanently fixed structures associated with access and use of the site for tourism and recreation.
Recreation	To provide for intensive recreational use and enjoyment of the reef, with structures limited to re-locatable facilities
Conservation	To maintain a large portion of the reef free from built facilities

	Infrastructure precinct
A	Swimming enclosure
B	Overnight commercial moorings
C	Coral viewing vessels
C1	Coral viewing vessels/diving and snorkeling
D	Diving and snorkeling
E	Seaplane main landing/take-off
G1 and G	Beach hire equipment limits

Where possible the pattern of zones should form a series of transitions in terms of restrictions or access (e.g. avoid placing a conservation zone beside a heavy industry zone; if possible try and separate the two with a buffer zone or recreation/commercial).

(b) Linking zoning with other coastal planning and management tools

Once zones have been established through a zoning plan, a number of related forms of management can be used in conjunction with the zones (Table 4.3). These other forms of management can overlay the zoning plan so that management can be fine-tuned for a particular area or resource.

The effectiveness of a zoning plan will ultimately rely on the community's acceptance of this plan and the government's commitment to provide the resources to implement it. Studies have shown that where the public has been actively and meaningfully involved

Table 4.3 Coastal management tools linked with zoning

Tool	Purpose
Time partitioning	Restricting access to an area or resource to specific times of the year
Facility/infrastructure restrictions	Specifying the type of gear which is used or what type of infrastructure can be constructed
Permit/licence quotas	Restricting the number of users accessing an area, using a resource or being allowed to undertake an activity
Production quotas	Restricting the amount of the resource which can be harvested
Licence – quota combinations	Limiting the number of users accessing the area, as well as controlling harvesting levels

in the planning process there is a greater acceptance of the plan, its regulations and their implementation (Savina and White 1986; Stone 1988; Ehler and Basta 1993; Kelleher 1993). Techniques for involving the community in planning and management are discussed in Chapter 5.

A number of activities are undertaken to implement a zoning plan, with communication, education, EIA and enforcement playing major roles. These activities are discussed in this section. The implementation of zoning plans is similar to other plans and is discussed in Chapter 5.

4.1.4 Regulation and enforcement

Regulation and enforcement are often perceived by the community as simple and easy options for achieving compliance with management initiatives. The basis for this simplistic view is that the majority of the community by its very nature tends to comply with the law and assumes that the rest of the community is the same. Clearly there is a sector of the community which, for a number of reasons, including a lack of understanding of the purposes of management initiatives, blatant disagreement with them, or economic motives, does not comply. For this sector of the community, regulation supported by enforcement is used along with other mechanisms such as awareness and monitoring.

(a) Regulations, permits and licences

Acts of parliament provide the broad legislative basis for managing particular resources and activities, but often do not provide detailed prescriptions which can be used to implement an Act's provisions. Regulations, permits and licences commonly provide implementation mechanisms by specifying what actions are acceptable under the Act, and the penalties for breaching it. Because regulations, permits and licences are easier to amend than an Act, they provide a flexible mechanism for managing the coast. However, as will be shown below, regulations, permits and licences only remain effective when sufficient resources are provided to enforce them and, in the long term, when implemented in combination with education and communication programs.

   Permits and licences are written approvals from government to conduct specified activities in specified areas. Commonly permits are used in conjunction with zoning plans as a means of enacting a zone's specifications and/or restrictions. The processes and criteria for issuing permits are generally controlled either by policy directions, regulations or specified in legislation.

   Permits can be used in a range of activities to assist in day-to-day coastal management activities, as shown by their use for the management of the Great Barrier Reef Marine Park (Alder 1993) (Table 4.4).

(b) Enforcement

Enforcement is a management tool used to effect compliance with Acts, regulations, permits, licences, policies or plans with a legislative basis. Enforcement is a management activity that is highly visible, and generally outcomes are achieved in a relatively short time when compared to other management mechanisms such as education programs. As a consequence, the public and politicians often perceive enforcement as 'the answer' to compliance. Enforcement is one of many mechanisms available to managers to encourage compliance with legislated management provisions, but it is generally temporary and short term. Research has shown that as long as the 'big stick' of enforcement is applied by an enforcement agency having a high profile in the community and actively patrolling the area, there will be compliance. Once the big stick is removed, however, many members of the community will revert back to their undesirable activities. But research has also found that when enforcement is used in combination with other management tools, long-term compliance can be realised.

   The various regulations, licences, permits and legislative tools used in coastal management are sometimes not worth the paper they are written on because they are not enforced. Of course, there can be a myriad of reasons for the non-enforcement: a lack of resources, not just financial but also staff; staff may lack the expertise needed to undertake various enforcement activities, or it may be culturally difficult to act as an enforcer; there may be a lack of political support to prosecute offenders and previous efforts to prosecute may have been unsuccessful, resulting in a reluctance to undertake further enforcement activities. The most common reason is simply a poor understanding of what it actually takes to effectively enforce the various 'rules' imposed by governments.

   Enforcement programs can also be very expensive and time consuming, and can be stressful for the enforcers. The constant reinforcement of an essentially negative message

Table 4.4 Permitted activities and examples on the Great Barrier Reef Marine Park (Alder 1993)

Permitted activity	Example
Exception to normal activities	Harbour works
Variable by their nature and need to be addressed on a case-by-case basis	Tourist programs
Subject to potential conflict between allowed uses	Mariculture ventures
New activities with unknown impacts – once their impacts are understood they can be classified as either allowed or prohibited	Establishing structures on the reef

---

Box 4.5

Enforcement of a marine reserve in the Philippines

In 1980 a marine reserve was established around Apo Island, Philippines. The marine reserve was established to assist in enhancing and maintaining fisheries resources for the local community of about 700 persons. The initial management of the reserve, however, was constrained by outside fishers who entered the area and not only over-harvested fish resources but also used destructive fishing methods such as illegal nets and explosives. Then, in 1985, an intensive community-based conservation program started on Apo Island under the guidance of Silliman University (a Negros Island based institution with a history of community outreach programs). This two-year program formally established a fish sanctuary on one side of the island and assisted the community to develop a management committee for full-time surveillance and protection of the sanctuary and reserve surrounding the island. This community-based enforcement, combined with an extensive education program and other initiatives in livelihood, has resulted in a significant increase in fish catch to island residents over the last twenty years. Today, the Apo Island coral reef and community groups are the focus of numerous educational field trips from communities with similar interests in other parts of the Philippines. It continues to be a success story twenty years later. The community has also developed a successful tourism industry, which provides benefits of approximately US$500 per hectare of reef per year to the local economy (Maypa 2003).

Today there are over 100 community-based coastal resource management projects (targeting fisheries, mangroves and coral reef resources) in the Philippines (Pomeroy *et al.* 1997). These are now mostly co-management projects whereby the community is working together with the local government unit (municipality or city) to ensure implementation (White 2004). Although Apo Island started as a community based Marine Protected Area, it is now being managed through collaboration of the local government and the community, as well as the national government since it was declared a national protected area under the 'National Integrated Protected Areas' Act.

– 'you are not allowed to do that' – by enforcement officers can erode their morale and also lead to long-term inefficiencies in program delivery. Hence, the trend in the effective compliance of coastal programs is to integrate enforcement with communication strategies aimed at pointing out those who breach the rules, what the consequences of their actions are, and more importantly why the rules were established to begin with. Communication and enforcement are now seen to go hand-in-hand acting to support each other.

Experience with enforcement programs in marine parks has shown that most people in the community want to comply with regulations, permits and licences (Alder *et al.* 1994). For this sector compliance is quickly gained once they are aware of the rules. The various regulations used in the management of the Great Barrier Reef Marine Park require an active enforcement program. The effectiveness of this program is described in Box 4.6.

Box 4.6

Enforcement program of the Great Barrier Reef Marine Park

Between 1985 and 1991 the Cairns Section enforcement program consisted of air surveillance and vessel patrols. Air surveillance was designed on an annual basis to survey particular areas of the Section at certain frequencies based on a stratified random sampling scheme. Vessel patrols were also designed to cover specific areas at a certain frequency, but weather and staffing constraints limited the statistical basis for the patrols. In either program, breeches of the Great Barrier Reef Marine Park Act, regulations or section zoning plan were recorded; these records were then used to examine changes over the six years of the study.

Total infringements declined steadily until 1988–9 and then remained constant (see figure below). This pattern was also evident for infringements related to zoning compliance, which declined from seventy-four in the 1985–6 financial year to eighteen in 1988–9 and remained at that level. Other types of infringements, however, were variable over the same time. The total number of infringements detected, and zoning plan infringements, were not significantly correlated ($p > 0.05$) to the amount of staff time or funds spent annually on enforcement (Alder 1994).

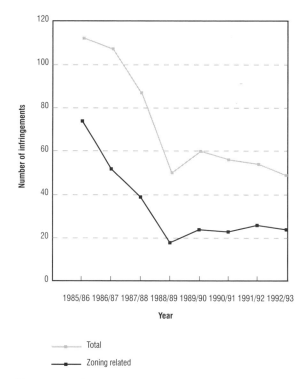

Zoning related and total infringements detected in the Cairns section, Great Barrier Reef Marine Park (based on Alder 1996)

continued...

Box 4.6, continued

In the corresponding time frame an extensive awareness and communication program was implemented. The program focused on raising user awareness of the park and that there were areas (zones) where certain activities were not allowed. To simplify users' understanding of zoning, all visual material for each zone was colour-coded, e.g. green was a national park zone which meant 'look but don't take'; blue was general use zone which allowed fishing; and pink was preservation – 'no-go'. Offices within the management agencies would also refer to the colour system when they explained the zoning system. A subsequent survey of the effectiveness of awareness and communication programs indicated that the zoning information was disseminated throughout the community and that there was support for management of the park. It would appear that awareness programs contributed to reducing zoning infringements (Alder 1994; 1996) (see Box 4.8).

The case studies shown in Box 4.5 and Box 4.6 highlight the need to include enforcement as a component of any coastal management planning program. Enforcement programs can be undertaken in a number of ways. Staff within an organisation can be designated as inspectors/officers and therefore have the power to enforce the provisions of an Act, or a plan if it has a legislative basis. Although one organisation may have responsibility for management, it may delegate enforcement activities to other organisations, as is the case in the Great Barrier Reef (Box 4.6). If the expertise does not exist within an organisation or affiliated institutions, the use of private security officers or subcontracting out the program is an option. Which option to use depends on a number of factors such as funding, expertise, support from politicians and support from the community.

In summary, whatever option is used to enforce permits, licences policies or plans, the long-term effectiveness of enforcement programs is enhanced when they are designed and integrated into other programs. This is especially so when enforcement is integrated with communication and education programs.

4.2 Social coastal management and planning techniques

The social dimension of coastal planning and management is often dealt with as an afterthought. Technical and scientific aspects can be emphasised, sometimes because it is easy to hide behind their 'objectivity'. The emotions, spiritual links and community values, aspects which are less predictable/objective, are easier to avoid or merely give them cursory consideration. As emphasised throughout this book, managing the coast is inextricably linked with managing society's use of the coast and therefore the social aspects must be an integral part of any management or planning program.

4.2.1 Customary (traditional) practices

Traditional knowledge is being lost very rapidly as its possessors die. Recording it is thus a truly urgent matter. Allowing it to vanish amounts to throwing away centuries of priceless practical experience. To record it with care and in the interest of its possessors – not just for the economic benefit of industrialised societies – is essential (Johannes 1989: 9).

This section outlines traditional resource management practices of non-western cultures and discusses how they relate to the planning and management of the coast. Customary resource management practices as they relate to the coast drawing are introduced first, drawing on general literature in the area (McCay and Acheson 1987; Johannes 1989) as well as some excellent texts written specifically about the coast (Ruddle and Johannes 1983; Johannes 1984; Smyth 1991, 1993). How these factors relate to the development of formal coastal programs is then discussed.

Cultural factors play a central, if not the central, role in the successful management of coastal areas. As described in Chapter 3, the cultural norms of a coastal nation will shape the boundaries of a coastal program, often long before notions of the exact details of program design have been considered. Much of the content of this book focuses on the development and implementation of coastal planning and management systems which are essentially founded on the cultural norms of western developed countries. These western norms include the basic rules of data collection and analysis, and consideration of alternatives within essentially Christian values of the relationship of humans with their environment.

However, much of the global coastline is inhabited by people of cultural groups having their own cultural values and religious beliefs. Often these do not conform to western Christian values. The result can be that these non-western views of the relation-ship between people and the coastal environment can be viewed as somehow diverging from the western 'norm'. Of course, this view is misleading – all cultural settings require unique management and planning solutions, including western cultures.

Consideration of cultural factors in coastal management is driven to a large extent by the re-vitalisation of indigenous cultures since the reduction of colonial powers over the last 100 years or so. The gradual withdrawal of European and North American influence from Asia, Africa, South America and the Pacific has seen a re-emergence and formalisation within government systems of indigenous cultures. This is coupled with attempts to reconcile colonial and indigenous cultures in the 'new world' of North America, Australasia, southern Africa and South America.

Like the other tools described in this chapter, using traditional knowledge and practices to assist in coastal management is a specialised activity. As such, relevant experts, such as sociologists and anthropologists trained in culturally appropriate com-munication techniques, should ideally be used. The authors have both witnessed attempts to elicit traditional knowledge in clumsy, inappropriate ways. This can often lead to those engaged in traditional practices to tell outside researchers what they think the researchers want to hear. Sometimes, locals can be mischievous, deliberately misleading outsiders who do not go about things in the right way, or can refuse to grant access or interviews to subsequent researchers.

*(a) Types of traditional knowledge and practice in coastal
management*

Traditional knowledge and practice in coastal management can be broadly divided
into knowledge of the biophysical characteristics of the coast, and of the various
management practices developed to manage the resource. The former focuses on
traditional understanding of elements of the coastal environment of direct use to local
populations, including an understanding of local oceanographic factors (tides, wave
refraction patterns for navigation) and to help predict the movement of fishery resources;
and knowledge of biological resources, most commonly linked in the coastal
environment with the exploitation of fish, crustaceans and other marine fauna. An
understanding of the schooling habits of a particular species of fish, for example, may
be used to design more efficient ways of catching those fish with available technology.
The use of so-called 'traditional ecological knowledge' has been documented in hunter-
gather cultures from the Inuit of northern Canada to Australian Aborigines and Solomon
Island fishers (Smyth 1991; Johannes *et al.* 2000).

Interwoven with traditional knowledge of the biophysical factors in the exploitation
of coastal resources are customary rules and decision-making hierarchies. The social
structure of traditional groups, such as extended families and tribal groups, determines
to a large extent how traditional knowledge of the biophysical environment is applied.
For example, Cornforth (1992) demonstrated the importance of customary decision
making in Western Samoa to day-to-day coastal management. In Western Samoa, and
some other Pacific nations, villages 'hold tenure' over coastal lands and waters, including
lagoons and nearshore reefs. The traditional basis of this is that villages communally
gain access to all the potential resources on an island, from hilltops to the ocean
(Crocombe 1995). Indeed, traditional customs include the use of management tools
described elsewhere in this chapter, including zoning, quotas on fish catches, development of regulations and policy (rules) and enforcement mechanisms (punishment and
shaming). An example of such approaches from Indonesia is given in Box 4.7. The use
and application of these techniques in the Pacific is also well documented (for example
Zann 1984).

The third important factor in traditional coastal management is the role of religious
or spiritual beliefs. In many cases these beliefs are intimately linked with cultural systems
and decision making, so that for all practical purposes they are one and the same.

Examples of traditional cultural values being followed, but with assistance on
introduced technologies, are fairly common. Again with reference to the Pacific, religious
ceremonies or visits from high-ranking members of neighbouring families may require
the presentation of 'sacred' foods, such as a turtle or prized reef-fish. The importance
of such occasions can outweigh day-to-day resource management considerations to
the extent that dynamite, poisons or other destructive actions may be used in order to
satisfy the cultural protocols.

Spiritual beliefs may also extend to restrictions on the taking of certain species of
marine life, such as where they may be within the 'totem' of a family group; while
other species may have special significance to particular age groups or genders. In the
Gilbert Islands of the Pacific, for example, no clan (extended family) member is allowed
to eat its totem; thus 'porpoise callers' cannot eat any crustacean, eel, octopus or
scorpion fish (Grimble 1972, cited in Zann 1984).

Box 4.7

Sawen traditional resource management in Indonesia

Sawen is a traditional resource management institution in west Lombok (Indonesia) that was lost in the mid-1960s under the rule of Soeharto and has been recently revitalised with Indonesia's recent reforms towards local autonomy (Satria 2004, in press). Originally sawen integrated the management of forests, the sea and farmland using cognitive aspects (local knowledge and resource management), regulatory aspects (codes of conduct) and normative aspects (world views and belief systems). A marine authority (*mangku laut*) identified the areas for closure and the timing of the closures and openings, as well as other practices such as the felling of trees. The authority prohibited the felling of trees onto the beach believing that trees had supernatural powers, but in fact the restriction prevented beach erosion. Members of the authority were determined through family lineage. *Lang-lang* (coast guards) enforced the rules and there was a strong moral obligation within the community to follow the sawen.

From the mid-1960s several factors contributed to the loss of sawen: the influence of revived Islam (implementing pure rituals) which did not support sawen practised by traditional Islam, which was influenced by local cultural norms; and the centralistic government of the 1970s and 1980s replaced traditional practices with fisheries laws and regulations. The centralistic regulations were difficult to enforce and led to destructive fishing practices, conflicts among fishers, loss of property rights and a loss of marine cultural identity.

In 1998 the reform era empowered local residents to manage local resources. In 2002, Kayangan, a coastal community in west Lombok, revitalised sawen to assist the community in addressing issues of over-exploitation, access rights and a lack of enforcement of fishing regulations in their offshore waters. The local government authority supported this initiative. A marine authority (*mangku laut*) and *lang-lang* were formed with several fishing rules established along with environmental rules banning such activities as sand and coral mining.

While some of the characteristics of the sawen have returned other aspects have not. Fines have replaced the moral and the spiritual aspects, which motivated local residents, and the understanding (appreciation) of the integration of the forests, farmlands and sea has yet to emerge. Nevertheless the restoration of sawen has benefited the area by:

- returning marine cultural identity;
- protecting small scale fishers;
- providing insights (i.e. local knowledge and wisdom) for implementation of local fisheries management;
- creating legitimate institution of community-based fisheries management.

(b) Balancing traditional and western approaches to coastal management

The prevailing view of the use of traditional approaches to coastal management is that it should be viewed in the same analytical way as any other approach (Johannes 1989), a view that has evolved from opposing positions on the efficiency of customary practice. Some view customary approaches as being the most efficient and equitable methods of exploiting natural resources, being honed over hundreds, and sometimes, thousands of years. Others point to the view that such practices were only sustainable due to low population densities in the past, and are now inefficient and unsustainable. Both views point to examples drawn from around the world. However, these views are used here to describe two ends of a spectrum (which has considerable 'grey' areas) which balances traditional and western approaches, as summarised by Johannes (1989: 7).

The truth lies between these extremes; wise and unwise practices coexist in many, if not most, cultures. The existence of the latter practices does not diminish the importance of the former.

Achieving a balanced view between the use of traditional and 'outside' approaches is one of the biggest challenges to effective coastal management in many nations today, especially in light of recent decisions to recognise indigenous rights over resources in coastal areas. Tensions between traditional and introduced management techniques may reflect larger tensions related to colonial influences and/or long-standing cultural differences. Nevertheless, the potential for harnessing traditional knowledge and integrating this with western approaches is enormous. Again, Johannes (1989) states with reference to biological information:

> ... the potential for the application of traditional environmental knowledge ... is quite simply, vast. Such information must not only be collected and verified. It must be balanced with more technical forms of biological research – population dynamics, pollution genetics ..., before it can be put to use.
>
> (Johannes 1989: 7)

(c) Integrating traditional knowledge, practice and beliefs into coastal management programs

How then, can traditional knowledge, practice and beliefs be integrated into some form of structured coastal management program? As has been alluded to above, the answer will depend on the scale and intensity of coastal management problems and the respective opinions and power of traditional groups and formal government organisations. The interplay of these factors can lead to a range of program types. For example, where coastal problems are not severe, and there is joint desire by governments and traditional groups to retain traditional customary management, there may a decision be taken to develop a 'minimum intervention' strategy. Thus, the coastal program simply formalises customary coastal management practice.

In cases where coastal resource degradation is significant, there is often the requirement for government intervention to employ western techniques to assist and/or overarch traditional approaches. In many cases the use of outside techniques is required because of the accelerated damage to coastal resources through the integration of western technologies with traditional practices. For example, the use of outboard motors on fishing boats has extended their range and speed, while using nylon fishing lines, nets and imported hooks has increased the fishers' efficiency, leading to overfishing in a number of areas.

The degree of traditional and government integration will depend to a large extent on the degree of local decision making and empowerment agreed to by those within the central and the traditional systems of governance (see Chapter 3). For example, governments may wish to formally recognise major parts of customary practice and management through the development of community management programs. A way to work out the relative use of western and traditional management approaches is through consideration of them in coastal management planning. Through the use of the participative management planning process (described in Chapter 5) the customary importance of an area to its stakeholders can be discussed, and the various roles, responsibilities and management actions required agreed upon. The result can be the clarification of the use of customary knowledge, practices and spiritual values.

4.2.2 Collaborative and community-based management

> ... sustainable development requires a long-term perspective and broad-based participation in policy formulation, decision making and implementation at all levels.
>
> (UN 2002: 2)

Collaborative and community-based management are powerful tools which have the potential to help address coastal problems at the local level. Both are capable of effecting socioeconomic changes, modifying people's activities at the source of the problem in a way which can ultimately help to meet management objectives. Poverty, for example, is often the reason for environmentally inappropriate fishing practices in many coastal areas. Managers will therefore often focus on improving the people's income, in doing so bringing about a shift from inappropriate to appropriate methods.

Collaborative and community-based management can also assist in integrating environmental and resource management activities into people's everyday lives: where a community makes some resource management decisions that affect their activities, management becomes a part of their lives. Furthermore, this type of management contributes to the socioeconomic development of the community. As mentioned above, problems are not just environmental, and all aspects of the community context must therefore be addressed. Partnerships between people and nature can be strengthened by actively involving the community in management. A sense of stewardship and responsibility for managing resources is often an outcome of collaborative and community-based management (Drijver and Sajise 1993). Various governments are aware of the benefits of collaborative and community-based management; the challenge for managers is to facilitate these forms of management. The next section describes collaborative and community-based management, and their role in planning and managing of coastal areas.

(a) Background to the development of collaborative and community-based management

Collaborative and community-based management in marine and coastal areas evolved from a convergence of several advances in protected area management, rural development and fisheries development during the 1980s. The 1980 World Conservation Strategy and 1982 Bali World Congress on National Parks emphasised the linking of protected area management with local area economic activity (Wells and Brandon 1992). This concept was further developed in the late 1980s to link conservation with sustainable development, and led to the establishment of Integrated Conservation and Development Plans (ICDP). These plans focused on balancing the conservation needs of an area with the socioeconomic development of the community which is dependent on the area. The ICDP approach has been developed in agricultural and forestry projects, which have advanced community involvement in the management of land-based protected areas. Community involvement in managing marine and coastal areas has, however, lagged behind land areas due to the issue of managing shared resources in multiple use areas.

The role of the community in coastal management is wide ranging and depends on a number of factors such a geographic scale, issues to be addressed, governance context, community motivation and capacity, and policy processes (Hale 1996). The community has several potentially important roles which contribute to planning and managing in coastal areas.

PARTICIPATORY COASTAL RESOURCE ASSESSMENT (PCRA)

Coastal dwellers and users are knowledgeable about local resources and can provide some of the biophysical information needed to make appropriate resource allocation decisions. Similarly, users can provide socioeconomic information more efficiently and effectively than most agencies. Through this PCRA maps and environmental profiles can be produced. Management costs (time, staff and funds) can be substantially reduced as a result.

PARTICIPATION

Stakeholders within a collaborative or community-based management program are generally more accessible if communities are organised. This provides more opportunities for managers and stakeholders to discuss key issues and to interact with each other. It also ensures prompt feedback from both groups which leads to more efficient resolution of issues and faster integration of stakeholders in the planning and management of an area.

DECISION MAKING

Stakeholders bring ideas, judgements and perspectives which can lead to substantive results and a final product of high quality (Baines 1985). This is particularly important since stakeholders are usually the groups that bear the majority of impacts related to access and resource use within an area. They are the users who generally have further restrictions enforced on their use of the area's resources and must bear the financial and social consequences. The design and implementation of programs and management prescriptions are more readily supported by the stakeholders and the general community when they play a major role in decision making than in the absence of participation.

INITIATING ACTION

Stakeholders can readily identify needed management actions. This provides a better incentive to suggest, initiate, and implement or support the needed actions. Again, this makes efficient use of limited resources.

PROGRAM EVALUATION

As discussed in Chapter 3 and at the end of this chapter, stakeholders have a vital role to play in the formulation and establishment of evaluation criteria for management, and to be active participants in implementation of program evaluation studies. Stakeholders can provide valuable insights and lessons about the design and implementation of a management program. This information is otherwise likely to remain unknown (Wells and Brandon 1992).

Managing agencies are also aware of the role the community has in planning and managing the coast, and many are shifting towards greater community involvement. This shift is increasingly being linked to broader trends in resource management towards a greater awareness of the relative roles of the community and lobby (or special interest) groups (Smith *et al.* 1997). Collaborative and community-based management are two approaches available to managers to increase the level of community and interest group representation in decision making, and are described below.

(b) Making the choice: collaborative or community-based management?

Collaborative and community-based management are the two major forms of effective community participation in coastal management programs. Which to pursue depends on the factors which affect the community's role, as discussed above. As an example, collaborative management is better suited to Sri Lanka's form of government and social structure, while in the Philippines community-based management is more of a possibility since local authorities have jurisdiction in coastal waters (White and Samarakoon 1994). The differences between the two forms of management are discussed below.

Collaborative management, as the name implies, involves all stakeholders in the management of resources. In this form of management the aim is to achieve mutual agreement among the majority of stakeholders on the available options. White and Samarakoon (1994) note that collaborative management has a number of common elements: all stakeholders have a say in the management of resources; sharing of management responsibility varies according to specific conditions but government assumes responsibility for overall policy and coordination; and socioeconomic and cultural objectives are an integral part of management. Collaborative management is well developed in fisheries management (Jentoft 1989; Lim *et al.* 1995) and a set of common characteristics is emerging (Table 4.5).

Table 4.5 Characteristics of collaborative and community-based management

Characteristics	Collaborative management	Community-based management
Initiative	Decentral	Local
Organisation	Formal	Informal
Leadership	Participant	Mutual adjustment
Control	Decentral	Decentral
Autonomy	Some	Yes
Participant	Yes	Yes

Source: Based on Jentoft (1989).

Community-based management uses a holistic approach to management by incorporating environmental, socioeconomic and cultural considerations in decision making by stakeholders. It is based on the concept of people empowered with responsibility to manage their resources. That is, the community together with government, business and other interested parties share an interest in co-managing resources with some decision making devolved to the community. The characteristics of community-based management are listed in Table 4.5.

In both approaches, consensual planning as discussed in Chapters 3 and 5 is the ideal process to formulate a plan of management. Community-based management, however, is rarely achieved since governments are reluctant to devolve power. Communities are often viewed as unqualified or unskilled to take on responsibility for managing, or communities are reluctant to take responsibility for decision making. Nevertheless, community-based management represents a set of ideals that many communities and their managers might usefully adopt.

Collaborative and community-based management represent the top end of Arnstein's ladder of citizen participation (Figure 3.6). Collaborative management itself is not at the top of the ladder because it retains an element of government decision making. In a well developed community-based management program local decision making is undertaken by community representatives, as shown in Figure 4.4 from the Seychelles. This form of management represents the top of the ladder. Examples of community-based management from the Philippines (White *et al.* 2002) and the Caribbean update (Smith and Homer 1994) demonstrate the effectiveness of this form of management in meeting management objectives. In the Philippines many islands and their surrounding reefs are planned and managed by the local community with the assistance of non-government organisations. Because the community, especially the fishers, have determined the management regimes, there has been a wider acceptance and compliance resulting in improved fisheries resources. In many similar cases community-based management is intimately linked with government and traditional cultural groups joining to develop culturally appropriate coastal management systems.

Collaborative and community-based management are not just a developing country phenomenon; they are also being developed in other countries such as Australia, Japan, Norway and the United States. Collaborative management has developed more widely than community-based, with a number of partnerships being established with resource management agencies. Collaborative arrangements can be based on either a sector or a geographic basis.

Either of these management approaches provides a framework for governments to

Figure 4.4 Marine management workshop participants, Seychelles

work with indigenous cultures in the joint management of coastal resources. A good example is the Great Barrier Reef Marine Park Authority, where aboriginal communities and the Queensland Fish Management Authority have agreed to ban gill-netting in the southern section of the park to address the problem of declining dugong populations (Anon 1997). Community-based management is effective in involving urban and urban-fringe residents in on-the-ground management activities. Coastcare is a joint Australian federal and state funded coastal management initiative in Australia which includes a major component focused on involving communities in on-the-ground management. More than 2,000 projects were funded under the Coastcare program between 1995 and 2002 (Clarke 2004). Under this initiative, community groups are encouraged to assist in dune, reef, mangrove and beach management, through activities such as the construction of dune access ways.

(c) Developing collaborative and community-based management programs

Community participation usually begins with a bottom-up approach involving major stakeholder groups. The process is initiated through a government commitment to devolve some power to the community, and the community's recognition of the need to manage local areas. If the commitment is made and stakeholders are aware of the need to manage, then community-based management begins to evolve in the community. Subsequent actions and developments by government and the community determine the progress towards full empowerment. The development and implementation of community-based management programs has been rapid in some countries such as the

Philippines, and slow in other countries. In countries where there has been a strong paternalistic or government dominated approach to management, collaborative management is more likely to be a possibility, with slow progress towards greater involvement and empowerment following.

There are five common principles in developing community-based management as identified by Drijver and Sajise (1993):

- a process approach: (similar to a bottom-up approach) managers and stakeholders agree on overall objectives, and then develop ideas and activities step-by-step towards achieving these objectives;
- participation: all participants have some form of power in all phases of planning and management;
- conservation and sustainable use: developed in partnership with all sectors of the community so that sustainable use programs are socioeconomically acceptable;
- linkages: between local management prescriptions, and regional or national level policies and strategies; and
- incentive packages: (or readily observed tangible benefits – social or economic) an integral part of any community-based management program; stakeholders must perceive some benefit from participating in the planning and management of an area.

A community-based management program has a number of components: community organisation, education, non-government organisation (NGO) involvement, social benefits, government support and institution building. Initially, community organisation is undertaken. It involves creating committees with representatives from various sectors of the community so that particular issues can be discussed and programs planned and implemented. Here the NGO component is included since they can assist in community organisation and education. Education is an important component; it informs the community on the resources they are using, their value, and how they can be managed through a community-based management program. Education programs also explain how management of the area or resources will benefit the community, and how they can be a part of the planning and management of the resources.

Real or perceived personal or social benefits, including ownership of resources or the management of those resources for sustained use, must be integral to the program. This can only be achieved if there is government support which ensures that the legal mechanisms allow for some of the management responsibility to be given to the community and financial support for particular development programs. Once community-based management is initiated, institutional development is another component. It is focused on supporting and training community groups so that they are given the skills and resources for long-term management. Support is often maintained through networking with other communities so that they have a support system to call upon.

The framework for community-based management is described well in Figure 4.5. Application of the framework and associated processes in the Philippines is illustrated in Box 4.8.

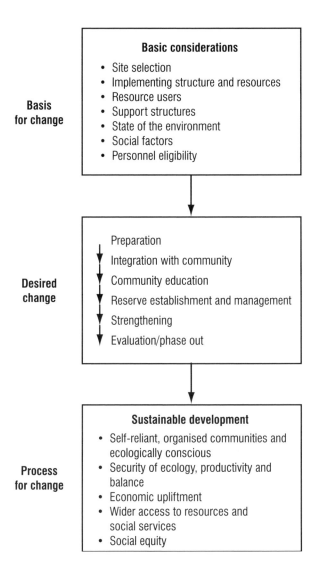

Figure 4.5 Framework for developing a community-based management program (based on White *et al.* 1994)

Box 4.8

Community-based fisheries management in the Philippines

Many Philippine islands suffer the same problems of deteriorating marine environments due to destructive fishing practices. The consequences of these activities are reflected in declining fish catches and correspondingly reduced disposable income derived from sales of valuable fish. Increasing poverty forces people to use more efficient and destructive fishing methods (White *et al.* 2002).

The Philippines government recognises the need to actively involve the community early-on in the development of management programs of marine and coastal areas to stem the decline of coastal resources. This recognition, combined with legislative changes giving local governments jurisdiction over 15 kilometres seaward of the low water mark (Rolden and Sievert 1993), and the early work of academics in the area of community development, set the scene for effective community based management of marine and coastal areas in the Philippines (Ferrer 1992).

Such community based management has enabled many local government supported marine protected areas (MPAs) to spring up around the country. The earliest such community and local government MPA was on Sumilon Island in 1974 through the assistance of Silliman University. A second well known island marine reserve is that of Apo Island, initiated in 1980, also by Silliman University. The Apo Island Marine Reserve was designed to enable local communities to protect and/or enhance their marine resources. This set the model for many more similar management regimes in the Philippines, so that now there are more than 400 locally managed MPAs in the country (White 1986; White and Savina 1987; White *et al.* 2002).

Apo Island symbolised an experiment in coastal management which has proved effective for coral reefs surrounding small islands and along some large island shorelines. The reserve model included limited protection for the coral reef and fishery surrounding the entire island and strict protection from all extraction or damaging activities in a small sanctuary normally covering up to 20 per cent of the coral reef area (White 1988a, 1988b). This reserve and sanctuary approach is providing real benefits to local fishing communities through increased or stable fish yields from coral reefs which are maintained and protected (White 1989b; Alcala and Russ 1990; White and Cruz Trinidad 1998).

The approach used to facilitate community-based management through small MPAs now is a fairly standard approach in the Philippines and is composed of six major activities.

1   The local municipal or city government develops a 'coastal resource management plan' for their shoreline and municipal waters. This plan generally includes several 'best practices' for coastal management which almost always includes one or more small marine sanctuaries or MPA. At this point some investment is made in a resource assessment and activities begin with the actual coastal community concerned.

continued...

Box 4.8, continued

2   Usually, there is integration into the community whereby one or more field workers live and work in the community for a period to assist the community to assess its problems and collect baseline data (environmental, socioeconomic, demographic and resource knowledge). This information enables further planning of the MPA.

3   Education: a continuous activity on an non-formal basis using small groups or one-on-one contact to assist in planning for the MPA.

4   Core group building: using existing community groups or facilitating the formation of new groups. Core groups provide guidance on how the project should be implemented and suggest potential solutions. Groups often reflect the interests of community members, for example the marine management committee (MMC) was formed by members interested in the reserve.

5   An MPA plan is developed and formalised through a municipal or city ordinance that legalises the MPA for implementation. At the same time, the ordinance helps to formalise and strengthen the implementing organisation providing ongoing support to the core group and its management efforts.

6   Some form of outside mentoring of the management organisation is maintained by the local government unit, a non-government organisation or an academic institution or some combination thereof. Training may be done for managing the MPA, collecting user fees, setting up buoys and signs or other needs of the particular site management plan.

The results of MPAs on Apo Island and many other sites are now substantial:

- municipal legal and usually some financial support for the MPA exists;
- demarcation using buoys and signs is usually in place;
- MPAs are managed by coastal or island resident committees which also patrol the area;
- municipal ordinances for the area are posted in the local language;
- moral support from the Philippine police is sometimes given;
- a community education centre may be established and be the focus for meetings and training programs in some areas;
- dive tourism is increasing in those areas where it is desirable to the community and local government;
- reef fishery resources and diversity have increased; and
- community satisfaction with management generally improves remarkably.

The success of small MPAs in the Philippines has generally followed the Apo Island approach. These cases highlight the need to combine community, environmental and legal approaches for a particular site with long-term institutional support from government, non-government groups and academia to set the framework for effective community based management, and to support it over time.

(d) Conclusion

Examples from the Apo and Sumilon islands readily illustrate the success and lack of success of community-based management in maintaining local reef fisheries. Apo Island has maintained a strong community-based management program of reef resources since 1985. At Apo Island management included reducing destructive fishing methods and closing a section of the reef (up to 20 per cent) to extraction and other damaging activities (White 1996). The community agreed to the closure. Sumilon Island management commenced in 1974; however, it has had a weak, intermittent program and less municipal support. Comparing fishery stocks of the two reefs shows the island with a strong community management program has increased its stocks significantly, while the other community has seen only a minor increase (Russ and Alcala 1994).

The success of collaborative and community-based management bear many similar features to consensus planning and implementation of such plans. Indeed, community-based management is one form of consensus management. Both are flexible management tools which can be applied to a range of social and cultural environments. They are flexible enough to meet the legislative requirements set by governments as well as incorporating traditional practices within the same management program. Collaborative and community-based management are recent planning and management tools which are being embraced by many nations. The challenge facing planners and managers is to improve the effectiveness of these tools and to broaden the scope of their use on the coast.

4.2.3 Capacity building

The ancient Chinese philosopher Lao Tse said:

> Give a man a fish and he will eat for a day.
> Teach a man to fish and he will eat for a lifetime.

The fishermen in the Central Visayas region of the Philippines taking part in a community-based fisheries management program Alix (1989) modernised this proverb to:

> Give a man a fish and he will eat for a day.
> Teach a man to fish and he will eat until the resource is depleted.
> Teach a community to manage its fishery resources and it will prosper for generations to come.

'Capacity building' is a term used to describe initiatives which aim to increase the capability of those charged with managing the coast to make sound planning and management decisions (Crawford *et al.* 1993). The term is used commonly by international organisations, especially the United Nations in its various programs. Capacity building is also increasingly used by national governments when new programs or initiatives are introduced and there is recognition that relevant expertise among the participants can be enhanced. This rather sweeping term, then, can be used to encompass a great number of apparently different activities, all of which are focussed on supporting and improving coastal management decisions. The focus of these activities is on the 'human capacity' of individual decision makers and coastal managers as well as the 'institutional capacity' (Crawford *et al.* 1993). The latter refers to the coastal management capacity of businesses, governments, non-governmental groups and communities.

The distinction between human and institutional capacity is a useful one: human capacity building is centred on training and professional development, while the other aims to improve institutional arrangements for coastal management. There is a blurring of the boundaries between the two in the discussion of research and data management, which require both the building of human and institutional capacity, as is discussed below (Table 4.6). Institutional arrangements were discussed and analysed in Chapter 3, and hence are not discussed further here.

Human capacity building can include anything from providing written training material, videos, facilitated meetings, or workshops to extensive long-term formal education programs, partnerships and mentoring schemes. The common themes of all these activities are training, professional development and improved expertise. They are not just restricted to administrative types of activities but apply to other areas such as strengthening the research capabilities of individuals or organisations. Collectively these various activities contribute to strengthening individual or institutional capability to plan and manage the coast efficiently and effectively.

An emerging component of the way in which coastal programs are developed and implemented is through the use of communication and marketing tools. These tools are increasingly fulfilling a variety of roles in coastal programs, such as promoting the use of a particular policy, law, plan, management tool, or the application of a particular institutional design (Table 4.6).

The nature and scope of capacity building programs will vary with the range of staff functions of the organisation. If the organisation's primary functions are administrative, then a capacity building program will focus for example on improving skills in various administrative functions, policy formulation and strategic planning. An organisation which is technically or operationally focused will have a capacity building program to strengthen field operations to improve their surveillance and enforcement capability, or develop skills in resource assessment and community development. Similarly, if

Table 4.6 Example components of a capacity building program

Capacity building group	Example tools
Human capacity	• Education • Training • Professional development
Institutional capacity	• See Chapter 3
Communication	• Marketing • Education • Training • Information dissemination • Database management
Research	• Data collection and analysis • Database management • Results dissemination

participants are experienced bureaucrats from either government or industry, a capacity building program will be very different from one which is used to improve the community-based management skills of local coastal residents who have had limited exposure to decision making. Irrespective of the administrative level or management focus, individuals and institutions need the knowledge, skills and confidence to participate in decision making. Capacity building programs play a critical role in providing this.

(a) Communication, education and training

This section introduces the use of communication, education and training techniques to assist in coastal planning and management initiatives. Communication is used here to describe the general act of imparting information in such a way that understanding is achieved and ultimately behaviour and attitudes change (Ferguson 1999). Within this broad umbrella are a range of approaches including programs in education, training and corporate-style communication. The term 'communication' is used to describe these, unless specifically referred to otherwise.

Five strategies can be used alone or in combination to influence behaviours and attitudes to ultimately achieve compliance with coastal plans and strategies (Global Vision 1996):

- technological, employing new methods or equipment such as the use of moorings rather than anchors; or economic incentives or disincentives as discussed in Section 4.3.4;
- enforcement, as is discussed in Section 4.1.4;
- social marketing, which draws on marketing and communication techniques; and
- education, to raise awareness and understanding.

Communication has several functions in coastal management, including:

- reduction of social conflicts and resource impacts;
- gaining support for management practices;
- reduction of management costs;
- the potential for increasing users' experiences of the coast; and
- contributing to the development of community-based management.

In contrast to the use of regulations and enforcement, implementation of communication programs can be inexpensive. In Australia, for example, it was estimated that an effective education program targeting fishers could be implemented for 2 per cent of the cost of enforcement (Bergin 1993). Alcock (1991) also noted that education costs less money and effort than enforcement. Communication programs, however, take time and require a long-term commitment of staff and funds before benefits are evident. Communication, however, can affect long-term behaviour changes thereby reducing management costs over time.

A major factor limiting the funding and support of communication programs is the time taken for their benefits to be realised (Saeed et al. 1998). The impact of enforcement activities is immediate and publicly visible, while the effects of communication programs are less obvious to the community and politicians; and managers are reluctant to assign adequate funding for them since it is difficult to measure the benefits. This issue was studied in relation to the management of the Great Barrier Reef Marine Park in Australia by Alder (1994) (Box 4.9).

Box 4.9

Changing awareness and attitudes of Cairns (Australia) residents towards management of the Cairns section of the Great Barrier Reef Marine Park

A long-term study (Alder 1996) of changing awareness and attitudes of Cairns (Australia) residents towards management of the Cairns Section of the Great Barrier Reef Marine Park has demonstrated the effectiveness of education programs in several areas. Their value in changing awareness and attitudes was evaluated using face-to-face surveys of Cairns residents in 1985 when management of the section began, and in 1991 prior to the review of the zoning plan. The results of the first survey were used to focus education and awareness programs on informing the community of the existence of the Great Barrier Reef Marine Park, its values, issues and management regimes.

The six-year study highlighted changes in community awareness and attitudes. Awareness of the park's existence increased significantly. Although the understanding of zoning (the basis for park management) increased, it was not significant; a detailed knowledge of zoning, however, decreased (see table below). A total awareness score was formulated for the 1985 survey. The median score increased from 3 to 4 in the period 1985 to 1991. In addition, support for restricting or encouraging specific activities in the park such as resort development, shell collecting, and commercial, spear and recreational fishing, remained high for both surveys. Support for encouraging fishing competitions and island camping remained unchanged and support for floating hotel development declined significantly. Support for park management remained high, but most respondents (46 per cent) were undecided about how effective management was.

The education and awareness programs contributed to improving community participation in the formulation of zoning plans. Although it did not increase the level of participation significantly, it enabled participants to focus on specific issues rather than broad general concepts.

Changes in percentage responses in community awareness to the Great Barrier Reef Marine Park

Variable	1985		1991	
	%	N	%	N
Park's existence	10	348	45	454
Zoning concept	65	34	70	201
Zoning details	46	22	19	24

Communication programs can be developed to involve stakeholders in aspects of coastal management ranging from facilitating participation in the management planning process, including defining goals and objectives, and developing policy and drafting action plans, to involvement in monitoring programs. Motivation and involvement of stakeholders is maximised when they can perceive the relevance of their participation. Again, communication programs can address this issue (Box 4.10). In the Caribbean, the recreational diving community is involved in monitoring coral reefs using simple methods that require a minimum of training (Smith and Homer 1994).

Box 4.10

Indonesian communication strategies for coastal management

Designing and delivering communication programs in Indonesia, as in any developing country, is not an easy task. Several constraints, other than the chronic ones of limited resources and expertise, need to be addressed in the development of any communication program at the national or regional level. Indonesia has 583 languages and dialects (Department of Conservation and Land Management 1983) and a diversity of cultures. Although Bahasa Indonesia is the national language, only those people who have completed high school studies understand and use it. Coastal dwellers, the most intensive users of marine resources, mostly speak their own local dialect, and therefore any communication program must include native speakers. Similarly, the literacy rate for coastal residents is considered low (Ministry for Population and Environment 1992); consequently communication programs must use alternatives to print-based media. Cultural and religious differences should also be incorporated at the local area level. Particular attention should be given to the different status of women since they are often the major exploiters of near shore coastal environments in Indonesia.

Act No. 5, Article 37 of the Conservation of Living Resources and their Ecosystems Act (Republic of Indonesia 1990) specifies that education is a part of the management of protected areas in Indonesia. Clearly the government of Indonesia recognises and supports the role of communication in protected area management. Reviews of publications and reports on the development and progress of Marine Protected Areas (MPAs) (Soegiarto 1981; Haeruman 1986) indicated the use of communication programs in their development, suggesting that until recently communication had had a low profile in MPA management.

In 1992, World Wide Fund for Nature Indonesian Program (WWF-IP) developed a communication strategy (1991–5) which focused on raising awareness among key agencies to address marine conservation issues and strengthen information, education and communication at the park level (Schoen and Djohani 1992).

Communication programs at the national and MPA level are underway in Indonesia. Current initiatives in MPA communication in Indonesia are either focused on specific issues or areas. Specific issues include dugong and turtle conservation, coral reef management, and mangrove management. Outputs from such programs include posters, brochures and comic books. These media are

continued...

Box 4.10, continued

usually inexpensive to produce and easy to distribute. Their effectiveness, however, depends on the education level of the recipients and how relevant the messages are to them.

The COREMAP program (Box 5.19) included a major communications strategy that was undertaken between 1998 and 2002. A mix of social marketing, public relations, local entertainment – education events, media advocacy, collaboration with NGOs and school-based activities were used in order to:

- increase government political and financial support for coral reef management programs;
- educate the general public, especially the younger generation about coral reef issues;
- motivate and foster community initiatives in local coral reef management;
- increase community and human resources invested in coral reef management; and
- create and enhance collaboration among stakeholders on coral reef management programs (Storey and Whitney 2003).

An evaluation of the impact of the COREMAP Communication Program found that:

- The two mascots which symbolise the importance of coral reefs have helped to deliver coral reef management messages. The mascots appeared on a popular Indonesian television spot three weeks after their launch in 2000 and 38 per cent of the audience recognised both mascots (Storey and Whitney 2003). Similarly recognition of the project's logo – SeKarang – is high among residents in the provinces where COREMAP is implemented.
- By December 2001, 63 per cent of the general public and 71 per cent of coastal residents in the COREMAP provinces were aware of the SeKarang logo, and 64 per cent of coastal residents knew of the COREMAP program, up from 3 per cent from early 2000 prior to the launch of the logo.
- A major impact was that, compared to people with low exposure, people with high exposure to the communication program had a high level of awareness of coastal reef issues and a willingness to act to address coral reef issues. However, people with low or no exposure to COREMAP communication were more cynical and pessimistic about coral reef management.
- COREMAP communication also facilitated increased discussions about coral reef management within coastal communities, and increased attendance at meetings on marine resource management (Storey and Whitney 2003).

The success of initiatives such as the COREMAP communications program suggests that communication has an important role to play in managing coastal areas and resources.

Other examples of communication strategies which can be easily understood by those targeted by a particular message in subsistance fishing communities in Indonesia and Papua New Guinea are shown in Box 4.11.

Box 4.11

Example cartoon books for communicating impacts of coastal dynamite fishing

Cartoon books for public education can be an effective way to communicate a message, especially in areas where literacy rates may be low. Examples of the use of such material to help reduce the use of explosives for fishing in Indonesia (Bason 1990) and Papua New Guinea (Hershey and Wilson 1991) are shown below. The first cartoon strip from Indonesia, using colloquial Indonesian, shows 'that evening … he … makes his first bomb'. This strip also shows the common technique for making such bombs, which is filling a bottle with explosives, lighting the fuse and then launching. Primitive fuses are used, often resulting in severe personal injury, as graphically shown by the second example from Papua New Guinea.

Indonesia

Papua New Guinea

Corporations use marketing strategies to develop products or services that will satisfy wants. They communicate the benefits of the products or services on offer to existing and potential customers, ensuring that demands are fulfilled to the satisfaction of the customer and the business (Armstrong 1986). This concept also applies to communication programs used in coastal management. Managers may wish to develop communication materials (products) and programs (services) which will alter specific behaviours or change awareness amongst users, which will satisfy management needs and users' wants. This focus on satisfying management objectives in the short term in order to benefit users and management in the long term distinguishes the use of education programs in coastal management from marketing in the business environments. Nevertheless, marketing concepts are becoming increasingly important in the development and implementation of communication programs in coastal planning and management. Examples of the use of marketing techniques that can be used in the management of marine parks are shown in Table 4.7.

Table 4.7 Marketing activities and their application coastal communication programs

Activity	Description	Application in coastal communication programs
Market planning	Setting of targets and markets based on corporate objectives and formulation of action plans	Establishing what changes are required in user behaviour and awareness to meet management objectives (e.g. reducing uncontrolled access to beaches by 50%)
Product development and planning	Developing new ideas and concepts, and testing the products to ensure they meet customer needs	Developing new education material and ensuring that it will work with intended audience before distribution (e.g. a TV ad to inform the community of the impacts of uncontrolled access)
Sales planning	Defining field or sales outlets	Defining the groups which will receive the program (e.g. surfers, fishers and local residents)
Marketing research	Collating information on actual and potential markets and users of goods and services	Determining the information wants and needs of the target audiences (e.g. where are the designated access points?)
Sales forecasting	Assessment of potential sales and market trends	Assessment of the potential short- and long-term impacts of the education program on the target audience (e.g. how long will it take to see a measurable reduction uncontrolled access?)
Analysis	Analysis of the product life-cycle	How long will the program be effective for? (e.g. how long will the TV have an impact on the target audience?)

continued...

Table 4.7 Marketing activities, continued

Activity	Description	Application in coastal communication programs
Target marketing	Formulating a more detailed definition of different groups that make up the market (segmentation) and determining where efforts should be targeted	A more detailed definition of the target audiences (e.g. surfers who belong to a club, surfers who are local residents, etc.)
Developing the market mix	Setting the blend of product, price, place and promotion to generate the responses the organisation wants in the target market	Balancing the available funding (price) for the program, with the intended messages (product), the most appropriate media (promotion), and target audiences (place) (e.g. balancing the cost of a TV ad and target audience within a limited budget)
Marketing and sales operations	Implementation of the marketing plan	Implementation of the education program
Marketing and sales control	Monitoring performance to ensure targets are achieved within the budget	Monitoring and evaluating the effectiveness of the education program
Feedback	Amending the plan as necessary	Revising the program to improve its effectiveness

Source: Based on Armstrong (1986).

Specialist education and training programs are becoming increasingly used as an integral part of coastal management initiatives. The development of training programs is increasingly being guided by Training Needs Analysis prior to program implementation. A Training Needs Analysis is a cyclic process of identifying training needs, developing strategies to meet the needs and then evaluating the effectiveness of the training and identifying subsequent training needs. Formal and informal education programs can be undertaken during program implementation to fill the technical and administrative gaps that staff might have. At regular intervals (usually annually) the Training Needs Analysis is updated and subsequent training and education programs are developed for staff. The Training Needs Analysis approach was used in the Indonesian COREMAP program with a high degree of success in both providing targeted training that met the needs of participants and their organisations as well as efficient use of a limited training budget which ultimately resulted in demonstrated improvement in the management of coral reefs (see Box 4.10).

Training programs are offered both in-house as part of the ongoing professional development of staff and by international organisations and tertiary training institutions. In recent years the fostering of regional centres of expertise in coastal management and planning has contributed substantially to the local delivery of education materials.

(b) Research and data management

Many coastal management decisions focus on complex issues of resource allocation and are therefore made with a degree of uncertainty. Managers attempt to deal with this uncertainty by basing their decisions on an analysis of the best available sources of information, including the opinions and perceptions of stakeholders. What, then, does a manager do if he or she judges that a large degree of scientific uncertainty remains?

Some existing information may have come from previous or current research programs. Often the planning process identifies information gaps or highlights the need for better or more appropriate information. If resources are available then research programs are undertaken to obtain the necessary information. If resources are scarce, the plan may recommend a range of research programs to provide that information, or it may recommend research programs (e.g. a coastal processes research program) to answer specific questions or issues. Whatever factors initiate the research programs, it is increasingly recognised that their outcomes, including the data and the management of the data, have an important part to play in reducing the level of uncertainty in any coastal management program (National Research Council 1995):

> This need (scientific information) is becoming more evident as the complexity of the relationships among the environment, resources and the economic and social well-being of human populations is fully recognised and as changes and long-term threats are discovered.

Before the 1990s there were few coast-specific research programs (e.g. the GESAMP – Joint Group of Experts on the Scientific Aspects of Marine Environmental Protection) developed to answer specific coastal management questions (National Research Council 1995; GESAMP 1996). Much of the research on the coast was focused on ecological or science questions, or to provide information for engineering projects or EIA programs at the site specific level. However, the passing of environmental legislation, including coastal, in the 1960s and 1970s (see Section 4.3.1) highlighted the need for scientific information for decision making (National Research Council 1995). In turn, this spawned a belief by decision makers that research programs were an essential prerequisite for decision making, and the views of 'uninformed' non-scientists were secondary. The result was a mountain of literature on various coastal research programs (scientific and social), including some very comprehensive analyses of the techniques for developing such programs and ways of managing the resultant data. The results were often collated and used opportunistically to provide scientific justification to reduce uncertainty in decision making. Despite this mechanistic view of research, it has contributed to improved decision making on the coast. The outcomes of coastal processes programs, greatly improving our knowledge and ability to define appropriate shoreline management strategies, and to improve the definition of coastal buffer zones, is a good example.

Global initiatives such as the 1992 Rio Summit and 2002 World Summit on Sustainable Development emphasised that decision making needs to be supported by a range of information, including social, from a variety of sources; and the need to link scientists and managers initiated a change in information requirements. As discussed earlier, purely researched-based systems evolved during the 1990s into a process by which the opinions of local or traditional knowledge and feelings of stakeholders are combined with scientific research through the principle of precautionary decision making. This evolution effectively recognised that research programs remain a crucial part of effective coastal management, albeit in a modified form.

Despite current research efforts, our understanding of coastal ecosystems and processes, social features and economic value remains poor at best. As a consequence, environmental decision making in the coast, as in most environments, is characterised by uncertainty. In this environment of uncertainty researchers can struggle since they are concerned with inquiry, description, and explanation whereas policy makers are concerned with reflecting societal values (National Research Council 1995). Decision makers, however, do not have the luxury of suspending their decisions until all the scientific information is collected and analysed because of a wide range of ethical and practical considerations (Latin 1993). Indeed, some have suggested that, from a management perspective (Welch 1991):

> While good science and information are important, inter-agency co-operation is the first prerequisite to sound management.
>
> (Welch 1991: 205)

The need to further evolve the linking of science and management, and heightened awareness of the importance of science, remains despite the advances made to date (National Research Council 1995). Latin (1993) suggests that conventional scientific norms may impede rather than promote reef conservation, because the response of science to uncertainty in the absence of considerable knowledge and reliability is no decision. This 'no decision' response by scientists also applies to the coast. The consequence of a no decision is significant since it maintains the status quo and does not contribute to problem resolution; in some situations existing problems may be exacerbated. However, constraints in the links between researchers and decision makers can be reduced if researchers relax their decision-making norms and managers involve researchers in all stages of decision making. Latin (1993) further suggests that scientific norms of knowledge and reliability must be relaxed if scientists want to facilitate better environmental management. The nexus between the gathering of information, the development of policy and decision making is shown with reference to Figure 4.6. An ideal view of this interaction is that decisions are made after a well ordered sequence, whereas in 'real world' decisions information gathering, policy development and impact assessment are all occurring at the same time, thus influencing each other (Kay et al. 1996b). The result is that information gathering, including science, policy and decision making, becomes interconnected and symbiotic (Feldman 1989).

The National Research Council (1995) has identified areas where researchers can be involved in policy formulation, including provision of internal advice in the form of a report, or through an internal advisory group using researchers within the agency or contracted services. Advice can be obtained quickly this way, and can be targeted. Using advisory groups external to policy making agencies can provide an independent evaluation of information, an approach which is useful when agencies require an independent review of internal mechanisms, and when it is cost effective to obtain the information outside of the organisation. Workshops are another forum to provide advice, but it is important that workshop participants include policy makers and scientists. Another approach is the use of informal policy advisory groups composed of

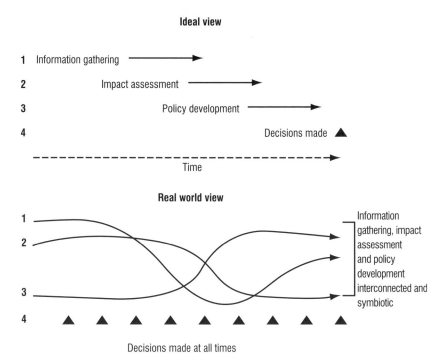

Figure 4.6 Ideal and 'real world' views of the interaction of information gathering, impact assessment and policy development for decision making (adapted from Kay *et al.* 1996b)

a range of internal and external researchers, stakeholders and decision makers to review published information and advise the decision making agency.

The question which arises from the above discussion is 'what coastal information do decision makers need which will also facilitate cooperation between researchers and managers?' The information needs will vary with the scale of planning. Information at the international level will be focused on large areas, summarised by country, highly qualitative and with limited precision. Information relevant to this level can include remotely sensed data, summarised demographic statistics and national economic analysis. At the site level, however, information will be very detailed and precise. Detailed site maps indicating individual plants, species lists with possible densities, and detailed geology and geomorphological characteristics would be typical information needed at this scale of planning. A survey of coastal managers in Australia highlighted twenty-eight types of information needed. The ten most important types provide an insight into the scope and nature of the information (Table 4.8). The table indicates that environmental information is a high priority, but information on social factors and other planning initiatives is also considered important.

The analysis of information for decision making is critical to coastal management, but it is often inadequately performed due to reasons such as limited time, funding and expertise (Bower 1992). Just as important is the reporting, since it must be in a form that is understandable to decision makers and others; and the methods of analysis must be well presented for peer review (Bower 1992). Therefore it is important to

Table 4.8 Ten most important information needs for Australian coastal managers

Top ten (of 28) information needs for coastal managers	Rank order of availability of top ten information needs (1 = best; 28 = worst)	Top ten gaps in information as identified by coastal managers	Barriers to information transfer (% frequency of respondents suggestions)
1 Ecosystems, habitats and species	14	Ecosystems, habitats and species	Inadequate information services (48%)
2 Environmental impact assessment	12	Development benefits and losses	Unclear locus of responsibility (15%)
3 Condition of rivers, estuaries, etc.	25	Recreation and tourism	Need for research and investigation (15%)
4 Recreation and tourism	6	Condition of soils and beaches	Absence of coordination and integration (6%)
5 Community priorities for coastal areas	24	Condition of rivers, estuaries, oceans	No access to local information (6%)
6 Strategic plans	17	Regulations and by-laws	Inadequate resources – financial, human (5%)
7 Condition of soils and beacher	16	Pollution indicators	Lack of clear policies on coastal management (2%)
8 Integrated resource management	26	Community priorities for coastal areas	
9 Public participation	15	Integrated resource management	
10 Coastal hazards	13	Public participation	

Source: Brown (1995).

include data analysis and reporting as part of the planning process and to include other interested parties who can assist in defining what information is needed and in prioritising information needs. Information needs will be guided by the issues identified, goals and objectives, program evaluation criteria and planning scales. The analysis techniques available will also influence the choice of information to collect and store.

Once the research component of a coastal management program is underway, it is imperative that the outcomes of research, including data, be adequately captured, stored, retrieved and reported (National Research Council 1993). This raises a number of issues regarding coastal data such as ownership, consistency and access. Data management options used will depend on how these issues are resolved and the sources of data. Large environmental datasets should be structured so as to be transparent, reliable, scalable, and distributed; where possible, data entry should be automated (Malafant and Radke 1995).

The use of Geographic Information Systems (GIS) to undertake these data management tasks is becoming increasingly widespread, especially when linked to the use of satellite remote sensing technology (Wright and Bartlett 1999). GIS can be extremely useful for coastal management, and especially coastal planning, because of the ability of such systems to store and analyse spatial data captured at a range of scales. For example, GIS was used extensively to produce much of the background information for the regional coastal planning exercises described in Box 5.13. GIS technology is becoming cheaper, easier to use, and more reliable (Harmon and Anderson 2003). This is especially so with the development of GIS interfaces that can be downloaded at no cost from the Internet, and in parallel the development of GIS applications that can be 'served' by organisations and accessed through normal Internet browsers (Ralston 2004).

The greatest constraint to the use of such techniques is rapidly becoming the quality of the data, and how it can be updated and improved, rather than computing limitations. While this problem is being resolved very quickly in developed countries through the development of structured, coordinated mechanisms termed 'Spatial Data Infrastructures' (Ralston 2004), it remains a critical problem in the developing world. Indeed, the issue of so-called 'digital haves and have nots' was recently the subject of the first phase of the World Summit on the Information Society (WSIS) held in December 2003 that included a Plan of Action in seeking to rectify these imbalances (WSIS 2003). The next phase of the WSIS is scheduled for Tunis in November 2005.

While it is tempting to become fascinated with information technology for its own sake, it is worth remembering Bower's (1992) 'Four Facts of Life' with respect to information for decision making:

1   No analysis for integrated coastal management can include all the information and analyse all the alternatives.
2   There are physical and psychological limits to a human's capacity as an information processor and decision maker. Too much information obscures the various trade-offs that are involved, which are the heart of the political process.
3   Only a limited amount of data relating to any given analysis can be presented at one time due to the complexities of coastal ecosystems, the complexities of decision making on the coast and the multiple use nature of the coast.
4   The format used to present the results will affect the amount of data that can be presented and will affect the extent to which the results are understood.

Management decisions must be made, and dealing with uncertainty is part of the decision-making process. Research, however, plays a critical role in reducing uncertainty and providing advice on a range of environmental, social and economic factors. Research, however, can only make a significant contribution when information sources, processes and outcomes are efficiently managed and shared effectively. When scientists balance their strictly scientific norms with pragmatic considerations, the effectiveness of management decisions is usually enhanced (Bower 1992). Recent initiatives such as the World Summit on Sustainable Development have introduced some of the needed changes including the ability for flexible partnership initiatives to support the development of management-focussed data management and knowledge sharing, such as the OneCoast initiative outlined in Box 4.12.

Finally, the use of the Internet in coastal management has exploded in recent years. Early evaluations of the impact of the Internet found that, while there was significant

---

Box 4.12

The OneCoast knowledge sharing initiative

OneCoast is one of over 250 'flexible implementation mechanisms' developed by governments, intergovernmental organisations, the private sector and NGOs to assist in the implementation of the World Summit on Sustainable Development by 'putting words into action' (UN CCD 2004). 'These "mechanisms", or more officially Type II partnerships, contrast with the main negotiated text of the summit, called the Type I text, by being much more action oriented' (Kay and Crow 2002).

OneCoast aims to provide an Internet-based information technology infrastructure that supports the diffusion of knowledge surrounding coastal management practice. The integration of information surrounding coastal management projects, the sharing of lessons learned, and open collaboration around emerging best practice, are all examples of how OneCoast plans to benefit the coastal community through using knowledge-sharing principles well established in the private sector (Seely Brown and Duguid 2000). It aims to establish services to coastal management 'communities of users' that provide a basis for diffusing capacity from both local and global contexts. It will support this through the capture and dissemination of the work of local communities of coastal management practice (who are using existing knowledge and creating new local knowledge) juxtaposed with the institutional experiences gained in program development and implementation (Wenger 1998) (see figure below).

Why now? The potential to develop a concept like OneCoast has emerged very rapidly in recent times through developing information technology standards that are focussed on information integration and interchange. The growing capability for governments, institutions and indeed individuals to interact with and contribute to globally accessible information sources in real time fosters a call to action for coastal communities of practice. The application of OneCoast to the classification and dissemination of coastal management projects in the Asia-Pacific region is shown in Box 5.1.

continued...

Box 4.12, continued

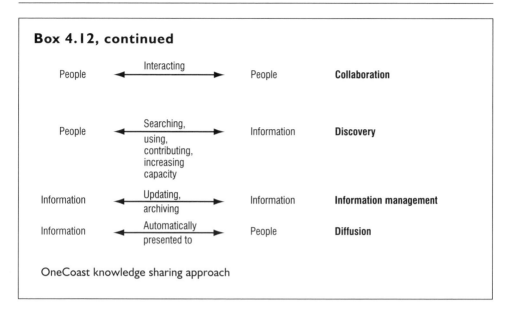

OneCoast knowledge sharing approach

potential for tangible benefits to coastal managers and planners in coming years, this has yet to be realised (Kay and Christie 2001). Although initially concentrated in developed countries where communications infrastructure is better developed and more reliable (Kay 1999), there is emerging evidence of the rapid adoption of the Internet in developing countries, especially among professional coastal mangers (CZAP 2002). The ability to quickly find and download information on coastal problems, experiences and techniques from around the world will add to the Internet's ability to bring like-minded people together (as shown in Box 4.12). However, searching through the enormous range of information available on the Internet increasingly requires the use of specialised sites to help 'navigate' through the apparent morass of data.[1]

(c) Section summary

This section has demonstrated the diversity of approaches to building the skills and professional and organisational infrastructure now acknowledged as essential for an effective coastal management program. Unlike other sections of this chapter which have outlined relatively clearly defined tools and techniques, capacity building remains an area of endeavour which requires extreme flexibility and cultural sensitivity.

As many coastal programs evolve from 'rule based' to 'participatory based' there is likely to be increasing demand for capacity building to play a central role in helping to deliver acceptable results; and perhaps for a corresponding evolution of coastal managers towards Olsen's (1995) 'ideal' coastal manager who 'besides being a good strategist and leader' will be equipped with the skills and knowledge for:

- conflict resolution;
- managing group processes;
- design and administration of transdisciplinary research programs;

- design and administration of public education and public participation programs; and
- program evaluation.

... and perhaps as well be skilled in the delivery of the coastal plans, tools and initiatives described in this book, including recreation and tourism management outlined in the next section.

4.2.4 Recreation and tourism management

The significance of tourism and recreation is often most evident in the coast. In fact coastal tourism is the most significant form of tourism, with domestic and international tourist flows in many countries dominated by visitors seeking the sun and the sea (Pearce 1987). The coast, with its beaches, dunes, coral reefs, estuaries and other coastal waters, has always been a natural playground. Coastal environments provide open space, the opportunity for leisure, relaxation, contemplation and physical activity. Changing recreation-oriented lifestyles in developed countries and the rapid expansion of tourism facilities in developing countries have placed considerable strain on coastal resources and in many cases intensified conflicting pressures on them.

Recreation and tourism are growth industries world wide. For many countries tourism is now a significant part of the economy (see Chapter 2). Indeed, in many coastal nations around the world, tourism is the most important single industry (WTO 2004) (Figure 4.7). The indications are that the growth in tourism will continue, with tourism in coastal zones being the major focus of that growth (Miller 1993).

Figure 4.7 Recreational pressures, Green Island, Great Barrier Reef (Source: John DeCampo)

This section includes a brief consideration of recreation and tourism planning principles; concepts such as recreational carrying capacity, tourism succession, and recreational planning methodologies focusing on the Recreation Opportunity Spectrum. This 'toolkit', described using case study examples, can be used in recreation and tourism planning at a range of spatial scales.

The terms 'leisure' and 'recreation' are used here in the sense of Patmore (1983) cited in Veal (1992):

> ... Leisure related to time, and the whole of non-work time in particular, and ... recreation related to the specific activities pursued in that leisure time. But the distinction is a convention, and its rigid application can occasionally stifle a full exploration of the values and satisfactions of the leisure experience.
>
> (Veal 1992: 45)

Following these definitions, the difference between tourism and recreation is defined by Kenchington (1993) as:

> tourism is the business of trading recreational opportunities for economic gain.
>
> (Kenchington 1993: 2)

Tourism can be generally considered to be the 'business' of recreation (Figure 4.8); but the distinction between tourism and recreation becomes less clear when it is acknowledged that the provision of recreational facilities on the coast by private industry and governments requires funding, and government funding is increasingly being gained through user-pays charges, such as park entrance fees. Given this trend, the distinction

Figure 4.8 Green Island beach hire, Green Island, Great Barrier Reef (Source: John DeCampo)

between recreation and tourism is judged here to be sufficiently blurred that the two terms are used interchangeably.

Recreational management and planning aims to enhance users' recreation experience of the coast while protecting and upgrading the coast as a recreation resource; in other words making coastal recreation more enjoyable and safe, without changing the coast in a way which actually reduces its attractiveness.

Recreation planning for a coastal area often aims to produce strategies that identify the appropriate: degree of naturalness, location, type and levels of access, type and extent of facilities, intensity of management, and level and type of recreational use.

Encouragement by governments of a tourism industry usually has multiple objectives which can be outlined according to scale. At the national level the aim of tourism may be to facilitate broadscale economic development within the nation's sustainable development strategy (Mowforth and Munt 2003). At the site level the aims may be to improve the local economy, maintain the area's cultural assets, improve local social conditions, and protect local coastal environments. Early tourism planning was focused on physical or promotional planning for the growth of tourism, but it has now evolved to using a balanced approach recognising the needs and views of tourists and developers as well as the wider community (Pearce 1989). This change in approaches to tourism management has also seen a call for tourism planning to be integrated with other forms of planning, and not to rely on tourism sector planning.

There has often been an absence of broader scale national and regional planning for the growth of tourism developments, with much of the focus of tourism planning being on managing the development of the industry within a defined area (Agardy 1990). In many cases around the world this local focus has produced a short-term economic gain, but long-term environmental degradation and resulting economic decline (Coccossis and Nijkamp 1995). Patterns of resort evolution have been described by Butler (1980) who outlines a six-stage evolutionary process: exploration, involvement, development, consolidation, stagnation and rejuvenation or decline (Figure 4.9). In this model of tourism it may be assumed that decision makers are seeking to reach the upper three outcomes shown in Figure 4.9 – rejuvenation, growth or stabilisation.

Planning for tourism development also seeks to address issues on the coast unique to the industry, such as visitors 'loving the environment to death', conflicts with other industries such as aquaculture and sand mining, and the strain on existing resources and infrastructure within the community (Mathieson and Wall 1982; Kenchington 1992).

The result has been the development of specialist recreational and tourism management tools and techniques (e.g. Lieber and Fesenmaier 1983; Kraus and Curtis 1986; Jubenville *et al.* 1987; Torkildsen 1992), some of which have been applied directly to coastal areas (e.g. Fabbri 1990; Wong 1993; Goodhead and Johnson 1996). These tools can complement, or be included within specific coastal recreation or tourism management plans or incorporated in broader integrated management plans as is demonstrated in Chapter 5. The most important of these management tools, and their underlying concepts, are described in the next sections.

(a) Concepts of recreation and tourism management

People choose to recreate. They choose to go surfing on weekends, walk on the beach after work, or go on holiday to coastal tourist resorts. Recreational choices include the

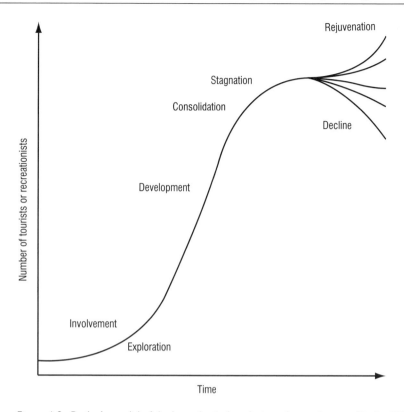

Figure 4.9 Butler's model of the hypothetical evolution of a tourist area (Butler 1980, as adapted by Pearce 1989)

type of recreational experience or activity sought, who to recreate with and recreational time and location. The notion of choice permeates the concepts of recreation and tourism management. By altering choice, through the provision of 'things to do' or 'things to see', or deliberately limited choice by restricting the provision of such choices, people's recreational experiences are being managed (McCool *et al.* 1985).

Recreational management concepts have taken the notion of choice and linked it to the relative impacts of different intensities of recreational uses on the environment, and on the recreational experiences of tourists themselves. One of the driving concerns of recreation managers after the Second World War (especially in North America) was an awareness of the problem of recreation 'succession' (Jubenville *et al.* 1987), referring to the evolution of a recreation site as more people become aware of its attractions. As visitor numbers grow at a particular site and outstrip the capacity of existing infrastructure, facilities are often upgraded and expanded, attracting even more visitors. If repeated a number of times, such demand-driven supply responses can significantly alter the nature of the site and the visitor experience and disenfranchise existing user groups. The danger, then, is that this positive-feedback spirals out of control with high intensity recreational uses inevitable, and a predictable 'sameness' of recreational choices. The succession model is similar to the models of tourist resort developments outlined above (e.g. Figure 4.9).

Concerns such as these are encapsulated in the concept of recreational Carrying Capacity, which focuses on the notion that there is a finite number of people who can visit an area before its capacity to absorb them diminishes. Once capacity is reached, a degradation of the environment or a reduction in the users' recreational experience occurs (Wager 1964). While this notion is instinctively appealing it has been of 'little utility for the manager looking for some rational reason for limiting use' (Jubenville *et al.* 1987: 29). However, despite these conceptual issues, carrying capacity has proved useful in the context of examining the impact of tourism on the coast and small islands in the Mediterranean (Coccossis *et al.* 2001). For example, in Malta (Box 3.3 and Box 5.17) a carrying capacity study resulted in the government placing a moratorium on the development of new hotels and instead focussing on increasing the revenue from existing tourists (Vella 2002).

However, as a general tool in recreation management, practical difficulties have led to a 'deceptively simple restatement of the problem' (Prosser 1986), the purpose of which was to examine explicitly the desired social and biophysical attributes of an area and how those attributes are to be effectively managed. Called the 'Limits of Acceptable Change (LAC)' concept, it focuses on the environmental and social conditions that are deemed to be acceptable, and the management actions required to achieve those conditions (Prosser 1986).

There are four main stages in the Limits of Acceptable Change planning process (Prosser 1986: 6).

1   Specify acceptable and achievable environmental and social conditions and defined by a set of measurable indicators.
2   Analyse the relationship between existing conditions and those judged acceptable.
3   Identify management actions needed to achieve acceptable environmental and social conditions.
4   Monitor the indicators of condition of an area and evaluate the effectiveness of management actions.

In practice the LAC concept is closely linked to a complementary recreational planning concept – the Recreation Opportunity Spectrum (ROS) (Clark and Stankey 1979). ROS considers recreation in terms of various settings and experiences available to different users. The ROS assists in recreational planning at a range of scales, but is commonly used at local and regional levels.

ROS recognises that different people look for different types and intensities of recreation, and that through the provision of a range of 'recreation opportunities' most users are accommodated. A recreation opportunity is defined as 'a chance for a person to participate in a specific recreational activity in a specific setting in order to realise a predictable recreational experience' (Stankey and Wood 1982: 7), which translates into planning for combinations of activities, settings and probable experience opportunities across a spectrum ranging from 'primitive' to 'modern'.

Thus, a recreational opportunity setting is made up of the combination of social, physical, biological and managerial conditions that give value to a place (Clark and Stankey 1979). This value can include those qualities provided by nature (e.g. vegetation and topography), those qualities associated with recreational use (e.g. use types and levels), and those conditions provided by management (e.g. facilities, roads and regulations). By varying these conditions management can offer recreationists a wide range of recreational settings and hence experiences ranging from modern holiday resorts to primitive 'back to nature' wilderness settings.

The value of the ROS as a planning framework is that it offers a conceptual tool for considering recreation as something more than simply different activities or areas. Instead, ROS highlights the issue of recreation and tourism management as being more than solely the provision of physical developments, such as resorts, campsites and walktrails, but rather providing a diverse set of recreation opportunities (Clark and Stankey 1979; Schmidt 1996). Beyond this value the ROS has specific application for (Clark and Stankey 1979):

- making inventories of, allocating and planning recreation resources;
- estimating the consequences of management decisions on recreation opportunities; and
- matching experiences people desire with available opportunities.

An example of the application of the ROS and LAC concepts is their combination for the purposes of national park management in the south coast region of Western Australia (Box 4.13).

Once criteria for management have been established for recreational management under the ROS and LAC system, tangible management steps can be designed. Examples of such steps for the management of the national parks shown in Box 4.13 are listed in Table 4.9.

Box 4.13

The recreation opportunity spectrum used for national park planning in the south coast area of Western Australia (adapted from CALM 1991)

The recreation opportunity spectrum is used by the Western Australian State Government Department of Conservation and Land Management (CALM) for planning ongoing management of national parks on the state's south coast (CALM 1991). CALM manage around 70 per cent of this extensive (approx. 1,500 km), sparsely populated and biologically rich section of coastline (WA Task Force 2002). The ROS concept has been applied to this section of coast to provide a management framework for recreation and tourism management.

National parks in the South Coast region, including those abutting the coast, were placed along the spectrum, as shown in the figure below.

Management prescriptions and recreational opportunities for parks at each position on the spectrum (see figure) are in the process of being developed. These management prescriptions will cover the range of factors listed below.

continued...

Box 4.13, continued

Recreational opportunities	Management issues
Access	How to get there; distance from nearest town; road types; access, proximity of parking to key features; parking capacity
Other non-recreational resource uses	Presence of western features (buildings, power lines, etc.)
On-site modification	Visual impact; complexity; facilities; disabled access; walks – with or without signs
Social interaction	Groups; availability of on-site information interpretation; appropriate use
Acceptability of visitor impact	Visitor impact
Acceptable regimentation	Visitor management; safety signs; management presence

CALM is working towards using each of the above management guideline prescriptions for each setting on the ROS (from primitive to modern – see figure). For example, the management actions for the factor of proximity of parking to key features for each ROS class could mean that users are expected to walk over 2 km from their cars in 'primitive' parks; whereas in the 'modern' parks they may be required to walk much shorter distances, in the order of 50–100 m, to reach recreational opportunities.

The ROS framework has proved to be successful on the South Coast and has been applied to other parts of the state requiring a regional perspective on recreation management.

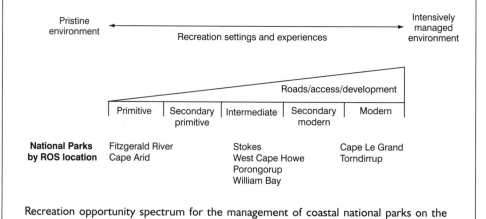

Recreation opportunity spectrum for the management of coastal national parks on the south coast of Western Australia

Table 4.9 Some measures to control the character of intensity of recreational use to meet desired management objectives in coastal parks

Type of control	Method	Specific control techniques
Site management (emphasis on site design, landscaping and engineering)	Harden site	Install durable surfaces
		Irrigate
		Fertilise
		Revegetate
		Convert to more hardy species
		Thin ground cover and overstory
	Channel use	Erect barriers
		Construct paths, roads, trails, walkways, bridges, etc.
		Landscape
	Develop facilities	Provide access to under-used and/or unused areas
		Provide sanitation facilities
		Provide overnight accommodation
		Provide activity-oriented facilities (e.g. camping, boating, swimming platforms)
		Provide interpretive facilities
Direct regulation of use (emphasis on regulation of behaviour; individual choice restricted; high degree of control)	Increase policy enforcement	Impose fines
		Increase surveillance of area
	Zone use	Zone incompatible uses (e.g. hiker only zones, prohibit motor use)
	Restrict use intensity	Rotate use (e.g. open or close roads, trails, campsites)
		Require reservations
		Limit usage via access points
		Limit size of groups, number of horses, vehicles, etc.
		Limit permitted duration of stay in an area (maximum and/or minimum)
	Restrict activities	Restrict building campfires
		Restrict fishing or hunting
Indirect regulation of use (emphasis on influencing or modifying behaviour, individual retains freedom to choose; control less complete, more variation in use possible)	Alter physical facilities	Improve (or not) access roads, trails
		Improve (or not) campsites and other concentrated use areas
		Improve (or not) fish or wildlife populations (e.g. stock or allow to die out)
		Advertise specific attributes of the area
		Identify range of recreational opportunities in surrounding area
		Educate users about basic concepts of ecology
		Advertise under-used areas and general patterns of use
	Set eligibility requirements	Charge constant entrance fee
		Charge differential fees by trail, zone, season, etc.
		Require proof of ecological knowledge and recreational activity skills

Source: Schmidt (1996).

(b) Recreation and tourism planning

Planning for recreation and tourism can be carried out as 'recreation and tourism only' exercise through sector-specific subject-plans, or through integration with other sectors. Often both are carried out, with broad scale national or regional tourist development plans concentrating on the requirements for the promotion of viable industries. There is an increasing trend for tourism planning to be incorporated into integrated planning initiatives: to recognise environmental thresholds that reflect the concept of carrying capacity and its application; to acknowledge the constraints for siting facilities (not every facility has to be on the coast); to incorporate sustainable principles into the design and construction of developments (e.g. silt curtains for marine construction); and to integrate social values into tourism developments, which is best done through community participation. Examples of such initiatives are described in more detail in Chapter 5, using examples from the Gascoyne Tourism Development Strategy and the Shark Bay Region Plan in Western Australia, and local tourism development planning in Sri Lanka.

Often the needs of the recreation and tourist industry, such as access to coastal land and good transport links, are planned for through integrated regional planning. As is shown in Chapter 5, these plans aim to bring together the various competing demands for coastal resources, including tourism demands, into an overall planning framework.

The steps common to most recreation planning initiatives follow the generic planning stages described in Chapter 5, especially as they relate to each scale of planning. However, recreation or tourism planning may require specific information and analysis techniques as part of integrated plans. The most commonly required recreation-specific requirement is an assessment of potential recreation demand. Recreation demand analysis requires the identification and analysis of existing patterns of use (i.e. how many people use each area, who uses which areas and facilities, when, and for what activities) and prediction of possible future changes in use patterns. A key issue in demand analysis is the difference between actual use of recreational opportunities versus the 'latent' or untapped demand.

An important issue in tourism and recreation planning is planning for the evolution of recreational opportunities and tourism products over time. A realisation that without deliberate planning a 'sameness' would creep into the style of recreational uses and opportunities was one of the driving influences in the development of the Recreational Opportunity Spectrum, described above. The same holds true for tourism goods and services. In a similar vein the evolution of tourist resorts and resort towns has become an important planning topic. Miossec's (1976, cited in Pearce 1989) model of tourist development, which conceptualises the growth of tourist regions through space and time, describes the staged growth of tourist areas and the resultant environmental and social impacts. This model was developed further and applied at Pattaya, Thailand by Smith (1992) (Figure 4.10).

Sustainability of the tourism industry is also reliant on local support, best obtained by encouraging local people to participate in all phases of tourist development. Without this support the industry may experience a succession of declining community attitudes towards tourism as described in Figure 4.11 (Doxey 1975). Again, that such issues of local community attitudes are now realised as important in tourism planning stresses the requirement for tourism to be managed through integrated area management programs (Pearce 1989; Smith 1992).

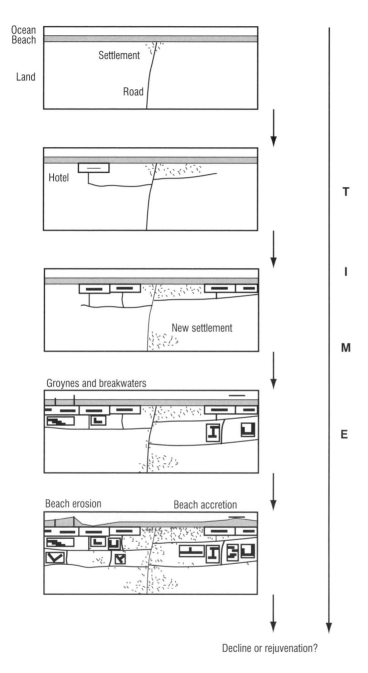

Figure 4.10 Tentative Beach Resort Model (adapted from Smith 1992)

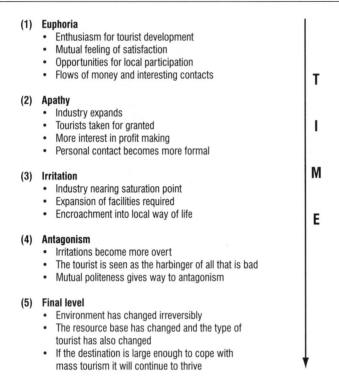

(1) Euphoria
- Enthusiasm for tourist development
- Mutual feeling of satisfaction
- Opportunities for local participation
- Flows of money and interesting contacts

(2) Apathy
- Industry expands
- Tourists taken for granted
- More interest in profit making
- Personal contact becomes more formal

(3) Irritation
- Industry nearing saturation point
- Expansion of facilities required
- Encroachment into local way of life

(4) Antagonism
- Irritations become more overt
- The tourist is seen as the harbinger of all that is bad
- Mutual politeness gives way to antagonism

(5) Final level
- Environment has changed irreversibly
- The resource base has changed and the type of tourist has also changed
- If the destination is large enough to cope with mass tourism it will continue to thrive

T
I
M
E

Figure 4.11 Example succession of community attitudes towards tourism (Doxey 1975, cited in Mercer 1995)

Miossec's model links the number of tourists with sustainable development. A decline in tourist numbers can often be caused by significant environmental degradation from poorly planned tourist developments, and from the impacts of tourist themselves on natural and social environments (Coccossis and Nijkamp 1995). Tourist developments built too close to sandy coastlines can, for example, cause chronic erosion problems, resulting in the loss of the beach that most of tourists came for in the first place! An example of such conflicts is clearly demonstrated by the rapid growth of tourist development on the Red Sea coast discussed in Box 4.14.

The incorporation of tourism development (including the Red Sea case study introduced in Box 4.14) into integrated coastal management planning is shown through a number of case studies in Chapter 5; for example, how the promotion of a locally sustainable coastal and marine tourism industry in Sri Lanka is integrated with other development goals is demonstrated in the analysis of the Hikkaduwa Special Area Management Plan (Box 5.18).

This section on recreation and tourism planning has demonstrated the subtle interactions between the coastal tourist or recreationist and their natural and social environment. The degree of this subtlety is being increasingly realised as the importance of providing a diverse range of recreational and tourism opportunities is understood. In addition, the use of management and planning tools, including environmental impact

Box 4.14

Tourism growth and management of the Red Sea coast

> Nestled between the desert sands of Africa and Arabia, the Red Sea has lured adventurous explorers for hundreds of years. Now it attracts thousands of tourists!
>
> (Hawkins and Roberts 1993)

The previously exotic and remote northern shores of the Red Sea within Egypt, Israel and Jordan are now the focus of coastal tourism developments. The area experienced rapid development from the early 1980s to September 2001 of the tourist centres of Hurghada and Sharm el Shihk (Egypt) and Eilat (Israel). There are now numerous coastal tourist centres along the shores of the Red Sea and these are expected to increase after stagnating between 2001 and 2003. New areas in Egypt and Jordan are also marked for development over the next decade as the countries and residents in the region seek to maximise the economic benefits of tourism. European holiday makers seeking new destinations were the main visitors, but residents holidaying within the region are increasing (Saad 2003).

A key drawcard for tourists is the well developed fringing coral reefs set in clear, warm waters. Scuba diving is one of the main recreational uses of these reefs. For example, of the 500,000 visitors to Sharm-el-Shikh (Egypt) in 1998, an estimated 150,000 were divers (Shehata 1998) (there are plans for diver numbers to grow to 300,000 per year to a projected 1.2 million visitors) (Hawkins and Roberts 1993). The actual growth of tourist numbers and their direct impact on the growth of marine tourist numbers is illustrated in the table below.

Growth of hotel beds and dive boats in the Sharm-el-Shikh (Egypt) region 1988–95

Year	No. hotel rooms	No. dive boats
1988	1,030	23
1989	1,276	25
1990	1,358	47
1991	2,906	60
1992	3,306	89
1993	5,190	120
1994	8,234	200
1995	11,384	240
1998	16,000	

Source: Anthias 1994, Shchata 1998.

The impacts of tourist growth on the coastal and marine environment have varied significantly around the Red Sea. In some areas, degradation of fringing reefs has been significant due to the direct effects of construction (infilling, sedimentation), indirect effects of tourist development (sewage, desalinisation, irrigation and rubbish) and the effects of tourists themselves (Hawkins and Roberts 1993) (divers and snorkellers damaging coral, anchor and mooring damage and the impact on local fisheries of increase seafood demand).

continued...

Box 4.14, continued

There is an increasing recognition in the region that tourism and its continued expansion can significantly damage the very coastal ecosystems that draw visitors to the Red Sea. A key component of the strategies to manage growth in the region is a marine protected area network. An example of one such protected area is the Ras Mohamed Marine Park at the southern tip of the Sinai peninsula (Sharm-el-Shikh, Egypt). The park, first declared in 1983, was expanded in 1989, 1991 and 1992, and two new protected areas were also added in 1992 and then linked to Ras Mohamed in 1994. In 1996 the remaining littoral areas were protected and in 1998 the Taba Natural Monument was established, bringing the totally protected area in the South Sinai sector to 12,000 square kilometres (Shehata 1998).

A simple zoning plan of open and closed areas, along with regulating the number of boats and divers/snorkellers, prohibiting anchoring, fish feeding and collection, and a policy of zero discharge are the primary mechanisms to manage day-to-day activities in the park. The EIA process is used to manage developments in and adjacent to protected areas (Shehata 1998).

The aim of management at Ras Mohamed and other initiatives in the area is to provide a strategic long-term view of tourist development and attempt to mix conservation and development to provide a sustainable future for both the reef ecosystems and the tourism industry.

assessment, economic analysis and risk management is becoming increasingly widespread in the tourism industry. The most important of these tools are described in the next section.

4.3 Technical approaches to coastal planning and management

Coastal planners and managers can choose to use a number of 'technical' approaches to plan and manage the coast. Many of these approaches are not specific to the coast; in many cases these approaches were developed for land based systems and then modified where necessary for use in the coast. The full range of technical techniques which can be applied in the coast is far too wide to include all of them here. Instead, this section discusses both the major and most common tools currently used in coastal management, and some which are not in widespread use but we judge to be on the brink of the mainstream.

4.3.1 Environmental impact assessment

Environmental Impact Assessment (EIA) is one of the most frequently used tools in coastal management. The use of EIA is globally widespread, and it is used in a variety of planning and management contexts, each of which is outlined in this section.[2]

The following clear working definition of EIA was developed in New Zealand during the reform of their resource management system (Ministry for the Environment 1988):

> Environmental Impact Assessment (EIA) is a process by which the impacts of a proposal (whether as a policy management plan or intended development) are identified early on in the decision-making process, so that these considerations are taken into account in the design and approval of the proposal.
>
> (Ministry for the Environment 1988: 6)

Within the EIA context, 'environment' refers not only to biophysical aspects, but also social and economic aspects. The general aim of the EIA process is to provide decision makers with the best available information which will help to minimise the costs (environmental and financial) and maximise the benefits of the proposed actions. Minimising environmental costs is often associated with managing environmental risk as discussed in Section 4.3.2. EIA is now an integral part of environmental planning and management of the coastal and marine environments of many coastal nations (Sorensen and West 1992).

The purpose of this section is not to detail the EIA process but to outline its use as a tool in planning and managing the coast. For more information and/or details on EIA there are several good texts, of which those of Glasson *et. al.* (1994), Gilpin (1995), Thomas (1996), Petts (1999) and Wood (2003) are drawn on in the following section. A number of international organisations have also produced EIA guidelines or manuals, including many international donor agencies (e.g. Asian Development Bank 1991). Many of these manuals and guidelines have been well summarised by Gilpin (1995: 74–87).

There is no coast-specific EIA, but the legislation or policies used throughout the world are either generic in their application, or 'the environment' is defined in them in terms such that it applies in the coastal zone. However, the special attributes of the coastal environment have been recognised through the publication of a number of guidelines and manuals written to assist in undertaking coastal EIA (e.g. Sorensen and West 1992; SPREP 1992; Vestal *et al.* 1995).

Some managers view EIA as both a political and a technical process (Gilpin 1995; Thomas 1996). EIA might be seen as a political process because it is based on society's value judgements, therefore decisions on whether a development should proceed are influenced by social politics. Furthermore, most governments have legislated for EIA, meaning that EIA decisions are also political judgements. Both social and political types of value judgements take place throughout the EIA process. Most governments have found that the best way to manage these judgements is through public participation. Depending on local legislative requirements, public participation can be an important element of many of the steps in the EIA process. The integral role of public participation in EIA also reflects its relatively recent evolution, and the simultaneous rise of citizen interest and subsequent involvement in decision making.

(a) Integrating EIA with planning

Close linkages between EIA and planning occur at a number of points in the EIA process. In general these links are becoming closer as the various forms of coastal planning described in this book (especially in Chapter 5) generally include consideration of environmental impacts.

In this sense, the classical EIA step-by-step practice has been linked to planning theories, most notably the synoptic (comprehensive or rational) theory (Smith 1993). Synoptic planning is based on the development of plans through a series of steps undertaken in a purely rational manner (Chapter 3). The synoptic model can be used to integrate impact assessment with planning. Thus, EIA can be used as method for looking at the impacts of 'planning' and plans on the environment by using a series of steps similar to synoptic planning (Smith 1993):

- identification of problems;
- defining goals and objectives in planning corresponds to the processes which assist analysing the potential impacts of goals and objectives;
- identifying opportunities and constraints in planning is similar to analysing the current environment and predicting the effects of planning actions;
- defining alternatives is similar in both planning and EIA;
- making a choice and implementing that choice is also similar in both processes.

At each of these steps impact analysis (similar to the above generic EIA process discussed above) is undertaken. The advantages and disadvantages of rational/synoptic planning as discussed in Chapter 3 also apply here. That is, the model usually only applies in the early stages of the process.

A 'manual planning' model is based on assessing the impact of 'the plan' using a process similar to the generic EIA described above or the use of a set of guidelines specific to assessing plans. In the manual planning model, the outcome of the planning process ('the plan') is subject to a single EIA process. If a synoptic approach is used, then an EIA is conducted at each step of the planning process. With either process, the potential impacts of a plan as part of the approval planning process are investigated.

The assessment of plans can be a statutory or non-statutory process depending on the nature of the EIA legislation or policy. If there are no statutory requirements for assessing the impact of a plan, governments will nevertheless assess the plan. Depending on the legislation and prevailing political climate, the public may or may not be consulted.

A project-by-project approach to EIA dominates the way that many developments and activities are managed around the world. This approach dominates primarily due to the legislative and administrative arrangements which generally apply to projects, and generally does not make provisions for how the proposal which is the subject of an EIA fits into the broader context. This limitation on the range of environmental measures managers can use within the EIA process has been criticised as a major failing of EIA (O'Riordan and Sewell 1981; Smith 1993). Individual projects may not have a significant impact, but several projects may in combination produce a 'cumulative impact'. Linear impacts can occur with developments that are spatially linear, such as strip development of continuous urban residential land along coastlines (Court et al. 1994; Vestal et al. 1995).

Planning process, especially the production of plans with a broad strategic focus, can enable the incorporation of measures to address some of the legislative and administrative constraints to the production of project-by-project EIAs. Strategic planning documents can be used to manage potentially cumulative environmental impacts. Planning processes may also trigger EIAs, through either statutory land-use planning systems or in the identification of potential impacts during strategic planning.

Plans or policies can limit the number or levels of cumulative impacts in several ways ranging from defining the number of facilities that discharge or contribute to environmental degradation, defining the total discharge into the environment for specific pollutants such as sulphur emissions, and specifying/setting the environmental standards for activities or structures that are allowed or permitted in a particular area. Plans and policies can also provide a mechanism for consulting the public on decisions regarding strategic land or water use, and other developments or facilities with the potential to add to the environmental problems of an area.

Similarly, cumulative impacts which occur in a linear fashion can be managed through planning by specifying where activities can and cannot take place. An example of how EIA is linked with regional strategic planning in the Great Barrier Reef is shown in Box 4.15.

The inclusion of EIA principles into coastal planning can ensure that the outcomes of planning decisions are similar to the desired outcomes for EIA. Similarly EIA can be used to ensure that plans will cause the minimum impact on the environment. Hence good coastal planning reduces the need for EIA, and when EIA is required, its scope can readily defined by pre-existing planning (Brown and McDonald 1995).

Box 4.15

Linking EIA and strategic planning on the Great Barrier Reef

The Cairns Section Zoning Plan for the Great Barrier Reef Marine Park is an example of how planning was used to address the issue of linear environmental impacts. In the re-zoning of the Cairns section in 1992, the community expressed concern with the number of applications for developing pontoons and other similar permanent structures on the outer barrier reefs of the section. These outer reefs form a line/chair of reefs and are noted for their aesthetics and good coral cover. To minimise risks, the zoning plan was modified to include a structure/no-structure subzone (see figure overleaf). The no-structure subzone (NSS) did not allow, or permit the establishment of, permanent structures on reefs. This ensured that linear developments along the outer reefs would be avoided.

In early 2004 a GBRMP wide zoning scheme came into effect and the No-Substructure Subzone that was used exclusively in the Cairns section was not included (GBRMPA 2003). Instead, the GBRMP regulations will be used to meet the objectives of the NSS. For other areas outside of the Cairns section, the new zoning plan makes provisions for the establishment of Remote Natural Areas, which have similar objectives to the NSS (GBRMPA 2003). Currently only the Far North section has established remote natural areas.

continued...

Box 4.15, continued

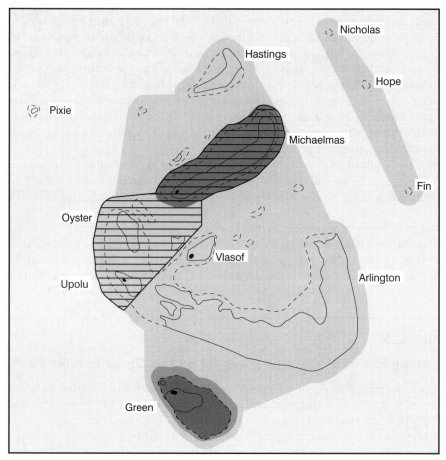

Source: GBRMPA.

Zoning classifications

☐ General use zone
▨ Habitat protection zone
▨ Buffer zone
■ National park zone
⊟ No structure subzone

0 ——— 5 km
N

Zoning map for the Cairns section of the Great Barrier Reef Marine Park

(b) Strategic environmental assessment

The focus of 'classical' EIAs on project-by-project environmental impacts, and their resultant inability to adequately address strategic environmental impact issues, as described above, led to the need for more strategic EIA tools. This need has become coupled with the increasing demands that sustainable development principles have placed on coastal planners and managers. One of the outcomes of this dual need has been Strategic Environmental Assessment (SEA), defined by the Australian Commonwealth Environmental Protection Agency as the (Court *et al.* 1994: 13) process of consideration of environmental impacts of policies, plans and programs applied to higher levels of decision making with the object of attaining ecologically sustainable development.

SEA can be considered as an EIA for programs, plans, and policies, but can also be considered as a planning tool in its own right. It allows the consideration of environmental impacts over a larger geographic area and development time frame. In addition, SEA enables subsequent EIAs on a project-by-project basis to focus on the details of the project and possible alternatives. Wood and Dejeddour (1992) consider SEA as the first stage of a two stage EIA system with traditional project-by-project EIA as the second stage. There are currently two main approaches to SEA: the first the extension of traditional EIA principles, and legislative procedures and requirements; and the second adopting a policy and planning rationale where environmental principles tend to be tailored in the formulation of policies and plans.

SEA assists in coastal planning and management in a number of ways. In decision making it enables managers to raise the importance of coastal concerns to the same level as the aspects of traditional development planning. SEA also facilitates consultation on a range of coastal issues between various organisations as well as the public. It can make EIA unnecessary, or reduce its importance, for specific activities if they are considered sufficiently at the plan or program level. Mitigation and compensation measures can be formulated for certain types of developments as a result of SEA as well as assisting in formulating or modifying codes of conduct.

A benefit of SEA is that it encourages the consideration of environmental objectives during policy, plan and program-making activities within organisations which traditionally avoided incorporating environmental considerations in their decision making. Some project EIAs may be redundant within an SEA. By reducing the need for or scope of EIAs, providing for a wider range of alternatives to be considered, and recommending suitable sites for projects, SEA allows managers to focus on specific aspects of EIA. Subsequent projects benefit when SEA facilitates the formulation of generic best practices. SEA enables managers to investigate other areas of impact assessment such as cumulative, secondary, long-term and delayed; as well as the impact of specific policies (based on Wood and Dejeddour 1992).

To summarise, the EIA process, like many other tools for managing the coast, continues to evolve to meet society's expectations and the needs of environmental managers, especially those in the coast. It is a well defined process which has been embraced and modified as a major environmental management tool in a number countries. Early in its history EIA focused on assessing developments on a case-by-case basis, which could consume considerable financial and human resources. However, as managers have come to understand EIA they have also explored how it can be modified and extended into a more cost effective and efficient process. Cumulative impacts and SEA are two major outcomes of this exploration. Through SEA the role of EIA in planning and managing coasts is now varied and applied at operational and strategic levels.

4.3.2 Risk and hazard assessment and management

> ... the concept of risk is neither entirely abstract nor wholly physical: it is socially constructed and so must be socially resolved through mediation and negotiation between the parties involved.
>
> (Gerrard 2000: 435)

There is no such thing as a zero risk. Each day we take risks, from the small risks like whether we will get wet while taking a beach profile, to the large risks associated with undertaking a coastal development in an erosion-prone area. The larger risks can be financial, such as the financial viability of building multi-million dollar marina developments on the coast; ecological, such as mitigating the risks posed to natural coastal ecosystems from heavy industries; or planning, associated with determining the proper location and design of structures to minimise the impact of natural hazards.

As the name suggests, risk and hazard assessment is concerned with assessing the probability that certain events will take place and assessing the potential adverse impact on people, property or the environment that these events may have (Newman *et al.* 2002). Coastal examples include failures of a chemical refinery on the coast causing damage to the plant itself, and to surrounding residents and the environment through the release of toxic chemicals into nearshore waters. Examples in the natural environment are analysing the potential impacts on a coastal region of severe storms or tsunamis. Not surprisingly, methods of managing risks once they have been assessed are called 'risk management' techniques. The importance of integrating risk management consideration into coastal management and planning has recently been brought into stark focus by the 26 December 2004 Indian Ocean tsunami. The regional and local-scale impacts of the tsunami and its implications for coastal management are discussed in Box 4.16.

This section considers all types of risks which influence coastal planning and management decisions. Before the details of risk assessment and management are described, some basic concepts of risk and hazard are introduced.

(a) Concepts of risk and hazard

Perhaps the most basic concepts of risk is that it is:

> a culturally framed concept which acts as a metaphor for individual feelings about loss of control, powerlessness and the drift of social change away from what is good for the Earth towards what seems to be bad.
>
> (O'Riordan 1995: 296)

This may seem to be a very big step from, for example, working out the impacts of a storm surge on a coastal community. Nevertheless, the concept that risk, and therefore

Box 4.16

The 26 December 2004 Indian Ocean tsunami: regional and local impacts and coastal management responses

At 00:59 GMT northern Sumatra experienced what was to be the fifth largest earthquake since 1900. The resulting tsunami started its devastating journey around the Indian Ocean and northwards into the Andaman Sea.

Within tens of minutes, the ocean was already rising along the Indonesian, Malaysian and Thai coastlines. It then took between 2–3 ½ hours for the tsunami to travel across the Indian Ocean to Sri Lanka, India and the Maldives. No warnings were sent to these countries and the elevated tsunami surged onto unsuspecting coastal communities. Unlike the Pacific (ITSU 2005), there was no tsunami warning system for the Indian Ocean. The extent of the area affected by the tsunami is shown in the map below.

An estimated 290,000 people were killed and 1.2 million were displaced or directly affected by the tsunami in the eastern Indian Ocean. The most affected by the combined effect of the earthquake (magnitude 9.0 on the Richter scale)

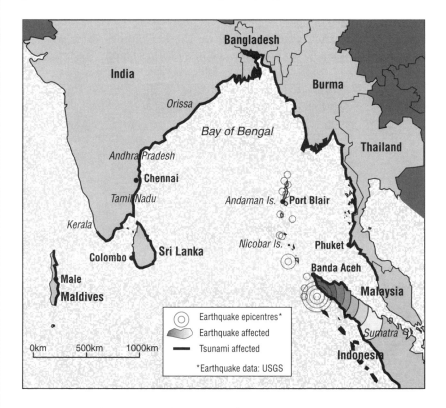

Extent of areas affected by the 26 December 2004 tsunami
(Source: Adapted from AusAid 2005)

continued ...

Box 4.16, continued

and the tsunami (at its greatest height it was 35 m (Japanese Research Group 2005) was the Aceh province of Indonesia with 98,000 confirmed dead and 132,000 people missing (OCHA 2005; USAID 2005).

The impacts on the coastal regions of Sri Lanka (Boxes 2.6, 3.10, 5.12 and 5.18) were 31,000 people killed, 15,000 injured and 500,000 people displaced (Sri Lanka Department of Census and Statistics 2005). The run-up of the tsunami was between 2.5–10 m in Sri Lanka depending on local topography (Japanese Tsunami Research Group 2005).

At the tourism and fisheries settlement of Hikkaduwa in South West Sri Lanka (Box 5.18) the impacts were restricted to a relatively narrow 1 km wide zone. Despite this localised impact at Hikkaduwa, over 1,500 people were killed and 60,000 displaced (Sri Lanka Department of Census and Statistics 2005). Personal accounts of the tragedy and videos taken by tourists in beachfront hotels show the scale of the impacts and the trauma endured by survivors (BBC 2005; Clarke 2005; Yahoo 2005). The impacts on the fishing industry were significant, with rapid assessments undertaken by the UN Food and Agriculture Organisation estimating the loss of approximately 50 per cent of all fishing boats in the country and varying degrees damage to 10 of the 12 fishing ports and related infrastructure in Sri Lanka (FAO 2005).

The biophysical impacts of the tsunami on the Hikkaduwa coast were less than the scale of the damage to property and loss of life suggests. The extensive

Tsunami damage reconstruction, Hikkaduwa, Sri Lanka (Credit: Sithara Atapattu)

continued ...

Box 4.16, continued

shallow fringing reef systems at Hikkaduwa – a major source of tourism revenue (see Box 5.18) – was still recovering from a devastating bleaching incident in 1998 caused by higher water temperatures from El Nino warming (Jinendradasa and Ekaratne 2000; Atapattu personal communication, 2005). As such, its already degraded state appears to have meant that tsunami damage was minimal. Early reports suggest that mangrove stands were relatively undamaged by the tsunami and acted to moderate the impacts on adjacent shoreland (Atapattu personal communication, 2005).

The impacts of the tsunami on Sri Lanka and other affected countries would have been significantly reduced by an integrated risk management approach used in developed countries around the Pacific rim prone to tsunamis. For example, the United States uses an approach that combines early warning, land use planning constraints, community education and evacuation preparedness (National Tsunami Hazard Mitigation Program 2001). In particular, the approach used in the Pacific North-West States of the USA is to embed tsunami warning, evacuation and education within broader coastal hazard management systems (Wood *et al.* 2002).

There is a commitment to establish an Indian Ocean tsunami warning system as a matter of urgency. There is also strong support at an international level to support national efforts to reduce long term vulnerability reduction to all future natural hazards throughout the ongoing relief operations and rehabilitation and reconstruction efforts (UN General Assembly 2005).

There are also encouraging signs in Sri Lanka and other countries that on-the-ground planning responses to the tragedy will build on their considerable history of coastal management (Boxes 3.10 and 5.12). However, it is important to note that despite the considerable coastal management activities in Sri Lanka there has been ongoing resource degradation and illegal development within the 'coastal zone' (see Box 2.6) defined under the Sri Lankan Coast Conservation Act (Figure 1.2). As such, it is likely that long-term coastal management responses to the tragedy at a national level will focus on implementation of existing strategies planned before the tsunami tragedy (Wijetunge 2005) to complement international initiatives.

The scale of the tragedy is a timely reminder to coastal managers of the critical importance of holistic risk planning and management within broader coastal management systems.

the management of risk, is deeply ingrained in how societies function is now widely accepted as the central tenet of risk management (Presidential/Congressional Commission on Risk Assessment and Risk Management 1997). The concept, which has taken many years to form, has mainly developed around the experiences of managing the risk of the nuclear industry or the impacts of extreme natural hazards such as tsunamis and cyclones (Box 4.17). Managers of hazardous industries have now realised that trying to convince a sceptical public that a particular industry is safe by bombarding them with high-powered science does not work; it can even be counterproductive. The focus today is on shared 'multiway' communication of risks (Figure 4.12).

Box 4.17

Cyclone risk management in Bangladesh (updated from Kausher et al. 1994, 1996)

The Bangladesh Cyclone Preparedness Program (CPP) provides an impressive example of the effectiveness of cyclone hazard mitigation measures. Cyclones of similar severity devastated the area in 1970 and 1991 (Box 2.11), but the loss of life in the later event was less than half of that in the former; and this despite significant population growth in the area during the intervening two decades. The difference in impact was generally attributed to the effectiveness of the CPP, set up in 1973 in response to the 1970 cyclone.

There are three components to the CPP: warnings, shelter construction and evacuation to shelters, and disaster relief. The system relies heavily on a grass-roots support system, based on more than 2,000 units of ten volunteers each. Volunteers have specific tasks such as cyclone warning, shelter management, rescue, first aid and relief.

The warning system, whereby volunteers supplement warnings over public radio by using megaphones mounted on rickshaws, is obviously a critical element of the CPP. Surveys undertaken after the 1991 cyclone revealed that 64 per cent of people took some precautionary measure on hearing the cyclone warnings, such as moving to cyclone shelters or well built houses or embankments (Hossain *et al.* 1992). How this figure compares to the total response of coastal residents to the warning will never be known.

Reinforced concrete buildings designed to save human life from cyclones are the second part of the CPP program. A five-year shelter construction program was initiated in 1972, but was abandoned after only 234 shelters were built, or an estimated 10 per cent of actual demand (Talukder *et al.* 1992). The program was re-started after a cyclone in 1985, and a further sixty-two shelters constructed, but few of the survivors of that event reported that shelters were available to them, stating among the reasons they had not sought shelter that the distance to the shelter was too great, that they felt their household goods would be at risk if they left them unguarded, and that there was a lack of proper facilities in the shelters. The post-1985 shelters also suffered from a design defect which had serious cultural ramifications. Having only one large communal room, unlike the three-storey shelters built after the 1970 cyclone, men and women were crowded together, something which is considered to be a violation of purdah for women (Talukder *et al.* 1992).

There are many lessons to be learned from the Bangladesh experience, including importantly that planning, even if in part flawed, can have outstandingly positive outcomes. But perhaps the most important lesson is the need to take a holistic approach to cyclone impact mitigation, embracing adequate (and culturally appropriate) shelters, evacuation routes, and improvements to the early warning system. Above all, however, an attempt has to be made to engender positive attitudes towards survival among those who, because their poverty, lack of housing choice,

continued ...

Box 4.17, continued

and their historical close relationship to the coast, have a tendency to face nature's ferocities with resigned fatalism.

The successes of the CPP have been built on in recent years with particular emphasis in embedding cyclone response planning into a broader coastal management framework (Centre for Water Policy and Development 2001). The development of an Integrated Coastal Zone Management Plan for Bangladesh was a 2001 initiative, using a multi-level (national to community) approach that sought to integrate cyclone protection into broader coastal management issues (de Wilde and Islam 2002).

(b) The risk and hazard management process

Developing a risk management strategy involves a number of distinct stages (Table 4.10). These stages are very similar to the basic phases used in strategic planning, including coastal management plans, namely:

- scoping and investigation;
- analysis;
- implementation (mitigation); and
- monitoring and review.

The traditional way to carry out a risk management program was to undertake each of the steps shown in Table 4.10 in turn, starting with hazard identification. This approach has now been found to be appropriate in only limited circumstances – mostly when there is little interaction with the public required, as in the case study shown in Box 4.18. In this example, a standard staged approach was undertaken, following the general phases described above. It demonstrates the 'technical' model of risk assessment

Table 4.10 Definitions of the stages of the risk management cycle

Step	Description
Hazard identification and prioritisation	• Determines what can go wrong by identifying a set of circumstances and established priorities of the most urgent hazards requiring policy action
Risk assessment and characterisation	• Evaluates how likely is it that a set of hazardous circumstances will arise and estimates their consequences • Determines subjectively the tolerable level of risk • Considers how risks can best be avoided, reduced to tolerable levels and controlled
Policy decision and implementation	• Chooses and implements a particular policy option in the most inclusive and open manner possible
Risk evaluation	• Provides feedback about the net effects of risk mitigation.

Source: Adapted from Gerrard (1995) and Soby *et al.* (1993).

Box 4.18

The management risks for the transportation of hazardous goods in UK ports (adapted from Gavaghan 1990)

Technical risk assessment was used in the United Kingdom to analyse the chances of a major accident occurring due to the transportation through ports and loading/unloading of hazardous cargoes such as flammable petroleum products, radioactive waste and potentially explosive fertilisers (Gavaghan 1990). The study used the normal steps in undertaking a technical risk assessment (see figure opposite) each of which is described briefly below.

Step 1 in the technical risk assessment is to study how a port is used in the movement of hazardous cargoes by analysing the various loading and offloading actions, storage procedures, and how the ships actually move around the port. This analysis is undertaken in such a way as to assist in describing possible accidents (Step 2), and specifically the frequency and consequences of these accidents (Step 3) (see figure opposite). Possible accidents studied included:

- the collision of two moving vessels;
- a passing ship striking a berthed ship;
- a ship running aground;
- a shipboard fire spreading to the cargo;
- the splitting of cargo during loading or unloading;
- ship failure due to construction defects;
- a ship being struck by falling aircraft;
- failure of shore-based equipment (pipelines, etc.); and
- the 'domino effect' of minor accidents combining to cause a major disaster.

The way in which each of these events is analysed is through 'event tree analysis', a standard method for such analyses in the cases of engineering process risks. One such event tree is shown in the figure opposite.

The figure also shows how the probability or frequency of such accidents is analysed (Step 3). At each branch in the event tree a probability is analysed for that particular event. For example, a ship struck at the jetty must be hit at an oblique collision angle for a spill to occur. The probability of this 'yes' case is 0.38 or 38 per cent (see figure opposite). Therefore, the probability the collision will not be oblique is $1 - 0.38 = 0.62$ (62 per cent). The results of this analysis show the overall probability that a gas carrier struck at a jetty will produce a gas spill is 0.019 or 1.9 per cent.

The next step in the technical risk assessment process is to analyse the consequences of each 'risky' activity. In this case study, the consequences to human life were studied. The results showed that a frequency of one human death every 2.5 years is likely to occur in British ports as a result of the movement of hazardous cargoes.

The final two steps of this risk assessment process are when judgements are made on the tolerability of risks. In the United Kingdom, the government's Health and Safety Executive (HSE) divides risk tolerability into:

continued ...

Box 4.18, continued

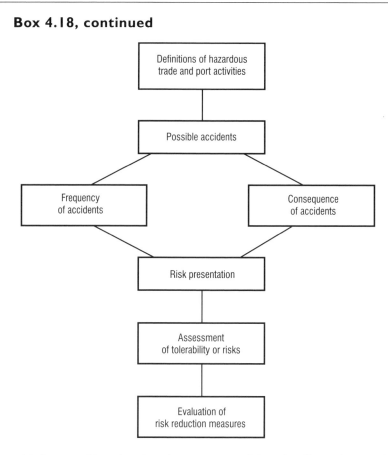

(a) Steps used in undertaking the assessment of the risks of hazardous goods transport in UK ports

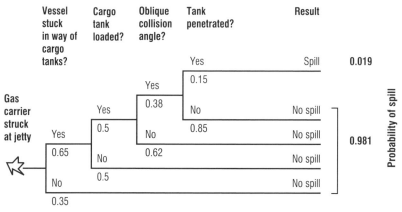

(b) Events and probabilities which must take place for a damaged tanker to spill its load

continued...

Box 4.18, continued

Negligible   acceptable to most people and does not require action.
Intermediate   industry must apply risk reduction measures, unless the cost
                    is grossly disproportionate to the lives saved.
Intolerable   requires risk reduction regardless of costs.

   The results of this risk assessment are currently being implemented in the
United Kingdom.

---

and management, that is with a focus on the technical analysis of risks coupled with
technical risk mitigation measures.

   Risk communication is not a large part of the risk management approach shown in
Box 4.18, mainly because the management of risks in this case is essentially an internal
management issue to do with the operational management of port facilities. However,
in situations conveying the message of risks, and how best to manage them, to others –
especially the general public – a different approach is required which places 'risk
communication' at its heart (Figure 4.12).

   The risk management cycle differs from a linear step-wise view of risk management
in that it emphasises the importance of feedback, so much so that the starting and
finishing points for risk management become blurred (Gerrard 1996). The key element

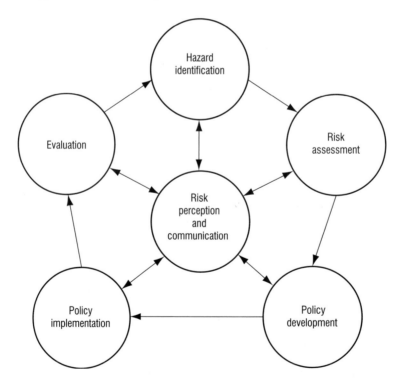

Figure 4.12 The risk management cycle (Soby et al. 1993)

in the risk management cycle is the central role of the risk perception and risk communication, which goes beyond the traditional view of 'public' (non-expert) perceptions, and the one-way communication of risk information from experts to the public. Applying this principle, the management of risks in the coastal zone which affect the general public, for example cyclone risk management strategies, requires the voice of the public to be heard at a very early stage in the process. Furthermore, their opinions should be incorporated into the decision-making process: managers can no longer afford to dismiss these views as uniformed or irrational. In other words, present day coastal risk management must be undertaken extremely sensitively and comprehensively (Gerrard 1994). These factors should also be taken into account if the study and management of risk is to be incorporated into planning procedures, including the production of coastal plans (Chapter 5).

A key part of the risk management process is the evaluation stage, which comes between assessing what the levels of risk are, and deciding what if anything should be done about them. A number of methods can assist in the evaluation process, the most common of which is Risk-Benefit Analysis. As the name suggests, this method is based on evaluating the economic efficiency of risk mitigation measures by adapting cost-benefit techniques (see Section 4.3.4). Other techniques are described in various specialised texts on risk management listed above.

A central concept in the risk evaluation process is that of 'tolerability' (O'Riordan *et al.* 1987). This is a carefully chosen word which says that people essentially 'put up with' (tolerate) risks, they do not necessarily accept them as was previously thought by risk experts. This notion then flows into the decisions which are made to reduce risks, depending on whether they are assessed to be 'tolerable' both by regulatory agencies and by the general public who will be affected by risk-related decisions. For example, in the United Kingdom much of the government's effort in risk assessment and management is focussed on the principle that efforts should be made to reduce risks only when they are either entirely intolerable, or sharing the cost of reducing them further is still reasonable given the extra risk reduction gained (Gerrard 1995). This is known as the 'as low as reasonably practicable' (ALARP) approach (Figure 4.13).

The ALARP approach yields three risk management areas, shown in Figure 4.13.

(c) Mitigating risks and hazards in coastal planning and management

Mitigation of hazards and risks is required where the risk management process has determined that risks are currently intolerable, as described in the previous section (Newman *et al.* 2002). The range of ways to mitigate risks is wide, and there are just as many ways to classify them, the simplest being to divide mitigation options into the following (not necessarily mutually exclusive) categories (Standards Australia and Standards New Zealand 1999):

- avoid;
- reduce the likelihood of occurrence;
- reduce the consequences;
- transfer; or
- retain/accept.

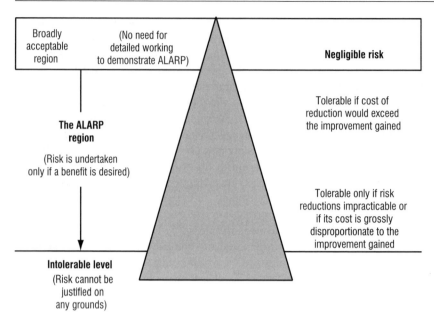

Figure 4.13 The 'as low as reasonably practicable' (ALARP) approach to risk management decision making (HMSO 1988; Gerrard 1995)

The wording of each of the above choices gives an indication of how they are applied, 'avoid' meaning that risky activities are either not undertaken, or that hazardous areas are identified and avoided. It is important to note that these options are generic, and will vary according to the types of hazards being addressed and the decision-making context. For example, in the context of managing drowning incidents on beaches, risk management means developing a holistic approach to the provision of safety services, education and skill provision (WHO 2003), while Table 4.12 shows the application of risk management in the context of coastal erosion.

Details of the background and application of the options shown in Table 4.11 are given in numerous studies of natural hazard impact management (e.g. Platt *et al.* 1992; Hewitt 1997; Chapman 1999; Mileti 1999; Heinz Center 2002). Each of the options

Table 4.11 Example options and measures for coastal erosion hazard management

- *Event protection*
  'Hard' engineering (e.g. sea-walls, groynes)
  'Soft' engineering (e.g. beach nourishment, dune enhancement)
- *Damage prevention*
  Avoidance (e.g. prevent development)
  Mitigation (e.g. relocatable or flood-proofed buildings, building codes)
- *Loss distribution (transfer)*
  Individual measures (e.g. insurance)
  Community measures (e.g. insurance, cost pooling, disaster relief and rehabilitation)
- *Risk acceptance*
  Do nothing

Source: Adapted from Kay *et al.* 1994.

can be applied by governments through a range of measures (Burby *et al.* 1991; Beatley *et al.* 2002), including legislation and the application of policy and economic instruments such as tax concessions or the removal of government assistance grants (see Box 4.25).

The key issue here for coastal planning and management is that there are many ways to address risk and hazard issues. Risk and hazard management can either be undertaken as a stand-alone exercise, as described above, or included as part of a coastal management planning process. Perhaps the principal advantages of the latter are: first, that if the planning process is undertaken in a consultative manner, the often difficult issues of risk and risk management can be openly discussed, and solutions found jointly by affected communities and decision makers; and second, the choices for risk and hazard management can be increased by inclusion into a management planning process. There may be good reasons for undertaking certain actions in the name of good coastal management practice, but which also contribute substantially to risk management. Management of coastal dunes in front of hazard-prone beach front property, for example, can be for the conservation of biological diversity, recreation and public access, and a way of providing a sand-buffer during erosive storms.

4.3.3 Landscape and visual resource analysis

One of the most commonly appreciated aspects of the coast is the scenery. Rugged cliffs, picturesque bays, idyllic beaches, rolling dunes, rocky headlands, marshes and tidal flats, forested hinterlands, breaking waves and expanses of water all help create scenery which is highly prized by people. The coast, as the interface between water and land, often has a richness and diversity of scenic features not found in inland areas.

The value of this coastal scenery hardly requires substantiation. It is a vital component of people's enjoyment of the coast. It is the setting for people's activities and is a strong influence on their sense of well-being and quality of life. Coastal scenery adds to property values and provides the settings, and often the attractions, for tourism. This economic value of scenery is increasingly being determined by environmental economists as part of the coastal planning process (Section 4.3.4). Coastal scenery is also regarded as an important value in its own right and in many planning cases it has proven to be the most important value in land use decision making. It is also a value that is an accepted component of many resource assessment programs.

Coastal scenic values are only one component of what can be called coastal 'landscapes'. The landscapes can be described and assessed in terms of values, which can include aesthetic value (based on the sensory perception of a place by the community), social value (based on the association between community, including indigenous people, and place), historic value (based on a connection with an historic figure, event, phase, or activity) (see Blair and Truscott 1989) and scientific value. While aesthetic values are often regarded as the most important component of landscape value, it is clear that all these common types of values often play an important role in the formation of landscapes. The relationship between landscape value and these related values is illustrated in Figure 4.14 (Cleary 1997, 2004). Landscape values are also closely linked to recreational values, but whereas the former is focused on values related to perception, understanding and enjoyment of the environment, the latter has been traditionally orientated towards the management of recreational activities and appropriate settings for these activities. Landscape values also underlie the tourism value of many places.

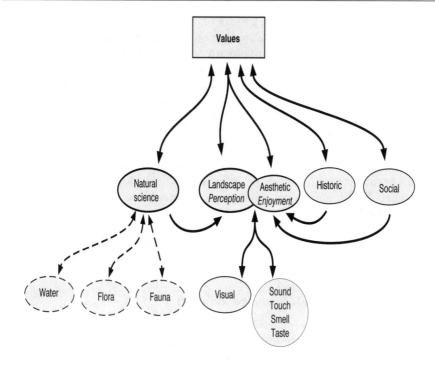

Figure 4.14 Relationship between landscape and visual values (Source: Cleary 2004)

There is a great need to assess and manage landscape values in coastal areas. The increased popularity of the coast as a place to live and visit, and the consequent development, has put immense pressure on many environmental values including landscape values. The will to protect landscape values generally stems from one of three areas:

- management agencies and individuals recognising the long term benefits of good resource management;
- lobbying by interest groups to protect the resource and to have their sentiments included in the decision-making process; and
- legislation and related legal cases requiring that landscape values be protected.

(a) Procedure development

Landscape assessment is a relatively new field and universally accepted methods of assessment do not exist. Formalised procedures for the assessment of scenic quality were first used in the 1960s and 1970s and developed rapidly as a response to legislation and related legal cases. In Britain and the USA, various Acts including The Countryside Act (1968) and the National Environmental Policy Act (1969) provided the initial impetus which ensured that aesthetic values were addressed in land management. One of the problems that these early procedures faced was the view that scenic values are relatively 'intangible' given that assessment results are based on 'subjective' judgements of observers (Williamson 1978). Despite being based on personal values or judgements, landscape

Box 4.19

Main components of assessment of values (Cleary 1997)

- Preference testing, which identifies the characteristics that are most important to the experience and enjoyment of people (see discussion on theoretical paradigms). Findings from specific studies are often applied in a general way to other studies. Holistic appraisals of places or developments should be restricted to evaluation rather than assessment and should be employed with caution (Castledine and Herrick 1995).

- Landscape character description and classification, which identifies and describes broad patterns of characteristics, allowing consideration of broad areas rather than individual sites. It provides an understanding of the diversity of places within a region, and the sense of identity of the whole region. It also gives a broad indication of appropriate land use and the basis for assessing significance, by providing an inventory of characteristics and by classifying their common patterns. Character is often classified under the categories of natural and land use character (land use is the character that most often changes).

- Assessment of significance, which uses the preferences of the local community, or established criteria that have been determined by research conducted in related studies, to identify the most important landscape features.

- Assessment of community use, which is a measure of the degree to which people experience the area. It identifies and classifies access routes and use areas. This information is vital to establish the degree of public interest and is commonly used to weigh other values.

- Assessment of views, which examines the main way most people receive environmental information as part of the perception process. It determines the general view experience and the key views available for people using an area. Selected key views may form the basis for visual simulations of the development, prepared as part of an impact.

- Assessment of wilderness quality, which is an indication of the naturalness and remoteness of places. This component helps ensure that new development does not diminish the extent of wild places.

- Assessment of the level of representation the values within defined scales (e.g. study area, regional, national). This information is used to weigh the assessed values. Values that are not well represented (i.e. rare) are given priority in land use management decisions.

values are no more intangible than any other values or study outcomes based on human preferences, such as the findings from market research or voting. As well, the standards for measuring the quality of any resource are based on subjective judgements, and that the issue is not subjectivity versus objectivity but rather the acceptance and consensus of standards and techniques used to measure quality (Brush 1976; Daniel 1976).

The trend in landscape assessment of including and integrating a wide range of aesthetic, social, historic and scientific values has occurred only in recent times. In the past, a large proportion of landscape assessments have been directed purely at visual values (Fabos and McGregor 1979). A number of reasons have often been put forward for this: there is a long history in many countries of people prizing the visual attractiveness of places (Johnson 1974; Laurie 1975; Fabos and McGregor 1979); people's perception of any place is largely visual (USDA Forest Service 1973); and there is a substantial body of legislation requiring management of visual values (Zube *et al.* 1974). Now, with the changing and broadening of definitions that has occurred in the landscape perception research field, it is recognised that the complex process of interaction between humans and the environment produces many outcomes other than measures of scenic quality (Zube *et al.* 1974).

KEY ASPECTS OF PROCEDURES

Excellent reviews of procedures by Fabos and McGregor (1979) and Ribe (1989) reveal a wide variety of procedures designed to assess a range of landscape values. The range of methods used shows variations in four key aspects: purpose, public input, paradigms, and techniques (Cleary 1997). The choice or development of any assessment procedure should consider these aspects. Following is a brief discussion of these taken from Cleary (1997).

PURPOSES OF THE PROCEDURES

The purposes of the procedures can be grouped into four main categories. The first purpose is to identify landscape values. This is usually undertaken over a large study area (regional or local scales as defined in Chapter 5) as part of ongoing environmental management programs. The second category of purpose is to identify impacts, usually undertaken as part of an Environmental Impact Assessment or management evaluation of development (see Section 4.3.1). The third purpose is to assist in the planning (rather than impact assessment) of development and can be specific to a development or apply to more general guidelines or policies, such as those within land use planning frameworks. The fourth category of purpose is to identify public preferences. This can be done to identify attitudes towards specific projects or to provide general preferences.

COMMUNITY INPUT

The second aspect of procedures, community input, varies from none to a heavy reliance on public preferences. Community interest has proven to be a key criterion in evaluating proposals in a statutory or legal context (McCloskey 1979; Castledine and Herrick 1995). Findings related to community interest have important implications for both assessing and protecting landscape values (Johnston 1989) and are the basis for a number of criteria which indicate the community interest value of a landscape feature. Key criteria are:

- there is a tested public preference for the feature;
- the significance of the feature passes a threshold;
- people can experience the feature; and
- the number of people experiencing the feature passes a threshold which indicates its scale of significance (e.g. regional).

There are four main possible inputs into the planning and design process:

Identification of values. Values and their protection are the focus of most planning frameworks. While there are well established assessment criteria for landscape values, there is merit in including the community in identifying values as it engages them in the process, creates better understanding of the process and the values, and in some cases (particularly for social value) adds to the values' knowledge base.

Influencing design. This is often the most desired (by the community) but least used avenue for community involvement in the process. This type of involvement (usually workshops and meetings) helps people to feel they have helped shape their future environment – that they have had a say. Even in the case where it may seem that a development has little design flexibility, changes can usually be made.

Influencing the decision-making process. This is characterised by lobbying, submissions/objections, use of the press, and strong local communications networks. If the first two parts of the community input process (i.e. 1 and 2 above) have neither been used nor been successful, a great emphasis is placed on this part of the process. In this case, there is often a feeling of disempowerment amongst the community. If the process has been successful then input usually reinforces the outcomes of parts 1 and 2 (i.e. there is much greater sense of ownership of the project design).

Influencing the planning framework. The various instruments within statutory and strategic planning frameworks are revised from time-to-time to reflect contemporary land use, values and community sentiment (Chapter 5). These changes usually happen over long time frames, not associated with specific projects, but occasionally the importance of an individual project will be a cause for changes to the planning framework.

THEORETICAL PARADIGMS

The main theoretical paradigms for procedures are based on two quite different views of the nature of landscape values (see Zube *et al.* 1974).

One view is that landscape qualities are inherent in physical characteristics of places and that assessment will reveal these qualities. This approach has resulted in a range of descriptive inventory or expert approaches, usually categorised as either a 'landscape approach' (McHarg 1969) or a 'formal aesthetic approach'. The underlying assumption of these approaches is that experts are the best people to judge which places or landscape elements are of the greatest significance. This view is becoming less relevant as land use decisions relating to landscape values place greater emphasis on public interest and on landscape assessments based on researched public preferences.

A second view is that landscape qualities are linked to a person's experience of the environment. There are a number of approaches based on this view: the 'psycho-physical approach' endeavours to identify the relationships between environment characteristics and a person's response; the 'psychological approach' endeavours to identify the reasons, particularly the psychological processes which make people respond in different ways to the environment; and the 'experiential approach' endeavours to gain an understanding of people's interaction with the landscape in terms of personal experience, feelings and meanings.

The psycho-physical approach, which involves eliciting, measuring and analysing public response to a series of environments, provides the most practical outcomes and is the most used (Zube *et al.* 1974). Even so, as Ribe (1989) points out, findings about the relationships between environmental attributes and perceived beauty tell us little about the aesthetic magnitude of particular differences. They also fail to tell us when the emotional nature of perceptual change is sufficient to constitute a new state, as between beauty and ugliness, or approval and disapproval. This information would be particularly useful for landscape impact assessment.

(b) Techniques

The fourth aspect of procedures to consider, and the one that reflects all the other aspects, is techniques. There is a wide range of techniques available for landscape assessment, which relate to the components illustrated in Figure 4.17, showing their general relationship to one another. These techniques are not specific to coasts and can be equally applied to inland areas. The understanding of the role of these techniques and the adequacy of their application in assessment studies and environmental impact statements is varied (Coles and Tarling 1991). The main techniques are discussed below and key points relating to process and standards are highlighted.

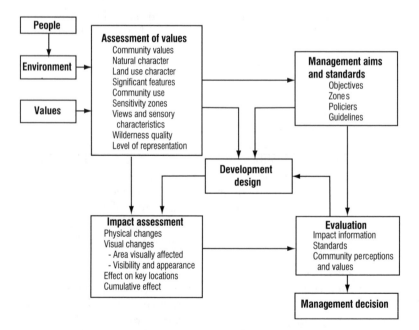

Figure 4.15 Main components of landscape management

ASSESSMENT OF VALUES

Assessment of values identifies existing landscape values (see Box 4.18), providing the baseline knowledge for a variety of further planning and management processes. Although there is considerable variation in assessment methodologies, the intentions of most are directed at answering three simple questions. What is there? What is most valuable, and how do people experience these? Having answered those questions, however, decisions have to be made about what weighting should be applied to the assessment components, variables and values. The processes employed to answer these questions generally have main components as described in Box 4.20.

The results from these components are sometimes combined to produce resource zones which, depending on landscape management and land use strategies, may form the basis for management zones.

These components, together with other considerations such as a sound theoretical basis, comprehensiveness and consistent application over large areas, have formed methodologies which have proven to be most useful for planning purposes and the most defensible against public scrutiny and court challenges. Further criteria relating to the development of procedures is provided by Fabos and McGregor (1979), Ribe (1989), Zube *et al.* (1974) and Castledine and Herrick (1995). One procedure that has been widely used in the past, particularly in the United States and Australia, is the Visual Management System (see Williamson and Calder 1979). While this procedure has a number of shortcomings, many of the techniques it employs have been used or refined in other assessment procedures.

IMPACT ASSESSMENT

Compared to resource assessment, impact assessment is a less developed, less understood and poorly applied process (Lange 1994; Institute of Environmental Assessment (UK) and the Landscape Institute 1995). It is essentially a process of comparing the conditions before a development with the projected conditions after, and then determining the corresponding effect on landscape values. It is not the purpose of impact assessment to determine whether a development is acceptable or not, but rather to produce a statement of effects. Acceptability is decided in the evaluation of the project. Typical failings in impact assessments are (Cleary 1997):

- there is no comprehensive reference to existing landscape values;
- changes are not comprehensively identified or are not measured;
- there is no statement of effects (sometimes erroneously substituted by a judgement of acceptability) or the statement of effects does not refer to comparable variables (i.e. landscape character, significance and access and views); and
- visualisations are used to substitute rather than support all of the desirable components highlighted in the above points.

Many impact assessments are relatively simple and all should involve the following few steps (Cleary 1997):

Box 4.20

Terminology in landscape and visual assessment

3D modelling is a technique, usually performed on a computer, where landforms and objects are accurately and mathematically defined in three-dimensional space. This allows the reconstruction of views of these landforms and objects from any location using a rendering process that applies textures to 3D forms to make them appear realistic.

Area potentially visually affected or *zone of visual influence* (ZVI) mapping is the total of the area that is within the seen area of all the points within the 3D form of the development. It should be treated as indicative only given it does not usually include vegetation cover.

Cross sections depict an object or area with part of the object or area cut away to highlight the profile or shape at the plane that defines the cut.

Evaluation is the process where assessment results are examined and used to make decisions about alternative futures, usually based on given standards.

Impact assessment is a process of determining how changes to the environment will affect landscape values.

Landscape is used by many different people for a variety of purposes, making it a rather ambiguous term. It has three main usages: the first refers to a scene (as in a landscape painting); the second refers to an area which has a common pattern of bio-physical features (as in a landscape ecology); and the third usage refers to a person's perception or image of an environment (i.e. it is a human construct) (Zube *et al.* 1974; Lowenthal 1978; Meinig 1979). The landscape architectural, environmental psychology and related professions to a certain extent use all definitions but specialise in an understanding of the latter.

Landscape value is the value they attach to a place based on 'their landscape'. *Landscape value* and *visual aesthetic value* are often used synonymously.

Seen area is a term used to describe the land surface that is potentially visible from a given point.

Sensory characteristics relate to the paths by which people receive environmental information (e.g. vision, hearing, etc.).

Values are measures of the importance people attach to things and typically stem from perception.

Visual characteristics relate to information received through the visual sensory path.

Visual effect is a term sometimes used to describe the likely visibility and appearance of a development (often described in terms of magnitude, contrast and duration) based on the visual characteristics of the development and variables such as landform (e.g. slope, proximity to ridges), vegetation pattern and height, and existing land use. Similar to *visual absorption capability*.

Visual aesthetic value refers to the visual aspects of aesthetic value. *Aesthetic value* refers to personal appreciation and enjoyment of things (e.g. objects, places, and processes). It can include beauty, functional and non-functional aspects of things, and does not necessarily include visual qualities.

- Determine the degree of physical changes to the site. Changes to vegetation and built form are very common and can be measured (e.g. area, height).
- Determine the visual changes brought by the development. Like physical changes, visual changes (expressed as visibility and appearance) are also easily measured, and can be expressed in terms of magnitude (often expressed as percentages of horizontal and horizontal fields of view), contrast (shape and colour), area affected (determined by seen area mapping (or ZVI)), and duration (timing of the changes described above).
- Determine the impact the physical and visual changes will have on the values of the area (i.e. landscape character, significance, access, views and wilderness quality) – which should have already been assessed. For example, a development in an otherwise natural area may be clearly visible from major lookouts, may permanently remove the natural character, may permanently remove a significant feature, may improve access, or may temporarily block views.
- Describe the impact from key locations, including from neighbouring properties.
- Determine the cumulative effect of the development. This may be done as a measure of distance between developments, percentage of a region affected by the development, or percentage of the main journeys in the region affected by the development.

Relevant supporting material from the steps above should be used to produce the effects statement.

AIMS AND STANDARDS

Management aims and standards set the desired future for landscape values. They often initially emerge in the parts of assessment reports dealing with management. Ideally these aims and standards will then be incorporated into plans, strategies or planning schemes which have statutory force as this will add to their defensibility (Castledine and Herrick 1995). In some cases the standards might be so high that some form of land reservation is recommended in order to protect values. Regardless of the statutory force of the standards, they should be widely promoted in the community to ensure that they are included in the early stages of development planning.

Typical ways to express these aims and standards are:

- objectives, which detail the desired future of the values in tangible terms;
- zones, which provide the geographic base for applying standards;
- policies, which are rules designed to fulfil management aims; and
- guidelines, which are usually recommended ways of achieving management aims.

EVALUATION

Evaluation is usually undertaken by the responsible planning or management authority in the lead up to making a management decision (which usually incorporates many other considerations). It is essentially a process of comparing the development's effects with the landscape management and land use standards of the area to determine its acceptability in relation to these standards. The determination may be either approval

Box 4.21

Case study – community involvement in the process (Cleary 2004)

Often we hear of consultation programs for major projects where the community feels frustrated with the process and often, as a result, concentrates effort into opposing such projects. At the same time, the project developers can feel frustrated by this opposition when in their minds a good consultation program was conducted.

Disempowerment aptly describes the feeling of the community in this type of situation – people may feel like they are not having meaningful input into shaping their future environment. These situations also often arise partly as a result of confused expectations. The developer needs to be very clear why they are undertaking consultation and involving the community, and the community needs to understand their role in the process. This is not to say that a developer can explicitly outline a very limited role for community involvement in a project and expect good community feedback. Experience from landscape and visual resource analysis shows that a high level of community involvement produces the best outcomes. Aspects of this involvement can be can be relatively straightforward – there is usually consensus on community values, and community ideas for minimising impacts are usually consistent with minimising impacts generally.

Landscape assessment provides one of the best opportunities for people to have meaningful input into the planning and design of development on coasts. There are several reasons for this, including:

1   Landscape assessment's primary role is to address community values. Many people in the community quickly recognise this and use the process as a conduit for expressing the feelings about a project, whether it be about their values or their approval or disapproval of a project.
2   Landscape assessment practitioners usually have the skills to act on people's expressed sentiment. With most projects they will play a key role in the planning and design of the project and given this, can provide immediate feedback to people and can carry the community input into the planning and design process. Part of this process is creativity, developing ideas into options and solutions that meet the needs of the developer and address the sentiment of the community.

A good example of the application of this approach is the redevelopment of the Smiths Beach Resort in Western Australia. This is a much-loved camping area in a picturesque part of the coast. The redevelopment was initiated by the current trends to built accommodation rather than camping and the pressure for higher returns, as a result of increased costs. Recognising that this was potentially a controversial project, the consultation program adopted a number of techniques to involve the community and foster a good working relationship. These techniques included:

continued...

Box 4.21, continued

- ongoing involvement by the community, commencing at the earliest possible stage;
- regular exchanges of information to ensure all parties were up to date;
- community values were identified and were explicitly addressed in the planning and design of the development;
- community ideas for minimising impacts were adopted as far as possible; and
- involving the community in the creative process of developing solutions.

As a result the project development application was lodged with a high level of community support, which in turn simplified the decision making process of the responsible authority.

---

or disapproval, with or without modifications recommended. The evaluation process is often weakened because of the lack of an adequate effects statement or inadequate management standards (Lange 1994). Without this information, the evaluation will revert to personal judgement, which will vary depending on whether they are a staff of the responsible management authority, proponent, planner, community member or relevant expert. In some cases, evaluators have been reduced to making judgements on landscape impacts using photographs of the site and descriptions of the development.

MODIFICATION

Modifications to the development to reduce landscape impacts can be recommended as part of the evaluation process (see Box 4.21). These will generally focus on one of the following.

- Site selection, which usually offers by far the greatest opportunity to reduce impacts. Techniques that can be employed for choosing sites with the least visual impact include mapping of use areas, slope, proximity to ridges, vegetation (pattern, density, height), soil (colour, erosion potential), and Seen Area Mapping which identifies the area seen from nominated sites and consequently the least seen sites.
- Design, which can reduce impact through appropriate layout, form, colour, texture, scale, and pattern of visible elements of the development.
- Screening, which uses ground modelling and planting to reduce the visibility of the development.

(c) Application of the techniques

Landscape assessment techniques have a number of strategic uses. They provide an understanding of the diversity within the region and the sense of identity across the region. Landscape character mapping provides a broad indication of appropriate land use (using landform, soil, vegetation, water and existing land use). Assessment of significance identifies areas of highest landscape value that might be best protected by land reservation or statutory controls. Mapping of various characteristics can help in recreation and tourism planning by identifying attractive features and key view points. It can also identify opportunities for landscape improvements such as vegetation corridors. Recommendations can be made as to the types of land uses appropriate for different areas based on the impact of various land uses on landscape values. Assessment also forms the basis for zones, policies and guidelines that can be applied over a broad area. It avoids *ad hoc* decision making, helps control incremental development, and allows planning for sustainable use of the resource.

At a development project level, assessment provides the basis for protection of landscape values and the creation of developments with sense of place. Some examples of application at this scale are:

- development can be kept away from areas with significant values allowing these values to remain an attraction for people, including those who might use that development;
- sites for development can be chosen which minimise the impact of the development on other users of an area while still retaining important features such as good views;
- design principles can be developed from local characteristics and can reduce the visual impact of the development while creating interest and a local identity;
- developments can be designed to offer a range of environmental experiences.

The application of landscape assessment to wind farm planning is shown in Box 4.22, which shows both the potential and challenges in its application in coastal management and planning.

Box 4.22

Wind farms and landscape planning

The development of wind farms on the coast has presented some interesting challenges for landscape planning (e.g. Gipe 1993; Metoc PLC 2000; Gasch and Twele 2002). Wind farms include very tall structures and are often sited in very prominent locations to capture high wind energy. The coast often has a high density of significant landscape features, by nature allows panoramic views, and forms the setting for a range of recreation, tourism and residential uses that focus on the landscape features and views. In addition, local communities have often expressed the notion that coasts are somewhat more precious then inland areas (usually based on personal opinion rather than assessment outcomes), suggesting that wind farms would be better located away from the coast.

The key to addressing these challenges is a good understanding of landscape values, determined by assessment. This knowledge allows values to be appro- priately protected, wind farms to be designed in a responsive manner, therefore achieving compliance with policies within the planning framework.

continued...

Box 4.22, continued

The wind farm at Albany in Western Australia (see figure) highlights some interesting questions, in particular whether values have been adequately protected, and the role community sentiment plays in the planning process. This wind farm is located in a reserve, undeveloped, with natural vegetation and is adjacent to significant coastal features, a walking track of national significance, and two national parks. At the same time, community support for the project has been high, even in this location. It can be argued in this case (and stated very simply here), that values have not been adequately protected and that community sentiment was too highly regarded in decision-making. The appeal of wind turbines or the level of community approval should not be confused with the protection of values. For example, levels of acceptance amongst people for wind farms is partly related to landscape and visual values but is determined by many additional factors, including: support for alternative energy sources; proximity to the wind farm; financial returns; perceived or actual effects other than aesthetic effects; general community benefit (including jobs and perceived positive or negative effects on tourism); and the effect on potential to develop adjacent areas. Partly because of this broad range of factors, the existing planning instruments, such as schemes, guidelines, policies, and overlays, are the most appropriate way to address levels of acceptance. These instruments have been devised partly to reflect the sentiment of the community of the day, and to accordingly provide direction and controls for new development.

Albany wind farm, Western Australia

(d) Conclusion

Landscape values are an everyday part of our lives and are increasingly being considered in coastal management programs. There are obvious benefits from landscape assessment such as better resource management, improved experiences for people using the coasts, protection of tourism assets, less land use conflict, stronger local identity for developments, and greater support for management authorities. As with other resource areas, the best results are obtained by specialists who understand the role of the various assessment techniques and who recognise the opportunities that assessment reveals. This is not to exclude the important role that the public plays in determining values and lobbying for their protection, for it is the response to public interest which has been the driving force behind most visual and landscape assessment work.

4.3.4 Economic analysis

Traditional economic analysis of coastal management problems has not always provided appropriate solutions and often confused coastal mangers by not supplying realistic guidelines for their actions. Economists analyse issues by using a set of economic rules which can appear to distort, or even misrepresent, real-world coastal issues. Indeed, many economists would now recognise that such distortion has taken place in the past, using apparent economic rationalism to lead to poor and/or short-sighted decision making.

However, during the past thirty years or so problems with the way 'classic' economic views the environment have been tackled by the rapidly expanding field of 'environmental economics'. Environmental economics attempts to provide valuations of the 'non-market' goods and services provided by the environment. This is the value 'of' and not the value 'in' the environment.

Environmental economists would be the first to say that their work is still relatively new. As such, its application has only recently moved into the mainstream of coastal zone management decision making, and only then mainly in Europe and North America. Nevertheless, as free-market economies spread, together with the community's growing awareness of the importance of environmental quality and its value, economic considerations in coastal management have also grown.

In addition, in an era of increasing accountability in government decision making and reduced budgets, the question of 'how much will this cost?', 'are you sure there are not cheaper ways of doing this?' or even 'is this coastal management program cost effective?' is never very far away. Coastal managers 'thinking economically' can help answer the 'how much?' question which could relate to deciding between spending money on a coastal management program and something completely different, such as building a new school or hospital. Or the question could relate to setting internal priorities within a coastal program, for example, which coastal wetlands in a rehabilitation program require urgent attention versus those which are of a lower priority. Thus, environmental economic analysis can assist in short-term decision making, such as the allocation of funds between government initiatives, and in the long-term decisions regarding the costs and benefits of specific developments along the coast.

Recently the values placed on the coastal resource have expanded to include the value of scenic views, beach access and other aesthetic qualities. Coastal land with ocean views and beach access creates high land values with the beach becoming more popular and environmental quality expectations more acute. Thus, there is a strong demand for uncrowded and good quality (clean) beaches and natural vistas. People are now willing to pay more to enjoy the coast, and are also willing to pay for its protection. A key issue, then, is how these non-market values can be quantified and incorporated within government coastal management decision-making processes.

As is the case with many of the coastal management tools described in this chapter, economic analysis, if not properly applied, can cause far more harm than good. The often complex procedures of economic analysis mean that it is really quite easy to present misleading or false opinions to decision-makers. Fortunately, there is a rapidly growing number of environmental economists and some very good texts on the subject. These include books specifically on applying economic analysis to coastal management issues including the excellent texts produced by Edwards (1987), the United States National Ocean and Atmospheric Administration (Lipton and Wellman 1995), the United Nations Environment Programme (Grigalunas and Congar 1995) and the Land Land–Ocean Interactions in the Coastal Zone initiative (Turner and Adger 1995). In addition there are more general introductory texts on environmental economics such as those by Gilpin (2000), Kolstad (2000) and Hanley *et al*. (2001) and more advanced texts such as that of Costannza (Costannza 1992) and Callan and Thomas (1996).

The increasing use of economic analysis tools in coastal planning and management has stimulated debate on how sustainability is factored into economically-based decision making. The debate is focussed on the choice of discount rates used to calculate net present values (Turner 1993). Discount rates are used simply because people prefer money now, rather than the same amount of money in the future. The ethical dimension to this argument is that by 'discounting the future', there is an implicit assumption that the present is worth more that the future. Also, arguments about discounting relate to how long into the future the economic analysis is undertaken. In other words, how many future generations are considered when looking at issues of intergenerational equity in sustainable development. The point to stress is that the choice to use particular discount rates in any economic analysis in coastal planning and management should be carefully made, either through reference to the texts listed above, or to an expert environmental economist.

Before the range of economic analysis techniques is described, some fundamental economic concepts, and how these are applied to coastal management issues, are introduced in the next section.

(a) Economic concepts

As Turner *et al*. (1995) summarise, economics is fundamentally concerned with the concept of scarcity and with the mitigation of scarcity-related problems. Viewed in this way, economic concepts have an important part to play in management decisions at the coast where the resolution of conflicts over sought-after space and resources is a fundamental element of coastal management and planning. Economics helps managers consider options for the most efficient allocation of resources, balanced by the needs of buyers (demand) and sellers (supply).

The classical economic concepts of supply and demand, and the cost of the opportunity forgone by society once a decision to allocate financial resources is implemented, played a central role in converting natural coastal areas into various 'developed' uses

Table 4.12 Examples of uses and environmental functions of mangroves

Sustainable production functions	Regulatory of carrier functions	Information functions	Conversion uses
• Timber	• Erosion prevention (shoreline)	• Spiritual and religious information	• Industrial/urban land-use
• Firewood	• Erosion prevention (riverbanks)	• Cultural and artistic inspiration	• Aquaculture
• Woodchips	• Storage and recycling of human waste and pollutants	• Educational, historical and scientific information	• Salt ponds
• Charcoal	• Maintenance of biodiversity	• Potential information	• Rice fields
• Fish	• Provision of migration habitat		• Plantations
• Crustaceans	• Provision of nursery grounds		• Mining
• Shellfish	• Provision of breeding grounds		• Dam sites
• Tannins	• Nutrient supply		
• Nipa	• Nutrient regeneration		
• Medicine	• Habitat for indigenous people		
• Honey	• Recreation sites		
• Traditional hunting, fishing, gathering			
• Genetic resources			

Source: Ruitenbeek (1991, 1994).

such as ports. The central economic rationalisation in these developments was to increase the speed of economic growth as measured by, for example, a nation's gross domestic product (GDP) and in doing so excluded reference to environmental costs and social amenity.

Classical economic analysis placed little value on natural environments, including those on the coast. Such places were seen as a bottomless sink, and therefore were valued at close to zero. Not until it was realised that the very environment enjoyed by people was becoming degraded to such an extent as to reduce amenity value, cause serious illnesses and deplete commercial fish stocks, did economists realise that the coastal environment had value; their classic economic models and solutions had failed to provide satisfactory answers (Smith 1996). Economists term such things 'market failures'. Following the increases in environmental awareness since the 1960s, this concept has been extended to include the economic goods and services supplied by an environment in its natural state – extending the fundamental economic concept that 'nothing is free'. This principle can be illustrated by looking at the economic valuation of mangroves (Table 4.12) (Ruitenbeek 1991, 1994; Tri *et al.* 1996; Balmford *et al.* 2002).

Perhaps the most important economic concept relates to 'economic value'. Its importance is summarised by Lipton and Wellman (1995) as follows:

> A fundamental distinction between the way economics and other disciplines such as ecology use the term 'value' is the economic emphasis on human preferences. Thus the functionality of economic value is between one entity and a set of human preferences.
>
> (Lipton and Wellman 1995: 10)

The characteristics of economic value are summarised in Box 4.23. Lipton and Wellman (1995) describe economic value through the different values placed on a polluted coastal area. In economic terms, this polluted area would only have less value than a pristine area because people prefer non-polluted areas to polluted ones. Clearly, an ecological view of the same issue would place different values on the pollution, especially if ecological damage had occurred – but how do you determine the monetary value of an ecosystem?

Box 4.23

Characteristics of economic value (from Lipton and Wellman 1995)

- Products or services have value only if human beings value them, directly or indirectly.
- Value is measured in terms of trade-offs, and is therefore relative.
- Typically, money is used as a unit of account.
- To determine values for society as a whole, values are aggregated from individual values.

An important problem in valuing the uses, functions and amenities provided by coastal environments is that many of these are provided 'free'. That is, no market exists through which their true value can be revealed by the actions of buying and selling (Pearce *et al.* 1989). Environmental economists find many ways to account for these values, as shown below. The important point to stress here is that different categories of economic value can be derived depending on the use and/or service they provide (Table 4.13).

An emphasis on quantifying the economic value of coastal resources is the centrepiece of all the economic analysis tools, the basic guiding concept being the distinction between the valuation of goods or services which are traded in a market (market goods), and those which are not (non-market goods) (Johnston *et al.* 2002). An example is when a commercial fish catch can be traded in the market place, whereas the natural coastal habitats which support the fishery cannot. A mixture of these values occurs when private coastal residential land is traded in the market place: the value the purchaser places on the land will be strongly influenced by the amount they are willing to pay to experience the benefits of living at the coast.

Both market and non-market valuation techniques rely on the economic concepts of 'consumer surplus' and 'producer surplus' to estimate the net 'willingness to pay' (Table 4.14). Without going into the intricacies of these concepts, they are fundamentally to do with the laws of supply and demand; that is, the more of a good that is supplied above what is demanded, the price is lowered to attract more buyers (Adger 2000).

Some economists argue that the easiest way to consider economic valuation is to think about the difference between 'value' and 'price'. In many cases price does not correspond to value. Consider the situation in the United Kingdom where sewerage sludge was dumped in the North Sea for many years. In that case the waste assimilation function of the marine environment was priced at virtually zero, despite its immense value in dispersing waste (Bateman 1995).

The most commonly used way to value resources traded in markets is the estimation of producer and consumer surplus using market price and quantity data. This is one of a range of well established methods which rely on the direct observation of market behaviour and value preferences. They are thoroughly described in most texts on resource valuation (e.g. Edwards 1987).

In contrast, the valuation of goods and services not traded in a market requires some assumptions to be made about how human preference for an environmental good or service is expressed. Three major types of procedures are available to measure these (Lipton and Wellman 1995): direct (including the Travel Cost Method and Random Utility Models); comparative (Hedonic methods); and experimental (methods which elicit preferences, either by using hypothetical settings, called contingent valuation, or by constructing a market where none exists).

Non-market valuation techniques such as these are essentially what separates environmental economic techniques from the classical approaches. However, in reflecting this difference, the application non-market valuation techniques is fraught with difficulty. This difficulty arose from problems of estimating economic values of non-market goods, especially in view of the recent gain in momentum of advancing environmental economics into economic decision making in industry, government and the community. Refinement of these methods continues. Meanwhile, it is as well to

Table 4.13 Categories of economic value

Economic value category	Description	Coastal example
Direct use value	Value obtained from direct, onsite use of a good	Recreational value of beaches
Indirect use value	Value obtained indirectly from a good, where you use another good that depends on the good in question	Coastal wetlands contributing to fish populations
Consumptive use	Good is consumed when used, such that the good is not available for others to use	Fish harvesting
Non-consumptive use	Good is not consumed through use, such that the good remains for others to use	Scenic coastal drive
Non-use value or passive value	Value obtained without the need to use the good	Knowledge of coastal biodiversity existence
Option value	Value obtained by maintaining the option to use the good at some time in the future	Conservation reserves

Source: from Grigalunas and Congar (1995).

Table 4.14 Economic valuation definitions

Consumer surplus	Excess of what consumers are willing to pay over what they actually do pay for the total quantity of goods purchased
Producer surplus	Excess of what producers earn over their production costs for the total quantity of a good sold

Source: Lipton and Wellman (1995).

note that the assumptions made in applying non-market valuation techniques can play a critical part in the results obtained.

Non-market valuation techniques and their underlying assumptions are too complex to be described here. The introductory texts of Lipton and Wellman (1995) or Grigalunas and Congar (1995) provide sources of further information together with the case study of valuing estuarine resource services in the USA of Johnston *et al.* (2002).

(b) Economic analysis tools

A number of general tools are available for economic analysis of coastal management issues (Lipton and Wellman 1995), ranging from relatively simple studies of the most cost-effective means of achieving a clearly defined goal (cost-effectiveness analysis and economic impact analysis), to analysing the regional costs and benefits of a range of interacting environmental, social and economic impacts (benefit–cost analysis) (Adger 2000). Benefit–cost analysis is the most widely used economic analysis tool because of its flexibility and broad applicability. Benefit–cost analysis normalises (usually to

Box 4.24

Steps in benefit–cost analyses (adapted from Lipton and Wellman 1995)

Step	Description	Issues
Describe quantitatively the inputs and outputs of the program.	Analyse the flows of goods and services.	Choice of discount rates.
Estimate the social costs and benefits.	Assign economic values to input and output streams. Use appropriate economic valuation techniques – market and non-market.	Choice of economic valuation technique(s).
Compare benefits and costs.	Total estimated costs compared with total estimated benefits. If the net value (benefit minus costs) or a project or action is greater than zero, then it is considered to be economically efficient.	Policy issues – is a net value greater than zero sufficient justification to go ahead with a project?

monetary values) the potential economic penalties and benefits of particular actions. It is frequently used in assisting coastal management decisions, most notably in Europe and North America, but is less applicable (as with all economic analysis tools) in places where detailed economic information is lacking (Kay *et al.* 1996a). As a result, economic analyses used in day-to-day coastal management are generally restricted to the developed world. Nevertheless, as Box 4.24 will demonstrate, economic analysis is being increasingly used in the developing world, albeit on a project-by-project basis at present.

Benefit–cost analysis compares the present value of all social benefits with the present value of opportunity costs in using resources (Field 1994). In its essentials, a simple benefit–cost analysis requires inclusion of time series data (e.g. depreciation and discounting of future day costs and benefits to present day values) regarding natural resources, environmental quality, and social and economic values. All benefit–cost analyses have three basic steps once the project has been fully specified, including the management options to be analysed (Box 4.24) (Lipton and Wellman 1995).

The three steps shown in Box 4.24 make actually doing a benefit–cost analysis look easy: it is not! Results of such analyses rely heavily on the choice of economic valuation techniques, discount rates and a host of other factors described in a number of texts on environmental economics (e.g. Folmer *et al.* 1995). However, as demonstrated by the case studies which apply benefit–cost analysis to coral reef management (Box 4.25), the technique can be a powerful weapon in the armoury of coastal managers.

Benefit–cost analysis can be carried out at a number of geographic scales, ranging from analysing the economic activities of a single activity to broad national or even international resource uses and environmental functions (Table 4.15).

Box 4.25

Benefit–cost analysis applied to the management of Indonesian coral reefs (after Cesar 1996; Cesar et al. 1997)

In order to evaluate the financial viability of protecting coral reef resources in Indonesia, an economic analysis has been applied by Cesar (1996) and Cesar *et al.* (1997). Cesar's analysis highlights individual economic benefit versus the economic impacts to society of various human activities which impact coral reefs (see table overleaf). These activities are from direct impacts (blasting/poisoning/over-fishing and coral mining) and indirect impacts from increased sediment loads from land based sources (urbanisation and logging).

Although Cesar admits that his figures are first estimates only, reflected in many cases by the large range in his estimates, they are broadly comparable with more recent studies in the Philippines (White *et al.* 2000; Burke *et al.* 2002). The economic analysis of Indonesian coral reefs enables a number of coastal zone management issues to be brought into focus. Perhaps the key issue is that all threats which could be quantified produced a total net loss to society greater than the net benefits to individuals; in some cases the net loss can be up to fifty times the net gain). The economic values presented in the above table show the net benefits of conservation for a single threat per km^2 of reef. Given the large reef area (75,000 km^2) and the multiple threats to which many of the Indonesian reefs are exposed, the potential sustainable economic benefits per year for Indonesia from coral reefs are US$1.6 billion (Burke *et al.* 2002).

Economic analysis was used by Cesar to assess who was responsible for the various threats to coral reefs shown in the figure below, and the economics of their actions. This analysis concentrated on the size of the economic stakes per person and the location of the individuals causing the threat. The size of economic stakes concentrated on the benefits which accrue to the individuals, families, boats and companies involved, including consideration of the economics of management of the operations, such as 'side payments' rents. The location of the individuals causing the threat focussed on whether they were local to an area (insiders) or came into an area (outsiders). Outsiders, for example, include large scale fishing operations. The analysis enabled the division of the threats to coral reefs to be classified, and policy responses designed.

The result of this approach is that the flows of money vary considerably for the different threats to coral reefs. For example, coral mining was found to be essentially a local activity with small economic stakes, and as a result a local 'threat-based' management approach, such as community-based management, is recommended. In contrast, poison fishing is a 'big-outsider' management problem with large-scale operations threatening extensive areas of remote coral reef. Here a national threat-based approach is recommended, with enforcement as its key element.

This approach has formed the basis of assessing the benefits and costs of the large Coral Reef Rehabilitation and Management Program, funded by the World Bank/Global Environmental Facility. The values determined by this program form

continued...

Box 4.25, continued

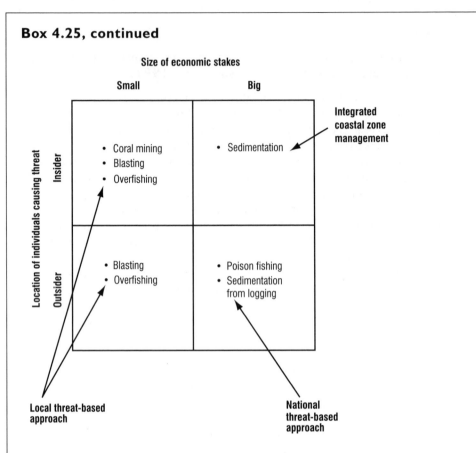

Policy responses to threats to coral reefs in Indonesia based on economic analysis of size of economic stake and stakeholder location (Cesar 1996)

the basis for the World Bank's decision whether or not to provide a loan to the Government of Indonesia for coral reef management. Only if the economic rate of return is high enough (above the so-called opportunity cost of capital), will the loan be forthcoming. This practice is standard in World Bank loans to traditional sectors (e.g. agriculture, infrastructure). However, the difficulty of quantifying environmental benefits has impeded the adoption of this approach to natural resource management investments. The coral reef analysis presented above is a good example of how the World Bank is extending the application of its economic analysis approach to coastal, marine and other natural resources management projects worldwide.

continued...

Box 4.25, continued

Total net benefits and losses due to threats on Indonesian coral reefs (US$1,000 per km^2)

Threat	Function							
	Net benefits to individuals	Net losses to society						
	Total net benefits	Fishery	Coastal protection	Tourism	Food security	Biodiversity	Others[1]	Total net losses (quantifiable)
Poison fishing	33	37	n.q.	3–409	n.q.	n.q.	n.q.	40–446
Blast fishing	15	80	8–170	3–450	n.q.	n.q.	n.q.	91–700
Mining	121	87	10–226	3–450	n.q.	n.q.	>67[2]	167–830
Sediment-upland activities	98	81	n.q.	192	n.q.	n.q.	n.q.	273
Overfishing	39	102	n.q.	n.q	n.q.	n.q.	n.q.	102

Source: Cesar et al. (1997).

Notes
1 'Others' includes loss of food security and biodiversity values (not quantifiable).
2 Estimated costs of logging for fuel in lime-burning kilns.
n.q. not quantifiable.

Assumptions: present value; 10 per cent discount rate; 20-year timespan; all figures are per km^2 of coral reef; intrinsic values not included.

Table 4.15 Scales of benefit–cost analysis (CBA)

Scope	Planning scale	Role of CBA	Mangrove related example
Single operator	Site	Optimise production	Evaluation of forestry profitability under different forest management options
Key resource uses	Site/local	Optimise joint production of two or more graded commodities	Evaluation of joint profitability of fisheries and forestry under different forest management options taking into account linkages between forest and fishery
Traditional production of local populations	Local/regional	Valuation of production	Accounting of physical flows of hunting and gathering and valuation of these flows at local and shadow prices
All resource uses and environmental functions in region	Regional/national	Optimise value of all uses and functions in region	Evaluate joint value of fishery, forestry, traditional uses and erosion control under different management options, taking into account linkages between forest, fishery and other ecosystem components
All resource uses and environmental functions (global scale)	Regional/international	Optimise value of all uses and functions	Evaluate joint value of fishery, forestry, traditional uses, erosion control and international benefits of biodiversity maintenance or climate control under different management options, taking into account linkages between forest, fishery and other ecosystem components

Source: Adapted from Ruitenbeek (1991).

(c) Economic instruments

Economic instruments deliberately intervene in the free market as a means of aligning production costs, as defined by the market, with social costs (total cost of production, including environmental costs) (OECD 1989). This notion of production cost adjustment as applied to mangrove management is shown in (Figure 4.16).

There are three main groups of economic instruments (OECD 1989):

1   Direct regulation.
2   Charges, taxes and subsidies.
3   Market creation.

Table 4.16 shows the three groups, together with a brief explanation of each and examples of their application to mangrove management.

The range of application of economic instruments in coastal zone management is enormous – from 'reef taxes', or levies placed on tourist visitors to coral reefs, to charging for fishing licences. As shown in Table 4.16 these approaches vary from using fines to enforce regulations to highly market-oriented approaches. As Ruitenbeek (1991) points out for mangrove management in Indonesia (Box 4.24 and Table 4.10), each approach has its advantages and disadvantages: direct regulation is the most administratively feasible, yet can be economically inefficient; whereas market creation is the most economically efficient, but is often the most difficult to implement. Between the conflicting goals of economic efficiency and administrative efficiency is the use of charges, taxes and subsidies (Box 4.26).

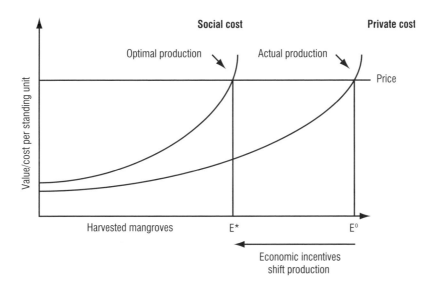

Figure 4.16 Rationale for economic instruments in mangrove management (from Ruitenbeek 1991)

Box 4.26

Example of using the control of subsidies in coastal zone management: the USA Coastal Barrier Resource Act (CBRA)

In 1982 the United States federal government enacted the Coastal Barrier Resource Act (CBRA). The legislation designated various undeveloped coastal barrier islands, depicted by specific maps, for inclusion in the Coastal Barrier Resources System. The Act removed certain federal government subsidies that otherwise encouraged development in the specified environmentally sensitive coastal areas. The Coastal Barrier Resources System now covers of 1.2 million acres comprising 1,200 miles of United States shoreline (GAO 1992).

COBRA's stated objectives were to (Beatley *et al.* 2002):

- reduce growth pressures on undeveloped barrier islands;
- reduce threats to people and property from storms and erosion and to minimise the public expenditure in such cases; and
- reduce damage to fish, wildlife and other sensitive environment resources.

The federal subsidies that were denied in CBRA areas included:

- new flood insurance policies (which also cover coastal erosion);
- federal monies for new roads, bridges, utilities, erosion control, etc.; and
- non-emergency forms of disaster relief.

However, a number funding measures not included in the Act were general 'revenue sharing' grants and a range of maintenance and repair grants, including those to existing public roads, jetties and infrastructure associated with energy resources (Godschalk *et al.* 1989).

These federal subsidies can be considerable, so their removal could be expected to have a marked effect on the level of development of barrier islands.

However, studies which analysed whether the Act had reduced development of undeveloped barrier islands showed it had not stopped the pressures, but had 'some effect' in discouraging new development (Beatley *et al.* 2002). For example, a recent study of randomly selected parcels of land on CBRA areas and non-CBRA areas found 19 per cent and 36 per cent respectively were developed (Salvesen and Godschalk 1999). The study found very different levels of success in controlling development between CBRA areas, with the conclusion being drawn that coastal management policies at state and local levels which were complementary to the objectives of the CBRA were important contributory factors (Beatley *et al.* 2002).

Evaluation studies also highlighted the complexities in using the modification of subsidies as a coastal management tool, and how they can be by-passed by using funds from other levels of government or the private sector. Also, the role of other federal subsidies, such as those offered by US federal government tax laws (such as tax breaks on second homes) and other subsidies specifically excluded from the original Act, was unclear in how they encouraged, or discouraged, development.

Table 4.16 The main groups of economic instruments and examples for mangrove management

Economic instrument group	Explanation	Example in mangrove management
Direct regulation	Direct alteration of price or cost levels	• Greenbelt regulations + fines • Trawling regulations + fines
Charges, taxes and subsidies	Direct or indirect alterations of prices or costs via financial or fiscal means	• Royalties and licence fees • Land or input taxes • Production or export taxes
Market creation	Creation of a priceable market in the use of an environment	• Auction of leases • Tradeable permits • Property right reforms

Source: Adapted from Ruitenbeek (1991, 1994).

Economic instruments are relatively blunt policy tools: their application is generally not directed at a specific geographic area, and can cover several issues (OECD 1989). Depending how the instrument is developed by a government, its application can cover the whole coast or be focussed on particular areas. But it is far less narrowly defined than, for example, a policy to require environmental impact assessments for new developments.

As well as those economic instruments that are deliberately developed by governments to improve coastal management, there are those 'inadvertent subsidies' which actively promote unsound coastal management practices, as in Indonesia, for example, where trade and regulatory distortions create inadvertent subsidies to the over exploitation of mangroves (Ruitenbeek 1991). In order to correct these distortions, Ruitenbeek (1991) recommended that the Indonesian government consider developing an economic incentives system to improve mangrove management. He recommended that five steps be followed in order to develop a charging system (Table 4.17) which must fulfil two key functions:

1   Providing adequate incentives to private operators to develop the mangrove system's sustainably.
2   Providing adequate funds to local or regional authorities to monitor, and where necessary, regulate mangrove development.

(d) Perverse subsidies

Economic subsidies that fail to account fully for social or environmental costs are known as 'perverse subsidies'. Although this term has been well known to environmental economists for many years, their enormous scale has only become widely known in recent years (Myers and Kent 1998). Current estimates of the global total of perverse subsidies range from a $950–1,950 billion per year (van Beers and de Moor 1999; Myers and Kent 2001), depending on whether the hidden subsidies of external costs are included (Balmford *et al.* 2002).

Table 4.17 Recommended steps to developing a charging system to assist mangrove management in Indonesia

Step	Explanation
1 Identify key resource areas	• Economic and ecological importance
2 Identify resource use conflicts	• Forestry vs. offshore fisheries • Tambak vs. ecological functions • Offshore fisheries vs. traditional uses
3 Identify inadvertent subsidies	• Low or unenforced fines • Low fees, charges or royalties • Trade distortions • Land tax or fiscal distortions • Input subsidies
4 Identify and evaluate corrective measures	• Reforms to existing system • Introduction of new measures
5 Implement preferred measures	• Pilot basis in critical areas • Broad basis regionally or nationally

Source: Ruitenbeek (1991).

The implications of perverse subsidies on the management of coastal resources is an emerging field. In the coastal fisheries sector, for example, the subsidies provided to develop national fishing fleets are viewed as one of the major factors contributing to the massive fishing over-effort globally (Milazzo 1997). Other examples include subsidies to drain wetlands to produce agricultural land (Secretary of the Interior 1994).

4.4 Chapter summary

This chapter has provided an overview of a number of tools used by coastal managers. Some of these are commonplace, whereas others are used only in specific circumstances and only in some parts of the world, but we believe are likely to become more widespread in coming years.

We chose to present a balanced view of current management tools available – not just the technical ones. Administrative, social and cultural management techniques are equally part of the modern coastal manager's toolkit.

Each of the management tools can be used individually, or in combination with others presented in this chapter and any other tools applicable to a particular situation. However, the effectiveness of these tools is often optimised when used as part of a coastal management plan.

Chapter 5

Coastal management planning

Coastal management plans can be very powerful documents. They can help chart a course for the future development and management of a stretch of coast and/or assist in resolving current management problems. This dual benefit is the greatest strength of coastal management plans: they can provide an eye to the future, but still be firmly based in the present.

Coastal management plans, and the processes used to develop them, can also be employed as part of any coastal program aiming to bring together (integrate) the various strands of government, private sector and community activities on the coast. As such, coastal management plans have the potential to play a vital role in the successful integration of various coastal management initiatives.

Finally, the processes used to develop coastal management plans can act as a kind of melting pot which helps blend together the various tools described in the previous chapter to deal with a range of issues. In doing so coastal management plans can assist in resolving conflicting uses and ensuring that management objectives are met. As will be shown below, this can enable coastal managers to tackle difficult and/or sensitive issues in a holistic, non-threatening way.

In order to present a structured discussion of the various types of coastal management plans the first section of this chapter presents a discussion of the different ways in which they can be classified. One of these classification types is then used to structure the description of coastal management plans – whether they are 'integrated' or 'subject' (non integrated) plans. Last, the processes by which coastal management plans are produced is described with special attention paid to designing a planning process which engenders not only a sense of ownership of the plan with stakeholders, but also a commitment to its implementation.

5.1 Classifying coastal management plans

Plans used in the management of the coast can be classified according to a number of criteria that form the basis of the terminology used to describe plan types in this chapter. The most common are the classification methods shown in Table 5.1.

Some of the classification methods in Table 5.1 are mutually exclusive, but most are not; indeed most coastal management plans produced today can be described according to one or more of the above criteria. Often a classification is required to accurately describe a coastal management plan by including information about its scale, focus and/or degree of integration. For example, a plan may be required in order to obtain

Table 5.1 Coastal management plan classification methods and plan types

Classification elements	*Plan types*				
Scope					
Geographic coverage	International	Whole-of-jurisdiction	Regional	Local	Site
Focus	Operational	Strategic			
Degree of integration	Subject	Integrated			
Statutory basis	Statutory	Non-statutory			
Reason for plan production	Required for funding	Required to clear statutory works conditions	Legislation which requires management plans	Direct response to management problem	Create commercial value
Process					
Participation	'Expert'	Participatory			
Flexibility	Fixed goals	Adaptive/action learning			
Worldview	Rationalist	Values-based			
Acceptance	Consensus	Directed			
Goal setting	Single goal	Scenario-based			
Context					
Cycle	New plan	Building on previous planning cycle			
Plan/programme	Stand-alone plan	Plan within programme			

funding, be integrated and strategic in nature, and cover a particular geographic region. This issue of plan classification is one that has recently been subject to analysis in order to develop an inventory of coastal management and planning initiatives in the Asia-Pacific region as outlined in Box 5.1.

Box 5.1

Project and program taxonomy for the coastal zone Asia-Pacific database

A Database of Coastal Projects in the Asia-Pacific region was developed for the 2002 Coastal Zone Asia Pacific (CZAP) Conference to: provide a resource for coastal managers, practitioners, researchers and others; facilitate the sharing of information and experiences; act as a means to identify gaps and avoid duplication of effort; and to enable the monitoring of trends and progress in coastal management.

The CZAP database contained 289 projects covering twenty-six countries. An analysis of the database found that most funding came from a few sources with an average project duration of five years, and a wide range of project types and implementing agencies (Chuenpagdee and Alder 2002). It was recognised at its launch that the CZAP Database was the first of its kind and, as a result, had limitations.

In preparation for the CZAP 2004 conference the CZAP database was redeveloped using technologies available through the OneCoast initiative (Box 4.12) in a form more accessible and integrated via the Internet than was previously the case.

In establishing this initial redesign of the CZAP Database, the development of a new way to categorise coastal management plans, projects and programs was developed. This approach used a 'topic map' approach (topicmap.com 2004). Topic maps provide a basis for knowledge representation and, through emerging Internet standards, enable systematic linking of information. The OneCoast reworking the CZAP Database utilising the topic map has enabled the initial development of the means to visualise how coastal management projects, plans and programs are classified and how they interact.

The approach uses an Internet specification that provides a model and grammar for representing the structure of information resources used to define abstract subjects (topics), and the associations (relationships) between them (Crow 2004). The specification is extensible, providing a basis for defining the characteristics of subjects and the contexts within which they apply. The intent is to allow the transparent analysis of coastal management and planning activity in the region and in so doing significantly improve the capacity for knowledge sharing. If successful, the new approach applied to the CZAP Database could be extended globally and in doing so significantly enhance the analysis undertaken by Sorensen (Box 3.9) on the global extent of coastal management activity.

Any one of the five methods shown in Table 5.1 could be used as the basis for structuring this chapter. Each has advantages and disadvantages. Choosing one classification method over any other could create an impression that one style of plan is more important than another; however, for purposes of clarity we have chosen the simplest classification method – by the degree of integration – to form the basic divisions in this chapter. Subject plans which have little or no degree of integration are described first, then integrated plans which attempt some form of integration are outlined. Within the discussion of subject and integrated plans the geographic coverage of plans is used as a way of structuring their analysis. However, before subject and integrated plans are discussed, it is worth discussing the other plan classification methods (Table 5.1) in more detail.

5.1.1 Coastal management plan focus

Coastal management plans can also be examined according to their focus on either strategic or operational issues (Figure 5.1). Strategic planning issues are concerned with the long-term future development of the coast, such as siting of ports or the location of future coastal urban developments. As described in Chapter 3, operational management issues are concerned with the day-to-day management of the coast, such as the issuing of permits, or on-the-ground management works, such as rehabilitation. Plans assisting in operational issues are usually called 'operational plans' or simply 'management plans'. The same terminology can be applied to plans which result from strategic management decisions (Figure 5.1), being termed 'strategic management plans'. There is also linkage between strategic management and operational planning. Strategic management decisions can set the framework for management planning in specific areas. For example strategic decisions on the siting, design and operations of tourist pontoons in coral reef areas will influence the day-to-day planning of those areas.

(a) Strategic planning

A strategy must be realistic, action oriented, and understood through all spheres of management. A strategy must be more than a cluster of ideas in the minds of a few decision makers, rather the concepts must be disseminated and understood by all managers.

(Thorman 1995)

Management		Planning	
		Strategic planning	Operational planning
	Strategic management	• Strategic plan • Strategic management plan	• Strategic operational plan
	Operational management		• Operational plan • Management plan

Figure 5.1   Coastal management plan types according to strategic or operational focus

Strategic coastal planning attempts to set broad, long-term objectives, and defines the structures and approaches required to achieve them. It is an ongoing process so that changing needs and perspectives of society can be accommodated, and as a consequence is often multi-dimensional and multi-objective. Strategic planning does not attempt to give detailed objectives, nor give a step by step description of all actions required to achieve the objectives. Strategic planning is the highest-order of planning; it attempts to provide a context within which more detailed plans are set and achieve specific objectives, as well as guide the development of government policy.

Strategic planning is a process in which the major elements determining the form, structure and development of an area, activity or resource are considered together and viewed in a long-term and broad perspective. The key functions of strategic planning are (AMCORD 1995):

- providing a long-term 'vision';
- planning, prioritising and coordinating; and
- providing broad regulation.

Strategic planning is an important part of management because it provides guidance in managing development within a longer-term framework than operational planning. Strategic planning is often on five to twenty-five year time frames, while operational planning is undertaken on an annual to triannual basis. Although strategic planning has long-term time frames, it is still an ongoing process so that changing needs and perspectives of society can be reviewed, generally at five year intervals. Strategic planning is also important because it is one of only a few frameworks which are multi-dimensional and multi-objective. Strategic plans can simultaneously focus on time and space while examining a range of competing issues and objectives.

The long-term, broad geographic focus of strategic planning and its position as the highest-order of planning, setting specific short-term objectives as well as the development of government policy, influences the use of other strategies within the planning hierarchy. It might seem from this that strategic planning is only appropriate at national, state and regional levels. However, while most strategic planning does occur at these levels, it does not preclude its application at the local or site level. Strategic planning is also relevant at these lower levels because local or site plans can incorporate a broad range of objectives such as sustainable development, improving access to the coast, and the sustainable use of particular resources. To achieve these objectives a long-term view is needed to produce fundamental changes in the local society's view of how areas or resources should be managed at all planning scales.

The long-term and broad perspectives taken in strategic planning facilitate a number of activities necessary for sound management (AMCORD 1995), which are also relevant on the coast. Strategic planning provides a channel for communication with the community and other stakeholders (e.g. steering committees, workshops). It enables managers and stakeholders to anticipate change in a well defined framework and to define a vision of how this change could be accommodated (e.g. tourism). In doing so, long-term objectives can be set and a framework for a range of long-term initiatives such as environmental quality can be established. Strategic planning provides a framework for other long-term or short-term strategies and policies for specific issues (e.g. fishing or tourism). Strategic planning through its multi-objective framework helps to identify long term action areas, establish priorities for action (e.g. structure plans or tourism development projects) and mechanisms to coordinate these actions. Along with prioritising, the resources needed to effect these actions can be identified.

Strategic plans generally deal with broad categories of management such as the appropriate uses of specified areas such as marine waters; particular resources such as fishes; development – economic, social and infrastructure; and environmental management. Again the multiple objective nature of strategic planning is highlighted. To accommodate these objectives in a planning framework, a strategy can be based on a number of mechanisms such as broad planning statements, policies, recommendations for existing and future programs or initiatives, a zoning scheme, or a combination of the above. Most of these mechanisms are discussed in detail in Chapter 4.

Like all planning initiatives, stakeholder participation is a fundamental component of strategic planning. Meeting the needs of all stakeholders through the multiple objective nature of strategic planning is difficult and there may not be agreement by all parties. Nevertheless, there needs to be consensus on a shared vision and acceptance on actions to realise that vision. This can only be accomplished through meaningful public participation as discussed in Section 5.5.1.

Strategic plans and resulting action programs can and should incorporate monitoring and evaluation to ensure that the strategy is working and that management can respond to changes in societal values and expectations.

(b) Operational planning

At the operational level, goals specific to the area's physical and socioeconomic conditions are formulated, and form the basis of the area's coastal zone management plan.

Goals or aims at the operational level will be guided by broad international, national or regional strategies as well as stakeholder participation in ways specific to local conditions. Area-specific goals may be to improve the livelihood of coastal residents through appropriate species and habitat management, or to maintain traditional-use opportunities.

Operational planning is concerned with how on-the-ground and on-the-water management actions will be realised. At the broader planning scale level this generally involves the allocation of financial and human resources, where necessary the formulation of statutory mechanisms, and the establishment or coordination of other organisations to undertake the activities required to give effect to the plan. Operational plans at the local or site level define the financial, infrastructure and human resource requirements needed to meet specific management objectives. This is usually done in the medium term (three to five years) to provide the time needed to budget for major capital works and projects, and the short term (annual) which enables agencies to implement the plan. The scope of these operational plans will vary with the available resources, administrative arrangements, and budgeting requirements for the agency responsible for managing the area.

5.1.2 Statutory basis of coastal management plans

The formal power of a coastal management plan as defined by its statutory basis has a large degree of influence on both plan contents and the approach to its formulation.

Some management plans, most commonly those associated with formalised land- or water-use planning systems, have the full force of law in their implementation. In contrast, other coastal plans may have been undertaken without such statutory force. These two groups of plans are generally called 'statutory' and 'non-statutory' respectively.

Statutory plans usually contain provisions regarding the use and management actions for particular areas of land or water. The most common of these are zoning provisions in statutory urban planning documents such as town planning schemes, and marine management zoning, for example related to marine protected area planning, aquaculture development, fisheries management or tourism planning (see Chapter 4 – Zoning).

Examples from Western Australia and the United Kingdom (Figure 5.2) illustrate the divisions between statutory and non-statutory coastal plans which influence coastal management. In some cases the division between statutory and non-statutory coastal management plans is blurred by legislation forming the framework within which they

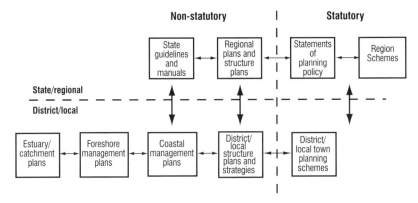

(a) Western Australia

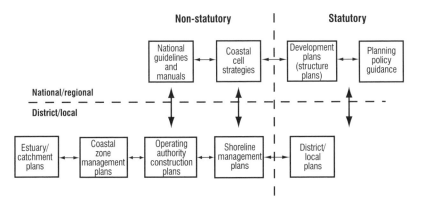

(b) England and Wales

Figure 5.2 Comparison of statutory and non-statutory plans influencing coastal management in Western Australia and the United Kingdom (from Kay *et al.* 1995)

can be developed; in other cases the division is specified by legislation which does not make plan preparation a legal requirement, but specifies plan contents. An example of this approach is the United States where the preparation of state integrated coastal plans is voluntary, but if the states choose to do so there are requirements specified in federal law (Chapter 3). These requirements are imposed to ensure that federal coastal management objectives are met.

5.1.3 The requirements of coastal management plans

The 'requirements' for coastal plans is used here to refer to the reasons why a plan is produced. This may seem rather obvious, in that coastal plans are produced to assist in addressing coastal management issues and problems (Chapter 2). However, this reason may be the direct cause of the production of coastal plans in some circumstances only. The direct cause and effect relationship (i.e. a problem produces a plan) can often be influenced by legislative requirements, influenced by inter-governmental relations, or be in response to community or political pressures. Coastal management plans may be encouraged, or sometimes a prerequisite, to obtaining funding for coastal management activities. The most frequently cited example of such a system is in the USA, where states must produce a Coastal Zone Management Plan in order to obtain federal government funding for various coastal management activities (Box 3.12). Another example of the linkage between the statutory requirements at a national level for coastal plan production at regional scales is provided by the New Zealand example shown in Box 5.15.

Other requirements for the production of coastal management plans include statutory provisions, such as those linked to environmental impact assessment requirements or planning approvals (Section 4.3.1). For example, in Western Australia management plans for foreshore reserves (site level plans) are usually required for the planning approval for some types of coastal urban developments. The requirement for such plans may also be linked to permit, licensing and other related statutory provisions (Section 4.1.4). In some cases coastal plans may not be a legislative requirement for the granting of permits or licences, but may be encouraged by the authorising government departments in order to provide a context for individual decision-making actions on the coast.

Finally, there may be direct legislative imperatives that require management plans to be produced in areas potentially subject to the impacts of coastal erosion and flooding, or for conservation areas such as national parks. Legislation which proclaims marine protected areas may require management plans to be produced ahead of proclamation, as is the case in Western Australian marine protected areas. In Indonesia a marine park can be declared without a management plan, but management actions cannot be initiated without such a plan. However, all Indonesian national parks (marine or terrestrial) require a management plan once declared. These approaches attempt to avoid the 'paper park' syndrome of declaring marine protected areas without providing a framework of resources to manage the area for its conservation values (Alder *et al.* 1995). Similar management planning requirements may be specified through legislation for terrestrial reserves protected for conservation purposes.

A key issue with coastal management plans which have some external requirements, be it funding, legislation or other reasons, is that these requirements place constraints on some aspect of the plan. Such constraints could include the contents of the plan, information needs, how the plan should be formatted, who should be consulted, the timeframe for plan finalisation, or the steps that must be taken to obtain approval (Box 5.2). The formulation of zoning plans for the Great Barrier Reef Marine Park is one example of how legislation directs the planning process as discussed in Box 5.1.

Requirements for plan production can have a profound effect on the overall shape of coastal plans. Clearly, plans must be produced to satisfy those constraints, such as being formatted correctly in order to obtain funds. If the constraints adequately reflect the practical issues of coastal management planning within a nation's administrative and political framework, this should not detract from management outcomes. However, where this is not the case, there is clearly a risk that satisfying the constraints imposed on the production of a plan can impede, or even, override sound coastal management practices.

An often overlooked requirement for coastal plans is community expectation. This is, after all, a major reason for undertaking coastal plans – that the community expects the best management of coastal resources. If the local community or stakeholder group

---

Box 5.2

Consultation requirements for zoning plans in the Great Barrier Reef Marine Park, Australia

The Great Barrier Reef Marine Park Act specifies that zoning plans will be prepared for sections of the park, meeting the following objectives (Government of Australia 1975):

- conservation of the GBR;
- regulation of the use of the park so as to protect the GBR while allowing reasonable use;
- regulation of activities the exploit the resource of the GBR region so as to minimise their effect;
- reservation of some areas for appreciation and enjoyment by the public; and
- preservation of some areas in their natural state undisturbed by man except for the purposes of scientific research.

The Act also specifies that the public are invited to make representation on two occasions: the first when it is decided to prepare a zoning plan, and after a zoning plan has been drafted. The GBRMP authority is required to consider any representation made and if it thinks fit, alter the plan accordingly (Government of Australia 1975). The draft plan is forwarded to the minister responsible for the GBRMP who either accepts it or returns it to the authority with comments for reconsideration.

Once accepted, the plan is laid before Australia's two houses of parliament for fifteen sitting days. If neither house passes a resolution to disallow the plan, it is passed and comes into operation on a date specified by the minister. If the plan is disallowed a new plan must be prepared, and the process begins again.

is not satisfied with the outcomes of a plan, they can actively work against it through lobbying, or by simply boycotting its implementation actions. The most commonly used method for avoiding this problem is to use a consensus-based model for producing the management plan, described in Section 5.5.1. An adaptive management approach (Section 3.1.4) can also avoid the situation of not meeting stakeholder expectations. If key stakeholders are a part of the planning process as well as plan implementation, they will have a better understanding of the constraints that the plan operates under as well as an appreciation of why actions taken under the plan are not necessarily meeting stakeholder expectations.

5.1.4 Degree of plan integration

Perhaps the main division in coastal management planning is between plans which attempt to assist in the management of issues through their integration with others, usually through the use of spatial management techniques, or managing issues through sector-by-sector prescriptions.

Plans which cover one particular aspect or sector of coastal management are termed 'sector' or 'subject' plans (Gubbay 1989). These include, for example, some natural resource management plans, such as a fishery management plan, coastal engineering, nature conservation plans and various industry-sector plans, such as a tourism strategy. Plans concerned with particular coastal management tools also fall into this category, such as the plans and strategies associated with the various coastal management techniques described in Chapter 4.

In contrast, plans that focus on the bringing together of various government sectors or management approaches, or attempt to address conflicts and the multiple use of a geographically defined area, are usually labelled as 'integrated coastal management plans'. The use of the term 'integrated' follows the sense described in Chapter 3 of generically joining together and does not imply the degree to which this joining occurs. Other words, for example coordination or harmonisation, could equally be used to describe such plans.

Integrated plans can also be called 'area plans' to denote their coverage of a specific area of coast. Area plans only equate with integrated plans where there is some element of integration attempted in the planning exercise. Without attempts at integration, area plans simply become subject plans which cover a particular area. An example of the differentiation between subject and area plans for the United Kingdom has been developed by Gubbay (1994) (Figure 5.3).

Nevertheless, subject plans can be included or accommodated in integrated plans at similar spatial scales. In some cases coastal management issues can be handled simply through a series of policy statements and initiatives, examples of which were described in Chapters 3 and 4. In these situations the level of integration is generally low, but agency coordination and cooperation is usually still required.

Clearly integration is the best planning option for many coastal management cases for a number of reasons: it has a holistic approach to solving issues; it is effective and efficient in its use of resources and easily handles multiple objectives. Another important feature of integration is its independence of spatial scales; that is, integration can be used at various planning scales. Nevertheless, there are numerous cases where a subject-by-subject approach is preferable. These cases are described in the next section.

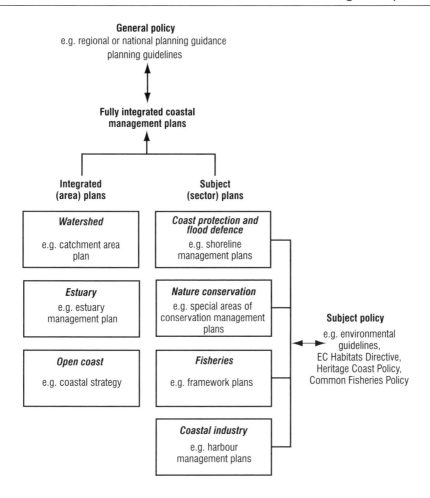

Figure 5.3 Components of integrated coastal management plans in the United Kingdom (Gubbay 1994)

(a) Coastal management subject plans

Subject plans are those developed to address a single issue, subject or sector and, as a result, may be deliberately non-integrative, or may be developed as a consequence of a recommendation of an integrated coastal plan. Subject plans can cover a range of topics – in fact any issue facing the management of the coastal zone described in Chapter 2. For example, they commonly include resource management plans (e.g. marine diversity plans) and industry sector plans (e.g. a transport or tourism strategy).

Subject plans are used for coastal management in a number of circumstances. Perhaps the most common of these is when they are used as a contribution to a broader approach to either an integrated coastal management plan or coastal management program. For example, in England subject plans are viewed by government as an important part of that nation's coastal zone management efforts (Gubbay 1994; Kay *et al*. 1995) (Figure 5.3). The United Kingdom government recognises that the effectiveness of these plans is optimised through their inclusion in a broader integrated coastal management programs that help to implement subject plans where feasible, and to ensure that subject plans do not conflict with each other, or with broader coastal management objectives and actions (Figure 5.3). It is unclear at present how the relationship between different subject plans, and between subject plans and integrated plans, will evolve in coming years as the UK government addresses the European Union recommendation for Integrated Coastal Management outlined in Box 3.8.

Integrated plans are described in more detail in Section 5.3, and subject plans in Section 5.4.

5.2 Designing a coastal planning framework

5.2.1 Design considerations

Before describing subject and integrated coastal plans in the next two sections, it is worth reflecting on how an overall framework for coastal planning can influence the approach and style of individual coastal plans.

A simple way of examining this issue is by considering the management of a typical coastal problem, such as the degradation of a coastal dune due to recreational pressure. There are a number of ways the problem could be addressed through direct management actions, but there are effectively only three approaches which involve the use of coastal plans (Figure 5.4). The first approach it to undertake immediate management actions, such as revegetation, access management, etc., without first producing a plan. In a situation where issues are few, or management actions simple and/or unlikely to cause conflict between different coastal user-groups, such direct action is the most appropriate approach.

The second approach is to write a coastal plan to guide management actions, then undertake those actions. This course of action may be the most appropriate where there are conflicting issues and/or users, or complex management issues.

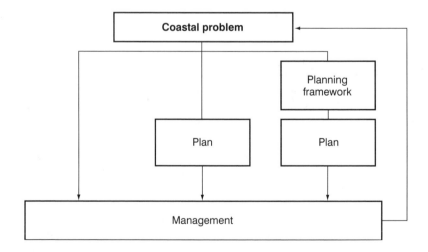

Figure 5.4 Options for coastal planning frameworks

The third option is to develop a coastal planning framework which considers the various types of plans available to address the particular management action and how the plans would interact with other issues and overall management objectives to assist in achieving desired management outcomes. Subsequently, a plan is produced and implemented by undertaking management works. Which option is taken again depends on available resources, legislative basis, social and cultural factors, and political priorities and acceptability.

It is important that coastal managers be able to distinguish between the different types of plans described in the previous section, and between different geographic scales of integrated plans outlined below. This way managers can make an informed choice regarding the need for a planning framework and which plans, or combination of plans, are the most appropriate for their circumstances.

The development of a coastal planning framework usually occurs when there is the need to resolve more than one issue or to formulate more than one plan. Thus, in the many and varied circumstances where management needs are greater than what can be addressed through a single *ad hoc* plan, a coastal planning framework is often designed. Such a planning framework is often part of, or closely linked to, an overall coastal management program. The form of coastal programs is usually dictated by the administrative, political, economic and social circumstances of particular coastal nations, as described in Chapter 3.

Assuming a coastal planning framework is required, the issues which require consideration in its design can be broadly grouped into four main areas (Figure 5.5):

* relationship with an overall coastal management program (including the type, number and intensity of management issues and problems) and other government policies, strategies and plans;
* choice of plan types and production styles;
* linkages between plan types; and
* scales and coverage of plans.

The most important factors influencing a coastal planning framework are the type, number and intensity of management issues and problems. This has a direct bearing on

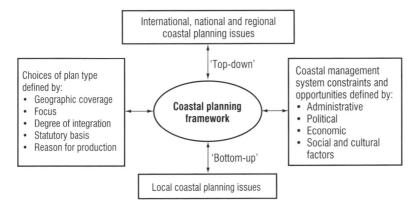

Figure 5.5 Major factors influencing coastal planning frameworks

the choice of particular styles of coastal plans and the tailoring of plans to fit particular objectives. These factors also have an indirect bearing on framework design through their influence on the shape and nature of an overall coastal management system. As discussed in Chapter 3, coastal management programs are constructed to reflect the management issues being addressed and the particular cultural, social, economic, political and administrative issues within individual coastal nations (Figure 5.5). Well designed coastal management programs emphasise the central role of coastal planning; therefore a coastal planning program as such often does not have a separate identity from an overall coastal management system. An excellent example of such an approach to the application of a coastal planning framework being embedded within an overall coastal program is provided by local coastal planning framework from the Philippines (Box 5.3).

Box 5.3

Coastal management planning certification in the Philippines

An innovative approach to promoting and assuring quality in coastal management plans has been developed in the Philippines (Courtney *et al.* 2001). It is based on the notion of developing specific benchmarking criteria and applying these criteria through an independent assessment process. This approach builds on the considerable body of knowledge in quality assurance (QA) as applied to environmental management issues through the International Standards Organisation (ISO) environmental management system series (ISO 14000), and has been adopted in the Philippines under Philippine National Standard, Environmental Management Systems – Specification with Guidance for Use.

The benchmarked certification approach is a direct response to the needs of the more than 8,700 coastal local government units (LGUs) in the Philippines. Certification criteria have been developed based on the LGUs' CRM mandate and internationally recognised best practices in CRM. The certification is voluntary and should be initiated by the local municipality. The approach is positive and proactive with multiple benefits as outlined below (DENR 2003).

- Encourage self-assessment by municipalities and cities through annual monitoring and evaluation of their CRM plans and programs.
- Encourage provinces to provide planning and information management assistance to coastal municipalities and cities and to serve as an information consolidation node for CRM.
- Encourage multi-institutional collaboration between local government and national government agencies at provincial and regional levels to achieve improved management of coastal resources.
- Validate results and benchmark local government performance in CRM through a multi-sectoral review committee.
- Provide a standardised system to evaluate progress towards achieving medium-term development plan targets of integrated coastal management adopted by 250 LGUs along 6,000 km of shoreline for the improved management of municipal waters by the year 2004.

continued...

Box 5.3, continued

- Provide recognition and priority funding status to certified municipalities and cities.

Importantly, the certification process provides a framework for benchmarking CRM performance of LGUs both as a basic service and as a road map for planning future directions and initiatives. The benchmarking criteria are used to benchmark LGU performance at three levels of certification: beginning, intermediate and advanced (see table below).

Benchmarking certification levels

Level 1 – Beginning CRM
Acceptance of CRM as a basic service of municipal/city government with planning and field interventions initiated (1 to 3 years) • Multi-year CRM plan drafted • Municipal Fisheries and Aquatic Resources Management Council formed and active • Baseline assessment conducted • Annual CRM budget allocated • Shoreline management planned • Planned CRM interventions initiated
Level 2 – Intermediate CRM
Implementation of CRM plans underway with effective integration into local governance (2 to 5 years) • Multi-year CRM plan finalised and adopted • Annual monitoring and evaluation of CRM plan and interventions conducted • Financial and human resources assigned permanently to CRM activities • Shoreline management guidelines developed and implemented • Planned CRM interventions implemented with measured success
Level 3 – Advanced CRM
Sustained long-term implementation of CRM with monitoring, measured results, and positive returns (5 years or more) • Multi-year CRM plan implementation fully supported by LGU and collaborators for at least 5 years • Regular monitoring of biophysical and socioeconomic impacts of CRM interventions • Annual programming and budget based on results of monitoring and evaluation • Shoreline management effective • Illegal acts stopped • Biophysical improvement measured • Socioeconomic benefits accrue to coastal residents • Positive perceptions of CRM interventions among stakeholders

The certification approach was first piloted in thirty-four coastal regions and then rolled out nationally. Guidance for developing a CRM plan is provided in a comprehensive manual (DENR 2003). As of December 2003, 111 coastal municipalities and cities have adopted the CRM benchmark system and achieved Level 1.

continued…

Box 5.3, continued

More than twenty municipalities and cities are now 'certified' through the certification system operating in three of the thirteen regions in the country.

An important part of the certification process is to celebrate success. An award ceremony is held to present an LGU with its certificate. There is also an award of a piece of equipment, depending on budget, which will help the LGU in its monitoring and enforcement activities, such as a handheld GPS for fisheries enforcement. In this way, local governments are encouraged to move forward to a higher level of certification.

Certification award ceremony (Source: Coastal Resources Management Project)

A coastal planning framework helps to choose between the wide variety of coastal plan types described in the previous section. The choice of coastal plans available to a coastal manager in any coastal nation will be constrained to a large degree by its systems of governance and, in turn, any overall coastal management system. This issue is particularly relevant to the statutory basis of coastal plans, the reason for their production and geographic coverage (Table 5.1). The latter is often constrained by the relative distribution of power, human and financial resources between levels of government and how these levels of government interact. For example, local-level planning may be constrained in countries where local government has small staffs and/or budgets. Similarly, the statutory planning systems in coastal nations, and how these powers are shared between levels of government, will largely constrain the choice of statutory or non-statutory coastal plans. A comparison of the coastal management plan types in the United Kingdom and Western Australia, shown in Figure 5.2, illustrates the point.

Legislative requirements may also dictate the approach to coastal plan production by defining, among other things, those who should be consulted. Where there are no such constraints coastal planners are free to produce plans using the various techniques described in Section 5.4.

Fitting together coastal plans that have been designed to have different scales, foci, degrees of integration, etc., can be compared to putting together a complex jigsaw puzzle. To take this analogy further, the task is made even more difficult by having neither a well defined picture to guide the assembly, nor well defined edges to the jigsaw, making its construction even more difficult. Pieces of the coastal planning jigsaw include how plans at one scale relate to those at another scale and how different styles of plans interrelate with each other in time, space and in the coverage of management issues. A nation's 1,000 km of coast could be covered by one overall national scale coastal management plan, ten regional-scale plans each covering 100 km of coast, and so on (Table 5.2); however, attempting to undertake 1,000 separate site-level plans covering 1 km each (Table 5.2) would clearly be a daunting exercise, even for the best resourced government; and even if resources are available they may not be able to address ecosystem-wide and offshore marine management issues that would be more effectively addressed at a broader planning scale.

However, attempting to cover long lengths of coastline with detailed management plans could in most cases be counter-productive unless undertaken in an extremely well structured, organised process over a long time period. The obvious danger in embarking on a large number of detailed plans is that the overall context of those plans is lost. There is also the danger of each plan attempting to produce similar outcomes for the coast; such as, for example, uniform types of coastal access which do not reflect site specific characteristics – the very purpose of site level coastal planning.

The opposite of attempting to cover a coast with a plethora of detailed plans is attempting to achieve detailed management outcomes with international, national or regional plans. In this case, the higher-level purposes of such plans, including identifying areas which require more detailed coastal management plans, become lost in an attempt to solve all management problems. This can also be counter-productive if there are different levels of government involved at the various management planning scales. For example, a national government may become embroiled in site-specific problems more effectively addressed by local governments or community groups, and vice-versa. The solution to the competing pressures for site-specific (operational) coastal

Table 5.2 Example coverages of different scales of coastal plans

Level of plan	Plan coverage (km of coast)	Number of plans required for 1,000 km of coast*
Whole-of-jurisdiction	1,000	1
Regional	100	10
Local	10	100
Site	1	1,000

Note
* For an example national coastline 1,000 km long.

management planning and higher-level strategic plans is to develop a structured program which identifies management priorities at regional, local and site level.

A hypothetical case of such a structured integrated coastal planning program is shown in Box 5.4 for a generic coastal nation with a 1,000 km long coast. In this example, whole-of-jurisdiction, regional, local and site-level plans are developed for priority areas under a five-year planning program.

Subject plans, as well as integrated plans, may seek to address a particular issue at a range of spatial scales. For example, a nation's fisheries management planning system may contain national-, regional- and local-level plans. Ideally the plans should be linked to each other as well as integrated into broader planning mechanisms focused on sustainable use of marine and coastal resources. The recommended management actions

---

Box 5.4

Integrated coastal planning program of a hypothetical coastal nation

Imagine a coastal nation with 1,000 km of coastline embarking on a coastal planning program. The various stakeholders in the management of the coast have decided that a multi-level integrated coastal planning approach is needed. They decide to develop national, regional-, local- and site-level coastal management plans which aim to assist in resolving issues of critical environmental degradation, conflicts over the current use of coastal resources, and future sustainable use of the coast. The international dimension to coastal management, such as relevant international treaties or assistance programs, is reviewed. The decision makers consider that a long-term approach with priority areas tackled in just five years is the best course of action. Annual reviews of progress will ensure rapid adaptation. After that, program priorities will be reviewed and the overall success of the approach evaluated.

The identification of priorities results in the development of a national-level plan, four regional, eight local and twelve site plans (see figure below). In some cases coastal problems are so acute and complex that a full 'cascade' of management plans from national- to site-levels will be developed (Location A on figure). In other areas, such as Location B, site-specific plans are warranted, but not a regional-level plan. Other areas, such as Location C, require local-level planning, but not regional- or site-level plans.

In this example, only the national-level plan covers 100 per cent of the coast (and by implication international-level plans), with progressively lower percentages covered by regional, local- and site-level plans. This was judged by the designers of the coastal planning program to be the most efficient mix of the various geographic coverages of plans within budgetary and human resource constraints.

It is interesting to speculate how this imaginary coastal management planning program would evolve after its first five years. Assuming that a major impact evaluation was undertaken (see Table 3.10), it may have been found, for example, that a particular level of plan was particularly effective, or that cascading plans

continued...

Box 5.4, continued

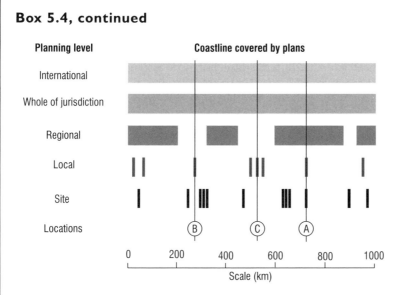

Coverage of integrated coastal management plans of an imaginary nation

had been found to be too complex in practice. This may have been especially so in relation to which levels of plan should be completed first – or deliberately paralleled with others. Other issues are likely to be whether the whole coast is covered with regional, local or site plans or whether existing areas are re-planned if the first round of plans did not meet their objectives or were unable to adapt to changing circumstances.

of such plans may be included in integrated coastal plans at the equivalent spatial scale (assuming that the integrated plans cover both coastal land and water). Of course, the opposite situation may occur with the outcomes of integrated plans being included into subject plans.

Which of these cases occurs is simply a matter of timing: the plan produced first will influence the second plan; the third plan will be influenced by the first and second plan, and so on for subsequent plans. This simple sequence assumes that coastal management issues have not changed over time – usually the exception – and hence the sequence of plans is likely to be affected by evolving circumstances, including the incorporation of previously unforeseen issues. Also, previous planning exercises may have uncovered new issues or problems which may have been considered unimportant, or were not considered at all. The result could be that plans are seen to exacerbate or even create coastal management problems, although in reality the plan merely brought the problems to the attention of planners and managers.

Of course, this sequencing effect will depend on the time elapsed between plans. If this time is long, say over five years, then previous plans may be out of date and of little relevance to subsequent planning initiatives. Plan sequencing is also determined by the statutory basis of any coastal planning framework. There may be statutory requirements to formally adopt the outcomes of previous plans.

In countries where some form of coastal planning has already taken place, the issue of which type of plans to produce first, in which areas, and to address which problems, may have already been addressed. In this case, it may be assumed that plan sequencing reflects coastal management priorities. Of course this may not be the case, with the order in which plans were produced reflecting other priorities, such as political imperatives or the need to satisfy funding requirements.

In contrast, in coastal nations where little or no coastal planning has previously existed, it is worth considering what may be the optimum sequence for plan production. A rather generalised answer to this question is that the most effective sequence will depend on the opportunities and constraints inherent in the governance of a coastal nation. The result is that a suitable sequence falls out from an analysis of governance issues, which in turn reflect social, economic, cultural and economic circumstances.

These sequencing issues also determine to a large extent the overall design of a coastal management program, of which coastal planning initiatives may be a part (see Chapter 3). The nation-by-nation approach to the sequencing issue is supported to some degree by analyses of the various national approaches to coastal management and planning listed in the bibliography. This literature is supported by the various international guidelines for coastal management programs outlined in Chapter 3 which stress a case-by-case approach to the design of coastal management and planning programs. Though this conclusion is the best available at present, it remains rather unsatisfying in that there is little general guidance through comparative analyses of coastal planning programs. Consequently, there are no definitive answers to the most effective overall design of coastal planning programs in general, and to the plan sequencing issue in particular. Clearly, this is an area worthy of future analysis particularly in the light of recent trends for great scrutiny of the performance of coastal planning initiatives outlined in the next section.

5.2.2 Plan monitoring and evaluation

Coastal planning practice has evolved in recent years, from considering evaluation as something that occurs as the last stage of the planning cycle, to a more holistic view of evaluation being a critical component of all planning stages – the most important of which being the design stage. As outlined in Section 3.4 this is an evolution from simple 'impact' evaluation to the development of evaluation frameworks that allow planning and management processes, and the basis for assessing their effectiveness, to be clearly articulated; and, in so doing, to be more accountable to stakeholders. Such stakeholders include funders/donors, decision makers, professional staff, and the coastal users and residents involved in the planning process.

Although it may be obvious, it is worth restating that the purpose of an evaluation is to draw conclusions and findings about the performance of the coastal plan or planning process being evaluated. Evaluation enables managers to determine if the plan is working to meet its objectives, for example to determine if ecosystem resources are maintained or improving, or that compliance rates are adequate. An evaluation of the plan also provides an opportunity to see if it is still relevant to current management challenges. This is not to say that the purpose of the evaluation is solely to make changes; the findings may be to reinforce an approach or set of tools currently being used – thereby strengthening the case for their continued support.

The first step in developing an evaluation framework for a coastal plan or coastal planning framework is to consider the evaluation 'logic' – especially if the evaluation form (see Table 3.9) has a focus on impact evaluation. Evaluation logic is defined as the thinking and development of indicators (or indicator) that are consistent with making a judgement about the 'worth' of a coastal plan or planning process. Fournier, cited in Owen and Rogers (1999), outlines the steps in defining evaluation logic as:

- Establishing criteria
  - In what dimensions must the indicator do well?
- Constructing standards
  - How well should the indicator perform?
- Measuring performance and comparing with standards
  - How well did the indicator perform?
- Synthesising and integrating evidence into a judgement of worth
  - What is the worth of the indicator?

Interestingly, the use of 'worth' in the above stages is quite deliberate in that it encompasses the concepts of intrinsic and extrinsic value within an evaluation (Owen and Rogers 1999). This allows for consideration of both the context of the evaluation and its internal merit. The terminology resonates with that used in environmental economics; Section 4.3.4 provides further exploration of this topic.

The practice of developing evaluation frameworks for coastal plans and planning programs is evolving rapidly, in part due to heightened awareness of the need for much greater emphasis on evaluating overall performance in coastal management systems, as outlined in Section 3.4. Emphasis on evaluation will become more common as the practice of developing meaningful methods for assessing progress towards sustainable development goals matures. For example, the approach to the development of sustainability indicators as part of the Malta Coastal Area Management Plan (Box 3.3), outlined in Box 5.17, is based on some of the most recent thinking on this issue (Bell and Morse 1999, 2003).

Two recent examples from the Philippines and the Mediterranean of different approaches to developing evaluation frameworks in the coastal planning are shown in Box 5.5 and Box 5.6 respectively. The Philippines example stresses that simple, understandable, yet rigorous evaluation mechanisms should be built into the planning process. Further, the approach taken in the Philippines outlines that the development of such success indicators, and the monitoring required to support them, should be undertaken in a participatory manner. It is an approach that echoes the participatory plan development approach outlined later in this chapter.

Experience in the evaluation of the Mediterranean Action Plan (see Box 5.8) coastal planning program is summarised in Box 5.6. The evaluation was the result of a deliberate approach to improving the effectiveness of coastal planning in the region and may lead to the development of the kind of plan-effectiveness framework that is successfully being applied in the Philippines.

Box 5.5

Monitoring and evaluation activities to support integrated coastal management planning in the Philippines

Monitoring and evaluation initiatives are a critical part of integrated coastal management planning in the Philippines. The Philippine approach to local-level coastal planning is an adapted quality management system that uses clearly defined benchmarks to define levels of certification (Box 5.3). Monitoring and evaluation activities are a central requirement to achieve (and maintain) certification. As such, there is extensive guidance provided to local governments on the concepts and practice of coastal monitoring activities and how these relate to the evaluation of plan performance (CRMP/DENR 2003). This guidance includes information on how to select monitoring indicators, monitoring methods and illustrative indicators (see the table below) such as:

- Municipal fish catch per unit effort (kg/fisher/day).
- Living coral cover and fish abundance inside and outside marine protected areas (per cent living coral cover, number of fish/500m^2).
- Mangrove area under effective management (hectares planted and managed).
- Household income in coastal bangarays (districts) (income/family).
- Frequency of CRM-related violations (daily, weekly, monthly).
- Level of stakeholder support for CRM plan and programs (percentage of stakeholders with knowledge of and supporting CRM best practices).

Importantly, the approach to monitoring and evaluation promoted in the local-level planning process is participatory. It also promotes using a mixture of quantitative and qualitative measures through extensive community and stakeholder involvement. The guidance given to local governments on the use of participatory monitoring in the development of their coastal plan is shown in the table below.

Guidance for the participatory evaluation and monitoring in Philippines coastal planning (CRMP/DENR 2003)

Misconceptions of monitoring and evaluation	Reasons for participatory monitoring evaluation
It is a worthless activity, which just wastes time and money.	It will guide your internal development and provide you with external accountability.
It is complex and technical and must be done by external experts, which makes it more expensive.	It keeps you focussed on one direction towards the attainment of your goals and mission.
Implementation is the important activity not monitoring and evaluation.	It occurs in an environment where you can honestly evaluate your own performance and that of those around you without fear of negative consequences.
There is a fear that unsatisfactory or negative results will cause problems and negative feelings of the group.	It is everyone's concern: everyone asks questions and shares and contributes towards the assessment.

continued...

Box 5.5, continued

Misconceptions of monitoring and evaluation	Reasons for participatory monitoring evaluation
It is usually imposed from the outside or top-down by provincial or regional agencies and staff.	It is a team-building process, which ensures all stakeholders put their heads together to arrive at the best decision for all.
Results are not used to improve implementation.	Evaluation must use both qualitative and quantitative descriptions to ensure that all relevant concerns are covered.
It is quantitative not qualitative.	All stakeholders have something important to contribute.
	It is an ongoing process which can be used to adjust, improve and finetune your activities.
	Nothing is perfect, there is always room for improvement.
	People working together to solve problems are much more effective than individuals working by themselves for the same goals.

A critical component of the Philippines approach is to consider monitoring and evaluation as part of the 'plan preparation and adoption' stage – not just following the 'action plan and project implementation' stage (see figure).

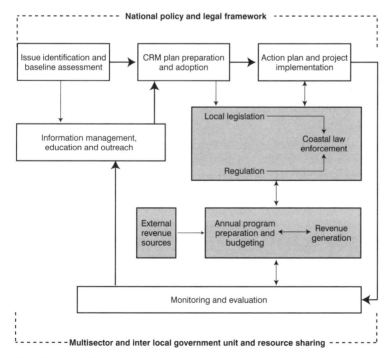

The Philippines local coastal planning program design

> **Box 5.6**
>
> **Lessons learned from the Mediterranean Action Plan Coastal Area Management Program**
>
> *Performance*
> - an evaluation mechanism has to be built in right from the beginning, while program monitoring must be linked to evaluation throughout project implementation;
> - fulfilment of project-level objectives in the planning phase does not automatically lead to implementation of recommendations, or of the plan;
> - fulfilment of project-level objectives does not ensure impacts beyond the immediate project area, unless results are widely disseminated and replicated elsewhere.
>
> *Integration*
> - environmental concerns must be integrated into the design and implementation of an initiative from the very beginning;
> - a program could be issue-oriented at the outset, primarily taking into account most of the factors contributing to these problem issues, but must become more comprehensive at later stages in order to deal with all complex linkages and to provide integrated solutions;
> - the interested national institutions, demanding and initiating the project, should be better identified at an early stage;
> - policy interventions must be closely linked to the objectives of the ICAM initiative;
> - without undermining the importance of technical capacities, it is advisable to ensure that the solutions to technical problems relevant to coastal environments are adapted to the local customs and cultural context.
>
> *Sustainability*
> - strong political commitment at all levels to the preparation and implementation of initiatives is the most important determinant of their sustainability;
> - participation of stakeholders and end-users from the design phase through to project implementation is of utmost importance;
> - a longer-term sustainability of the project should be secured, while greater importance should be accorded to an easier utilisation of project results by the institutions and those who benefit from these results.

5.3 Integrated coastal management plans described by geographic coverage

As the name suggests, integrated coastal management plans aim to bring together environmental, social, economic and political considerations which influence the use of coastal resources into a plan or plans which provide a coordinated direction for coastal managers. When integrated plans are formulated using these four considerations, the framework is set for effective decision making in the coast. Historically, decision making has been independent of these considerations, contributing to inappropriate or conflicting decisions about how the coast and its resources are managed.

Integrated coastal plans are now widely used as a mechanism to draw together disparate and uncoordinated decision-making processes of coastal resource management (Chapter 3). They can be developed in response to a number of coastal management issues, but the most common is simply conflict between different uses which cannot be adequately addressed by a number of single subject plans. These conflicts are often due to differing social, economic and environmental values held by coastal resource users. They can be managed in a number of ways, such as using legislation, policies, zoning provisions, and the many other techniques described in Chapter 4. However, in many cases the most effective and efficient management option is the formulation of an integrated coastal plan undertaken through meaningful stakeholder involvement. Integrated plans are currently the most widely used approach to addressing multiple and/or conflicting issues by providing a framework for focussing the efforts of those charged with managing the coast. This focuses managers towards a common goal, and in doing so assists in coordinating and integrating their actions.

As described in Chapter 3, integration is not a tangible management outcome, but rather a way of thinking about the designing of planning processes which use communication, negotiation and coordination skills to help stakeholders reach informed decisions about how the coast and its resources will be used. These methods are used to bring stakeholders together to open up and maintain dialogue, and to develop mutual understanding and commitment. Once established such an integrated planning framework can then focus stakeholders on discussing, analysing and prioritising coastal issues. Management prescriptions can then be agreed to, and a commitment made by the plan's authors to its implementation, ideally through a coordinated implementation system.

The different levels of understanding and awareness of what may often be disparate coastal issues can be addressed through integrated coastal planning designed to accommodate differing needs. Training, capacity building and information exchange (Section 4.2.3) can strengthen integration mechanisms such as collaborative and community based management, cooperation and coordination as discussed in Chapter 4.

Integration in coastal management planning can be between levels of government, coastal users and the community, or between different sectors of one level of government. It can therefore provide an important mechanism for coordination between one or more sectors, and/or levels of government.

Planning scale refers to the geographic coverage of plans; or, more literally, the scale of any maps produced as part of the plan. For example, a coastal zone management plan covering 1,000 km of coast would include small scale maps depicting the greater study area, whereas a much more localised plan would have much larger scale maps covering a small segment such as 1 km of coast.

It is important to note at the outset of this section that coastal management plans which operate at various scales (Table 5.3) are very different from each other. As will be shown, such plans can range from broad statements of intention by international organisations, to detailed site design plans developed by a community group. Nevertheless, all these plans, at whatever scale they operate, share the fundamental elements of planning: they define a future direction, and describe steps in order to

Table 5.3 Scales of coastal management plans

Level of plan (scale)	Key role
International	• Transboundary issues • Creating a common purpose
Whole-of-jurisdiction*	• Administrative arrangements • Setting national objectives and principles • Focus on priorities
Regional	• Translating international and national goals and objectives to local outcomes • Aggregate local needs and issues to formulate national and international priorities and programs
Local	• Community involvement in setting management options
Site	• Managing well-defined problems • Tangible results of all planning levels can be seen

Note
* National jurisdiction used here as an example

achieve that direction. At each scale of planning, the purpose and scope of planning differs. Which level of planning to undertake is determined by the issues and level of future planning and management of the study area; it is also strongly influenced by its location within a planning hierarchy shown in Table 5.3.

Coastal nations often choose whether to develop their coastal management planning approach with the geographic hierarchy shown in Table 5.3. Planning at the international, whole-of-jurisdiction and regional levels is generally strategic in keeping with broad guidelines and policies. Local- and site-specific plans provide more prescriptive guidelines and management actions for specific activities, development, infrastructure and use of marine and coastal resources. Whether to use such a 'cascade' of integrated coastal plans (Environment Committee 1992) will depend on the particular coastal issues being addressed, and the legislative basis of coastal management.

Integrated plans at different scales can be linked in a number of ways: through the deliberate flow of recommendations from higher level plans, or through the linking of common Guiding Statements. Another method is to encourage linkage through grant-in-aid schemes. The most widely cited of these is the United States, where state coastal management plans are encouraged by federal government grants. There are also common linkages between integrated coastal planning at the international and national level through the encouragement of by donor agencies of national coastal strategies which form part of international initiatives, such as the United Nation's Regional Seas program.

Higher levels of integrated plans can also actively encourage the production of similar styles of localised plans. This can be through recommended actions of the higher order plans, as a means to address the localised problems that higher levels of plans cannot address specifically, because of their more strategic nature. High level plans are often called special area management plans and are in common use throughout the world, Sri Lanka and Western Australia being examples.

A critical issue when considering scales of integrated coastal planning is how plans are to fit together. Ten local area coastal management plans each covering 20 km of coast do not achieve the same things as one higher order plan which covers the entire

200 km. There is a danger in thinking that a plethora of local area coastal plans will achieve the same objectives as higher order plans; they usually cannot. A related issue is that because local area integrated plans are often focussed on areas experiencing major localised problems, the 'local fix' approach can become endemic in a nation's approach to coastal management planning. The result can be that higher levels of coastal planning become neglected as coastal planning moves from crisis area to crisis area with the symptoms of coastal problems being addressed, but not their cure (Donaldson *et al.* 1995).

An example of a hierarchy of coastal plans used in Western Australia includes regional, local and site-levels plans. Characteristics of such plans – their scope; plan production processes; common issues; and typical nature and table of contents – are shown in Table 5.4.

There is no doubt that integrated planning is one option for managing the coast at any planning scale. Integration is not an easy concept to describe let alone translate into planning actions. There is no single recipe for effecting integration since, as described below, it varies with planning scales, and social and economic conditions.

5.3.1 International integrated plans

Coastal management planning at international scales is highly strategic and focuses on developing broad strategies and action plans to ensure common efforts between coastal nations. International-scale initiatives include global programs and those developed between groups of countries. National groupings can be dictated according to the regional boundaries drawn by international organisations such as the United Nations, or by economic or political groupings such as the Association of South East Asian Nations (ASEAN).

At a global scale, international organisations can play an active role in the development of international initiatives focussed on particular issues, or promote and coordinate the development of a particular coastal management tool or approach. Examples include governmental institutions like the International Maritime Organisation (IMO), the United Nations Environment Programme (UNEP) and the Organisation for Economic Cooperation and Development (OECD); and non-governmental organisations such as the World Conservation Union Nature (IUCN) and World Wildlife Fund (WWF). A recent instance of a range of international organisations working together is the global Millennium Ecosystem Assessment, which is assessing the world's ecosystems, including marine and coastal; examining possible future scenarios; and informing policy makers, communities and researchers on possible responses to improve the balance between ecosystems and human well-being (Box 5.7).

Global-scale initiatives can be effected through voluntary programs, such as the International Coral Reef Initiative, or through formal mechanisms such as memoranda of understanding, agreements or action plans – as in the case of the Regional Seas Program (Box 5.8 and Box 5.9).

The single most important global-scale plan of action influencing coastal planning and management continues to be Agenda 21 – the outcome of the United Nations Conference on Environment and Development (UNCED 1992). Agenda 21 is essentially a global plan of action for sustainable environmental management and economic development. Of the forty chapters of Agenda 21, the chapter addressing coastal and

Table 5.4 The hierarchy and characteristics of Western Australian coastal management plans (WAPC 2003)

	Regional plans and strategies	Local coastal plans and strategies	Site design plans
Scale and coverage	1:100,000 to 1:25,000 1,000 m–100 km of coast	1:25,000 to 1:1,000 100 m–1 km of coast	1:1,000 to 1:200 1 km–100 m of coast
Example types of plans	• Land use strategy • Natural resource management strategy • Coastal strategy • Conservation estate management plan • Issue based, e.g. tourism, transport	• Local planning strategy • Local coastal strategy • Rural strategy • Coastal management plan • Greenways plan • Foreshore management plan	• Revegetation plan • Coastal rehabilitation plan • Recreation management plan • Walk trail, road or car park design • Landscape plan • Signage plans
Scope of plan	Incorporates several local government areas, and may include inland as well as coastal components. Determines broad scale patterns of land use and management	Covers all or part of one local government area. Can provide more detailed planning for specific study nodes within the study area Guides development of local policies, planning decisions and management direction Designates development setbacks, coastal reserves and other areas of coastal utilisation. Also outlines rehabilitation areas, amenity sites and access ways for specific coastal nodes	Outlines specification to occur on the ground, such as detailed planning of infrastructure, landscaping and rehabilitation works within a foreshore reserve
Plan production	Usually state government in association with local government	Usually local government in consultation with state government, or a developer meeting subdivision or approval requirements of state and local governments	Land manager, local government, developer or community group

	Regional plans and strategies	Local coastal plans and strategies	Site design plans
Common issues	• Land use planning • Urban development • Tourism development • Port and industrial development • Conservation estate planning • Management arrangements	• Development setbacks • Assessing land capability • Identifying sites for rehabilitation • Foreshore width determination • Protection of sensitive areas • Vehicle and pedestrian access	• Resolving conflicting uses • Designating vehicle and pedestrian access • Appropriate species for revegetation
Nature and contents	• Broad scale patterns of land use • Designation of development zones • Transport networks • Coastal reserves • Recreation nodes and major conservation (park) areas	• Detailed designation of coastal nodes and areas for particular activities • Information on the coast • Allocation of areas for amenities and facilities • Assessment of recreation demand • Prioritisation of management resources, arrangements	• Location and design of infrastructure and recreation facilities • Detailed access provision • Detailed designs of amenities and facilities including landscaping • Detailed provisions for the protection of significant vegetation communities, wildlife, wildlife habitats and cultural heritage

Box 5.7

Millennium Ecosystem Assessment

The Millennium Ecosystem Assessment describes how the future might unfold from an ecosystem perspective based on four scenarios (in italic):

- Economic growth with policies that promote public good (*global orchestration*)
- Protected areas and reserves (*order from strength*)
- Local and regional adaptive governance (*adapting mosaic*)
- Green technology (*technogarden*).

The four scenarios were modelled for the future of the Gulf of Thailand marine ecosystem to 2050. The Gulf of Thailand is a shallow, tropical coastal shelf system in the South China Sea. It has been heavily exploited since the 1960s when a trawl fishery was introduced. Since then the area has been subjected to intense, steadily increasing fishing pressure (Pauly and Chuenpagdee 2003). The ecosystem was highly diverse with large long-lived species (e.g. sharks and rays) but it is now dominated by small, short-lived species that support a high-valued invertebrate fishery (e.g. shrimp and squid). The by-catch of the trawl fishery is used for animal feed.

The Ecopath with Ecosim modelling software was used to explore different futures by varying the mix of ecological, economic and social optimisation to the year 2050 (Christensen and Walters 2004). The modelling provided outputs on how the commercial fish catch of the system, as well as its value and biodiversity, will alter under four different policy scenarios as illustrated below.

Combining the narrative storylines and quantitative modelling makes it clear that how the future unfolds will vary greatly according to the alternative policy approaches taken over the next fifty years. All scenarios show a marked reduction in landings and resulting landing value after 2010 due to continued policies of maximising fishing profits. The extent of this adjustment varies markedly by scenario (Figures a and b). Then between around 2010 and 2030 the models suggest plateauing of landings given the state of the disturbed ecosystem and the effect of policy measures that consider social imperatives (e.g. jobs) as well as maintaining or rebuilding ecosystems. Around 2030, landings are modelled to increase as the ecosystem state recovers due to rebuilding of the system as a result of the adoption of management policies. However, while biomass (food) production increases, there are fewer (low diversity) short-lived and highly reproductive species (Figure c).

The bottom line is that future management will need to make trade-offs regarding maintaining diversity, biomass and profits. Management that focuses on maintaining biomass may result in a loss of diversity but could maintain profits. If the management objective is to improve profits from high value fisheries, then biomass may be maintained but biodiversity will decline substantially. A program to rebuild the Gulf of Thailand ecosystem (diversity) will be difficult and may only be possible by drastic catch reductions, which will result in a decline in short-term profits and employment.

continued...

Box 5.7, continued

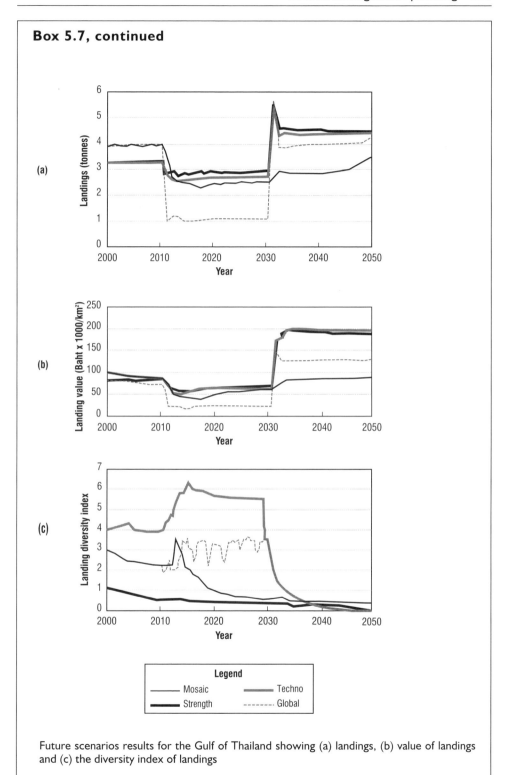

Future scenarios results for the Gulf of Thailand showing (a) landings, (b) value of landings and (c) the diversity index of landings

Box 5.8

The regional seas program

The United Nations Environment Programme (UNEP) initiated the Regional Seas Programme (RSP) in 1974. By engaging neighbouring countries in comprehensive and specific actions to protect their shared marine environment, the program aims to address the accelerating degradation of the world's oceans and coastal areas through the sustainable management and use of the marine and coastal environments.

There are currently 140 countries participating in eighteen regions. Fourteen regions were established under UNEP auspices: Black Sea, Caribbean, East Africa, East Asia, the Kuwait Convention region, Mediterranean, North-East Pacific, North-West Pacific, Red Sea and Gulf of Aden, South Asia, South-East Pacific, South Pacific and West and Central Africa; five partner programs for the Antarctic, Arctic, Baltic Sea, Caspian Sea and North-East Atlantic are planned to join soon and the Upper South-West Atlantic is under development (UNEP 2002). Sixteen regions have formally adopted regional action plans (in order by year of adoption):

- Mediterranean (1975) revised (1995)
- Red Sea and Gulf of Aden (1976) revised in (1995)
- Kuwait (1978)
- West and Central African (1981)
- Wider Caribbean (1981)
- East Asian Seas (1981)
- South-East Pacific (1981)
- South Pacific (1982) revised (2000)
- Eastern Africa (1985)
- Baltic Sea (1992)
- Black Sea (interim) (1993)
- North-West Pacific (1994)
- South Asian Seas (1995)
- Arctic (1998)
- North-East Atlantic (1998)
- North-East Pacific (2001)

Each action plan is written according to the perceived needs of the governments concerned to (Schröder 1993):

- link assessment of the quality of the marine and coastal environment and the causes of its deterioration with activities for its management and development, and the rational use of its resources; and
- promote the parallel development of regional legal agreements and of action-oriented programme activities.

continued...

Box 5.8, continued

A review of action plans highlights the common concerns managers have in managing marine and coastal areas and resource: coastal development; exploitation of fisheries; habitat loss and endangered species; land-based sources of pollution leading to eutrophication and increased health hazards associated with seafood; marine pollution and oil spill contingency planning; and sea level rise. The RSP recognises that each region is unique and that there is no one model which can apply to every region. Nevertheless a common suite of management prescriptions is used to recommend a range of initiatives to address these issues. Where possible these initiatives are at the regional level with each government coordinating its actions. Additional programs specific to the region may also be recommended. For example, the Red Sea and Gulf of Aden Action Plan recommends the formulation of national contingency plans for combating oil pollution (UNEP OCA/PAC 1986), while the East African Region Action Plan recommends regional cooperation in tourism (UNEP OCA/PAC 1982).

All RSPs have their own actions plans and financial mechanisms (trust funds), but only thirteen have associated conventions: Black Sea, Caribbean, East Africa, Kuwait, Mediterranean, North-East Pacific, Red Sea and Gulf of Aden, South-East Pacific, South Pacific, West and Central Africa, Antarctic, North-East Atlantic and Baltic Sea, South Asia, East Asia, Upper South-West Atlantic, Caspian Sea, Arctic while the North-West Pacific RSPs have no legal instrument (UNEP 2002).

In addition, UNEP also collaborates with a wide range of international organisations. UNEP's lack of implementation ability has been criticised as one of the problems with the RSP (Hinrichsen 1994). However, in program areas where countries have worked under the action plan framework and committed major funds, such as in the Mediterranean Action Plan contribution to national and regional planning (see Box 5.17), there has been greater support from member states.

A decision of the twenty-second UNEP Governing Council set out the elements of a new global strategy for the Regional Seas (UNEP 2003). The strategy is based on the central idea of the Regional Seas Conventions and Action Plans that, first and foremost, they should be used as regional instruments, contributing to sustainable development. The central elements in this new global strategy are:

- *Commitment:* the strategy calls for member states to develop an enhanced sense of 'ownership' toward their respective RSPs, leading to stronger political and financial commitment to their implementation.
- *Participation:* the strategy calls for new partnerships, inviting an increase in the participation of civil society and industry in the development and implementation of the regional programmes.
- *Sustainability:* the strategy invites member states to give their programmes sound and lasting financial support.

continued...

Box 5.8, continued

- *Partnership:* the strategy foresees the use of the conventions and action plans as a platform for the regional implementation. To help realise these goals, the strategy calls for continued administrative support from UNEP to the RSP, and foresees a number of more specific objectives, including:

  - increased horizontal cooperation between RSPs and action plans;
  - the strengthening of links with international organisations; and
  - participation in the Barbados Plan of Action on small island developing states.

- *Science-based and ecosystem-based management:* intensified monitoring and assessment activities facilitating a science-based decision-making system, including participation in the Global Assessment of the State of the Marine Environment and the Global International Waters Assessment; and promotion of the ecosystem-based management of marine and coastal environment (Adler 2004).

Box 5.9

The East Africa regional seas program

The coastline of the Eastern African region is an area rich in natural marine resources and breathtaking scenic beauty ... yet this is being seriously threatened by marine pollution, habitat destruction and the pressure of growing populations, urbanisation and industry.

(Iqbal 1992: 1)

East Africa is one of the regional seas programs, the framework for which is described in Box 5.6. The East African regional seas program covers the countries of Comoros, France (Réunion), Kenya, Madagascar, Mauritius, Mozambique, Seychelles, Somalia, Tanzania and South Africa (UNEP 2004).

The action plan was initiated in 1981–2 with the production of various baseline reports on the status of the region's coastal zones, summarised in UNEP (1982). The results of this work were to develop a draft regional action plan, recommend a number of priority actions in the region, and recommend that the draft action plan and two regional protocols be endorsed by the member governments. These came into force on 30 May 1996.

The first meeting of the contracting parties occurred in March 1997 and an outline set of operating procedures, including financial arrangements, was agreed to. The program is currently focussed on capacity building and public awareness raising on the integrated management of marine and coastal areas (R. Congar, personal communication, 1997).

continued...

Box 5.9, continued

A central component of any regional seas program is the contribution of governments into a trust fund to implement their own decisions. In the case of East Africa, the trust fund is also being contributed to by donor agencies.

The ratification of the legal instruments underlying the program has helped to revitalise coastal management in the region. The government of the Seychelles and UNEP established the Regional Coordinating Unit of the East African Region on Sainte-Anne Island. In addition, a Regional Centre for Coastal Areas Management was established in Mozambique. There are also a number of *ad hoc* expert groups established under the regional seas umbrella and a number of coastal management initiatives being undertaken by individual nations in the region (e.g. Moffat *et al*. 1998; Hale *et al*. 2000; Torell 2000) which are to be supported by the regional seas program.

Despite the long lead time required to establish the program, the East African regional seas program gained significant momentum to focus and prioritise regional coastal management funding and action. This is particularly important for directing the attention of donor agencies and international development assistance towards management problems that are seen as priorities by governments in the region. A recent analysis of the regional seas program found that the project implementation varied considerably from country to country. For example, Mozambique had fully implemented an overall coastal management approach in contrast to the fledgling initiatives in Madagascar (UNEP/FAO/PAP 2001). The RSP is also the regional focus for the implementation of global initiatives, such as the International Coral Reef Initiative (ICRI) and the Global Programme of Action for the Protection of the Marine Environment from Land Based Activities.

ocean management issues (Chapter 17 – Protection of oceans and seas) is the longest and most detailed. As described in Chapter 3, Agenda 21 effectively laid out a new paradigm, based on the principles of sustainable development, for the planning and management of coastal areas, that was essentially re-validated at the most recent Earth Summit that took place in 2002. The key focus of the 2002 Earth Summit for coastal management was a much greater emphasis on developing tangible measures for Agenda 21 implementation. Although a non-legally binding document, the global-consensus reached in its adoption has pervaded the coastal programs of many nations through the adoption of sustainable development principles into statements of coastal program goals and objectives (Chapter 3).

Global-scale initiatives are complemented by international efforts between groups of countries, or bilateral agreements between two countries. An example of the latter is agreements regarding the conservation of migratory species, such as the Agreement for the Protection of Migratory Birds in Danger of Extinction and their Environment, between the Governments of Australia and Japan (1974). The combined effect of global-scale, bilateral and regional-scale international initiatives on national and more detailed coastal management planning can be significant.

International initiatives are also important for building the capacity of coastal nations to implement coastal planning and management programs. Training, professional development, scientific research and data management (Chapter 4) are undertaken by a host of international organisations to assist coastal nations which may lack such facilities.

In summary, international initiatives, be they global, regional or bilateral, represent the broadest scale coastal planning initiatives. They can provide the context for initiatives at the national and sub-national levels outlined below.

5.3.2 Whole of jurisdiction integrated plans

The term 'whole jurisdictions' is used here to describe entire sovereign nations and those sub-national governments with significant legislative and/or budgetary powers. The most common type of such sub-national governments is state and provincial governments within federal systems. The defining issue is the ability of governments to choose between legislating to develop a whole jurisdiction coastal management approach or using an approach without the enactment of specific new laws.

Primary coastal planning focus at the whole-of-jurisdiction scale is on administrative arrangements for developing coastal planning frameworks, and articulating statements of goals, principles and objectives. Through the joint development of effective coastal planning frameworks and clear statements of what plans are attempting to achieve, more detailed coastal plans at regional-, local- and site-levels are provided with an unambiguous 'space' in which to develop.

The combined effect of developing administrative arrangements and guiding statements of direction for coastal planning commonly results in broad strategic whole-of-jurisdiction coastal plans and policies. Depending on administrative, political, economic and cultural circumstances such plans can establish requirements for the development of coastal plans in subsidiary jurisdictions, such as local or state governments. In some cases, these requirements may be prescribed within national legislation or policy, for example in the USA (Box 3.12) and New Zealand (Box 5.10). In other jurisdictions, the sharing of roles and responsibilities between levels of government may mean that national-scale coastal plans provide a framework to encourage, through non-statutory means, other levels of government to adopt national approaches (Box 5.10).

The general approach of combining administrative arrangements with the formalisation of guiding statements in the development of whole-of-jurisdiction coastal plans has been undertaken in numerous coastal nations. Here we focus on the nations used as case studies throughout the book in order to illustrate variations in this approach. The New Zealand example is used to show a legislative-based approach which specifies national requirements through a national statement of policy, backed by the national Resource Management Act (Box 5.10). This approach is contrasted to Indonesia's non-statutory national marine resource planning initiative (Box 5.11). Finally, the Sri Lankan national approach to coastal management planning, which is closely tied to the overall structure of its coastal management program, is described. These three examples also demonstrate how whole of jurisdiction planning can set the framework and guide lower order planning as described in the next three sections.

Box 5.10

National coastal planning in New Zealand

In the 1980s New Zealand embarked on a major process of legislative reform of its resource management legislation. This culminated in the passing of the Resource Management Act (1991), which remains the governing legislation for the management of New Zealand's land, air and water and was a very early national policy reform process to embrace sustainability principles. The Resource Management Act rationalised more than fifty Acts governing the coastal environment (Rennie 1993). The purpose of the RMA is the 'promotion of sustainable management of natural and physical resources' (RMA section 5).

The Resource Management Act authorises national policy statements, which can address any issue covered by the Act. As of 1991, no new planning instrument was to be inconsistent with national policy statements. The Act also contains provisions that allow both the Minister of Conservation and the Minister for the Environment to intervene in decisions when issues of national interest arise. The Act provides a framework for integrated management of the coastal environment, and requires the preparation of the New Zealand Coastal Policy Statement (NZCPS) by the Minister of Conservation – ensuring that integrated coastal management receives special attention in the RMA regime (Rosier 1993).

A formal hierarchy of coastal management planning in New Zealand has been established through the RMA, and is described further in Box 5.15. There are no specific guidelines for the production of sub-national plans and policy statements, but the specific requirements of the RMA have ensured that plans produced are similar in content and mode of preparation by the use of draft plans and extensive consultation and public hearings.

Draft New Zealand coastal policy statements were released for public comment in 1990 and 1992. The final NZCPS was released in 1994, for the first time including a Maori language version (Department of Conservation 1994a, 1994b). The coastal policy statement has a series of specific coastal policies, examples of which are shown in Box 4.2.

Policy 7.1.1 of the NZCPS requires that the Minister of Conservation monitor the effectiveness of the New Zealand Coastal Policy Statement in achieving the purpose of the Act by: (a) assessing the effect of the statement on all subordinate regulatory planning instruments, and (b) working with regional councils and other interested groups to establish a 'state of coastal environment' monitoring program. The Department of Conservation established an external peer review group in 2001 to advise on the development of a monitoring review process.

The recommended process included a series of regional workshops focussed on eliciting feedback from practitioners on the use and effectiveness of the policy. The workshops also sought to identify key issues to be addressed by the peer review group. Overall, workshop feedback indicated that most councils supported the retention of the NZCPS as a resource management method (Young 2003). Support was strongest at the regional level of planning. It was also agreed that a revised NZCPS should be shorter and provide more direct guidance on issues of national priority.

continued...

Box 5.10, continued

A host of other issues was brought forward in the workshop process, including problems with the NZCPS management of emerging issues (such as rapidly increasing pressure for marine aquaculture, biodiversity and bio-security management); institutional arrangements; the division of powers and funding between the various levels of government and Maori; and the use of non-statutory tools (such as guidelines).

The Minister of Conservation commissioned an independent review of the NZCPS in 2003 with the purpose of determining whether or not there was a need for it to be rewritten (Rennie 2004). The review included an analysis of central government actions in support of the NZCPS, and an in-depth analysis of polices, plans and permits issued by regional and district councils selected for their representativeness. Interviews were also conducted with local and central government officials and private sector practitioners, academics and additional key stakeholders. The review process was carried out at a time when the national level of government was preparing a framework for the management of the foreshore and seabed, drafting an oceans policy, and preparing legislation to reform management of the NZ aquaculture industry. A planned series of meetings with Maori had to be abandoned after the first two due to a major crisis in Maori–government relations following a court decision that implied Maori rights to retain customary rights to the foreshore and seabed (Rennie 2004).

The reviewer recommended to the minister that policies about protecting natural character of significant coastal landscapes and dealing with natural hazards needed to be more specific (Rosier 2004). Most of the public submissions raised the indiscriminate use of vehicles on beaches, so national priorities about managing public access to the coast also needed to be reconsidered. The outcomes of the review along with public submissions provided the basis for re-drafting the NZCPS.

In addition to NZCPS policies, proposed changes in central government policy mean many coastal issues may be dealt with in other national policy initiatives at a national level. Important policy programs include the completion of the New Zealand Biodiversity Strategy, the development of the national environmental performance indicators, and the ongoing preparation of an ocean's policy for New Zealand. Other reports provide information and ideas about how New Zealand may improve the management of the coastal and marine environment; for example, the Office of the Parliamentary Commissioner for the Environment completed a comprehensive report about current problems in managing the marine environment (PCE 1999).

Progress in revising the current NZCPS is delayed by a plethora of other national reviews relevant to management of the coastal environment currently being under-taken by various government departments. The most important of these reviews are the Marine Reserves Act, Marine Protected Area Program, Oceans Policy Review, Foreshore and Seabed Bill, State of Environment indicators program, and the Aquaculture Moratorium (pending completion of amendments to RMA) (Rosier 2004).

continued...

Box 5.10, continued

The passage of the Foreshore and Seabed Bill through Parliament will most likely delay action on any of the reviews in 2004. This is the most important review (Office of the Deputy Prime Minister 2004) affecting the future role and content of the NZCPS. The current proposal is to vest ownership of the foreshore and seabed in the Crown, integrating all rights and interests within existing systems for regulating activities in those areas. Maori customary rights and interests would be protected through their right to participate – not through ownership rights.

The continuing New Zealand experience in developing a national approach to coastal planning demonstrates the long-term commitment required to its development. The experience also clearly shows how closely coastal planning and management is interwoven with the fabric of a coastal nation's overall policy and legislative development process.

Box 5.11

Indonesia National MREP

As part of Indonesia's commitment to manage its marine and coastal areas, in 1993 it initiated, with funding support from the Asian Development Bank, the Marine Resources Evaluation and Planning Project (MREP) (Ministry of Home Affairs 1996). The project objectives were to improve marine and coastal management capabilities in ten provinces, and to develop and strengthen the existing marine and coastal data information systems. The project finished in 1998 with several successes, such as an improved capacity to plan for the coast, a planning framework at all levels of government, some coast plans at the regional and provincial levels, mapping of coasts, and development of geographic information systems (GIS).

The project was redesigned and launched in 2002 under its new name of Marine and Coastal Resources Management Project (MCRMP). The focus of this project is to improve the quality of marine and coastal environments and to prioritise the management of critical environments with the support of other government sectors. The project has also been extended to fifteen provinces, forty-three kabupatans and a handful special management areas (SMA), which can span more than one province. Each province will identify at least one Marine and Coastal Management Area (MCMA) and a range of coastal programs will be developed within each of these areas. Programs will include conservation area management, spatial planning of coastal and open ocean areas, integrated coastal zone management, pollution control in the coastal areas, restocking and enhancement of marine resources, and restoration of coastal forests and marine habitats (DKP 2003).

continued...

Box 5.11, continued

The ultimate aim is for all Indonesian provinces with a coastline to have a coastal area plan implemented. The overall hierarchical framework for these case studies is discussed in Chapter 3 (Figure 3.8). The range of issues to be considered by these plans is shown by those being faced in the province of Sulawesi Selatan (Box 5.14).

In each MCMA the tiered approach to developing a coastal management program will be used as described in Chapter 3. Each province will develop its own strategic planning based on a vision and goals to reflect provincial priorities. The goals and objectives will be translated into policy, and policies will be implemented in the MCMAs using a number of tools, including zoning plans as described in Chapter 4. Policy will also guide the formulation of zoning plans for large areas within the province. In addition, where necessary, site or subject plans can be used to address issues or problems outside of the zoning plans and provincial policies. Implementation of zoning plans and other specific planning will be achieved using many of the tools described in Chapter 4.

Local management plans are also being developed in key areas and for critical issues (Box 5.18).

Box 5.12

National coastal planning in Sri Lanka

Sri Lanka has made significant advances in coastal resource management in comparison with most other countries in the Asia-Pacific region. There is a need to consolidate the gains made through further refinement and sophistication of strategies adopted in coastal and marine resources management.

(IUCN 2002)

Sri Lanka has been undertaking formal coastal management planning since the 1981 passing of the Coast Conservation Act (CCA). Since then there have been at least two cycles of policy reform (Clemett 2002). The first cycle was the amendment of the CCA in 1986 to refine procedures and the 1990 adoption of a comprehensive Coastal Zone Management Plan (Olsen et al. 1992). The second was the adoption of the Revised Coastal Management Plan in 1997. The Revised Coastal Management Plan is founded on six national strategies (Coast Conservation Department 1996):

1   The coastal management program will proceed simultaneously at the national, provincial and local levels with the collaboration required to achieve effective and participatory resource management by governmental and non-governmental agencies.

continued...

Box 5.12, continued

2   Implement a program to monitor the condition and use of coastal environmental systems and the outcomes of selected development and resource management projects through the collaboration of national agencies.
3   Implement a research program to provide a better understanding of ecological processes and social and cultural information.
4   Implement a program to strengthen institutional and human capacity to manage coastal ecosystems.
5   Update and extend the scope of the master plan for coastal erosion management.
6   Implement a program to create awareness, by national and provincial government personnel and NGOs, of the strategies for coastal resource management and the issues they address.

Each national strategy (above) is accompanied by an explanation of why the strategy is important and a list of implementation actions.

An important addition to earlier coastal planning initiatives in Sri Lanka is an emphasis on coastal planning at regional and local scales in addition to the national level. The different topics and activities to be covered by each level of coastal plan are listed in the table overleaf.

Implementation of the Revised Coastal Zone Management Plan was staged across national and local levels. Although regional-level (provincial) coastal planning is suggested in the revised plan, this mid-level planning has not developed traction (Aeron-Thomas 2000). The use of local level plans (Special Area Management Plans), which are considered to be flexible enough to accommodate the major local coastal management issues in Sri Lanka, was an important component of the Revised Coastal Zone Management Plan. One of the special area management plans produced in Sri Lanka, at Hikkaduwa, is described in Box 5.18.

Sri Lanka is currently in its third major cycle of coastal planning and policy reform to consolidate the progress made to date, learn from experience, and to adapt to the changing socio-political realities of the country and evolution in international practice (IUCN 2002a).

It is clear that the 26 December 2004 Indian Ocean tsunami has placed renewed emphasis on the importance of coastal management planning in Sri Lanka at all levels from local to national (Wijetunge 2005).

continued...

Box 5.12, continued

Topics and activities addressed by Sri Lankan national, provincial and local coastal plans

National	Provincial	Local (special area management)
• habitat monitoring and management • implementation of guidelines for resource use • access and resource use conflicts • a change in the practical and legal definition of the 'coastal zone' • guidance, incentives, regulations and procedures for provincial coastal plans • decentralisation of permit procedures • formulation of procedures for local-level plans and their implementation	• assessment of trends, condition and use of coastal resources and land • identification of major regional coastal resource management issues, opportunities and constraints • mapping of areas of concern which may require local-level coastal planning • identification of areas suitable for conservation for development (e.g. roads, harbours, tourist resorts) • designation of green belts along the coast within which construction is prohibited or restricted	• range of local management problems, issues and concerns addressed through participatory plan production processes as described in Box 16.

Source: Olsen et al. (1992).

5.3.3 Regional scale integrated plans

> Regional-level planning and analysis confers a number of advantages that are absent from local- and national-level planning. At the regional level, it is possible to address and resolve problems faced by entire ecosystems. Very often these issues cross a number of jurisdictions and can only be effectively addressed with a regional geographic focus.
>
> (Jones and Westmacott 1993: 127)

Regional plans and strategies are used to address issues and problems which span a wide geographic range, generally covering more than a single local government authority. Typical lengths of coast covered by such plans are between 100 and 1,000 km. Some regions are defined in legislation; other regions are defined according to the issues being addressed. Integrated regional coastal plans establish a regional framework for on-the-ground or on-the-water coastal management, implement policy developed at the whole of jurisdiction level, and can provide the stimulus for the formulation of local- and site-level coastal plans.

The key focus of regional-scale coastal plans is to provide a bridge between whole-of-jurisdiction plans and policies, and local- and site-level initiatives. The regional level of coastal plans is often the first planning level which is sufficiently detailed to become spatially oriented. International or whole-of jurisdiction plans generally cover too much coast to translate broad economic, social, and ecological considerations into tangible management recommendations or provide practical guidance on matters such as locations and/or mechanisms to spatially separate conflicting uses of the coast. Regional coastal plans can address issues of urban and infrastructure development, resource allocation, transport, tourism, access and conservation.

The form of regional integrated coastal plans can closely reflect whole-of-jurisdiction plans in that regional goals and objectives, and in some cases regional planning principles, can be developed. Depending on the links between whole-of-jurisdiction and regional plans (determined by legislative, funding or other factors described above) there may be potential to develop a specific regional identity. Even if the constraints on the contents and form of regional coastal plans are restrictive, there is usually scope for the inclusion of regional planning and needs and issues.

The content of regional strategies and plans will vary according to the issues addressed, the needs of the region and the approach used to formulate the plan (see Box 5.13 and Box 5.15).

Regional coastal plans can often be the most difficult scale of coastal planning to develop. The primary reason for this difficulty is the 'bridging' role of such plans, situated as they are between tangible local issues, and more strategic initiatives at national or international levels. The challenge, then, is to develop regional coastal plans which are tangible enough to provide clear guidance to local and site-planning, and at the same time sufficiently strategic to assist in the implementation of national and international objectives.

There are several ways in which regional plan implementation can be achieved, including changes to town planning schemes, formulation of policy, or the drafting of specific detailed or sectoral plans. Implementation is usually a staged process which should be managed through a forum to ensure that the process is consistent and ongoing.

Ideally members of the steering committee charged with formulating a plan also participate in its implementation.

Five examples of regional coastal planning are presented below. Each illustrates how regional planning focuses on broad issues while providing guidance for local planning. The Central Coast Regional Strategy (Box 5.13) is a good example of spatially integrated planning; it also highlights how land and marine use planning can be integrated. The Sulawesi Seletan case study takes a much broader view but provides a well defined framework for subsequent planning (Box 5.14). The New Zealand example shows how a structured approach to coastal management plan is constrained by the national policies, legislation and long-term legislative reform (Box 5.15). In contrast, the example from the Strymonikos, Greece European Union ICZM Demonstration Project (Box 5.16) was an important precursor for the development of a Europe-wide approach to coastal management analysed in Box 4.2. Finally, the approach to regional coastal planning undertaken in Malta (Box 5.17) highlights the multi-layered coastal planning approach developed in the Mediterranean through the Coastal Area Management Planning program. The Malta example also introduces the overall approach to the use of Sustainability Indicators using the soft systems methodology described in Chapter 3.

Box 5.13

Central coast regional strategy (Western Australia) – planning context

The strategy is an example of a regional-scale integrated coastal management plan covering the 250 km of coast immediately north of Perth, Western Australia (Western Australian Planning Commission 1996). The issues promoting the strategy are shown in Box 2.4 and the principles used in its development shown in Box 3.13.

The strategy is a non-statutory 'bridge' between state-wide policies and local plans, both of which can be statutory and non-statutory. The document is strategic in nature. The strategy is focussed primarily on coastal land use, although it does address marine planning issues through the development of non-statutory marine 'precincts'.

The process for producing the strategy was based on a steering committee made up of state and local government officials, local councillors and community representatives. Consultation was extensive, with community workshops, the production a background information 'profile' report and a draft strategy for public comment.

The resulting coastal land use plan recommended a hierarchy of settlements along the coast, separated by existing or expanded conservation areas. Future tourist, industrial and residential development nodes were also identified (see figure opposite). Thus, the regional strategy provided broad guidance as to which land and marine uses were appropriate in which locations. Local coastal planning could then concentrate on helping to manage well defined problems or use

continued...

Box 5.13, continued

conflicts. For example, dune areas badly degraded by illegal 'shacks' can be transformed into tourist nodes. The Central Coast region has been very successful in attracting grant funding under the Coastcare/Coastwest program partly as a result of the strategic framework put in place by the strategy (Coastwest/Coastcare 2000).

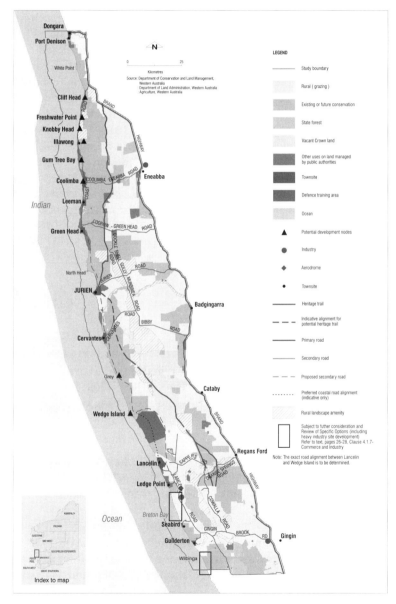

Central coast regional strategy land-use plan

continued...

Box 5.13, continued

In addition, the planning strategy has provided a mechanism to focus the activities of state government in the region. Through the process of producing the plan, and the mechanism for its implementation, the emphasis has been on developing an integrated view of coastal land and marine use planning to ensure the region's sustainable development.

A strength of the implementation process was the presence of a strong local champion who chaired both the steering committee and the subsequent implementation committee. This enabled a continuation of the momentum developed during plan preparation. An interim review of the strategy was undertaken in 2002 with the assessment of the extent to which the actions contained in the recommendations were completed. The review was triggered by the realisation that the majority of actions had been implemented (or were matters that were essentially long-term and ongoing in nature) and that the information on which the strategy was based was becoming outdated.

The strategy was developed as part of the regional planning programme of the Western Australian Planning Commission. As such, there are priorities attached to the regional planning programme. Although a full review is programmed within the next three years, this is subject to the broader priorities of government and the specific priorities of the planning commission. Widespread acceptance of the strategy has contributed to its successful implementation and has actually served to lessen the need for review.

Box 5.14

Regional coastal planning – Sulawesi Selatan province Indonesia

A provincial coastal strategy was drafted under the tiered approach of the National Marine Resources Evaluation and Planning Project MREP (Box 5.9) in 1998. The strategy deals with a number of issues through the setting of a vision and management goals (see Box 3.12). To achieve these goals the following indicative policies were suggested:

- all coastal planning and policy-making coordinated at the provincial level and regional level subject to the oversight of a provincial coastal steering committee;
- raise public awareness of the value of resources and processes so as to encourage responsible resource use;
- all planning efforts will contribute to the orderly implementation of the Sulawesi Selatan Coastal Planning System (adopted from ICZM);
- use the best available information when making decisions and improve the information base for decision making in relation to coastal resource management whenever possible;

continued...

Box 5.14, continued

- for kabupatens (head of district) which are not located in data-rich environments, use alternative approaches that are strongly reliant on consensus-building; methodologies such as participatory resource assessments, expert working groups, advisory groups, and concordance mapping, etc., will be used.
- priority attention will be given in planning efforts to identify strategies for poverty alleviation in coastal villages;
- there shall be no further net loss of mangrove in Sulawesi Selatan and where possible efforts will be made to rehabilitate existing forests and re-plant forests in suitable areas;
- all coral reefs in Sulawesi Selatan waters will be protected from unsustainable exploitation and damage due to human activity;
- marine and coastal tourism development will be actively encouraged and promoted, provided that such development is undertaken with due regard to the ecological and social carrying capacity of the development site; and
- new coastal aquaculture operations shall only be permitted where they can be proven to have no adverse environmental impacts as specified in the higher order plans.

The strategy provided the province with interim guidance for coastal areas and issues while the strategy is being revised through wider inter-agency consultation prior to provincial government approval. In addition, an oil spill contingency study for the Makassar Strait and a mangrove rehabilitation plan were prepared and await funding.

These policies will be strengthened under the Marine and Coastal Resource Management Program (Box 5.9) using integrated spatial (zonation) planning, resource management, conservation area management, habitat restoration including mangroves, and rebuilding of marine resources. In addition to conducting planning initiatives at the regional, local and site level, community consultation and formal and informal education programs are also included in the program.

The Sulawesi Selatan coastal strategy will continue to guide the development of the zoning plan for the Spermonde Archipelago, which was identified as a Marine Coastal Management Area under the MREP (Box 5.9) and is now a case study for the Coral Rehabilitation and Management Program (Box 5.18).

Box 5.15

Regional coastal planning in New Zealand

New Zealand's Resource Management Act (1991) requires that each of New Zealand's sixteen regional councils develop a regional policy statement and a regional coastal plan. However, an interesting twist in the requirement of Regional Councils to develop regional coastal plans is that they are only compulsory in the 'coastal marine area' – the area seaward of the mean high water mark at spring tide to the limit of territorial waters (see figure below), or the 'wet' component of the coast (Rennie 1993). The RMA allows a regional coastal plan to form part of a broader regional plan where it is considered appropriate in order to promote the integrated management of a coastal marine area and any related part of land in the coastal environment. Some regional councils chose to prepare a regional coastal environment plan in order to break down the artificial division between 'wet' and 'dry' components of the coastal environment.

Originally, regional coastal plans were required to be consistent with the RMA and could not be inconsistent with the New Zealand Coastal Policy Statement (Box 5.10). However, the RMA was amended in 2003 to require that plans give effect to the NZCPS, providing more priority for national objectives. Importantly, regional coastal plans are 'facilitative' plans, enabling development to occur under certain conditions, because the Act restricts most human activities in the coastal marine area unless provided for by a rule in a plan or a resource consent. On the other hand, district plans, prepared by territorial authorities for management of land in the coastal environment, are 'restrictive' plans, because development is permitted unless it is restricted by a rule in a plan or a resource consent. Now, district plans are also required to give effect to the NZCPS policies.

Initial evaluation of the success of the regional coastal plans shows that there have been some advantages and some disadvantages flowing from the legislative requirements controlling their preparation. Statutory timeframes requiring completion of regional coastal plans by 1 July 1994 to meet public notification requirements were unanimously condemned as unrealistic. Because the New Zealand Coastal Policy Statement was not finalised until May 1994, councils faced the dilemma of releasing proposed planning documents that would be incomplete and possibly inconsistent with the statutory requirements. The public consultation processes that followed the release of discussion documents and subsequent proposed plans were extensive for most plans. In addition, the outcomes of plan revisions resulting from the submission processes were often challenged in the environment court, bringing further delays.

Despite procedural problems, the statutory requirements of the Act have ensured that each region has undertaken some coastal planning. The strategic nature of these documents has also undoubtedly enabled decision makers to take a long-term view of the use and management of coastal resources. Proposed plans have also been useful in assessing and making decisions about resource consents through specification of matters to be addressed in any assessment of the environmental effects of an activity. The most recently completed provisions in

continued...

Box 5.15, continued

most plans have been those relating to aquaculture management areas (AMAs). Consequently, some regional councils have obtained approval for parts of their coastal plans, enabling them to come into effect while other parts are still before the courts. The consequence has been an extended 'transitional phase' and a national moratorium on new marine aquaculture applications to enable regional councils desiring aquaculture to specifically provide aquaculture management areas in their plans (Rennie 2004). By the beginning of 2004, the Minister of Conservation had approved only nine of the sixteen potential regional coastal plans (Rosier 2004).

All plans and policy statements must be reviewed at least every ten years after coming into effect. But delays in implementing the plans mean that few plans will be reviewed prior to 2008. Taranaki and Hawkes Bay Councils have had their plans reviewed. The major findings are that the plans were effective, especially in their role of facilitating development in the coastal marine area, which may have been prohibited in the Resource Management Act (Rosier 2004). Future plans will probably be more concise and better focussed on issues. Once the New Zealand Policy Statement is reviewed, the Minister for Conservation may require specific changes to be made immediately in sub-national plans and policy statements.

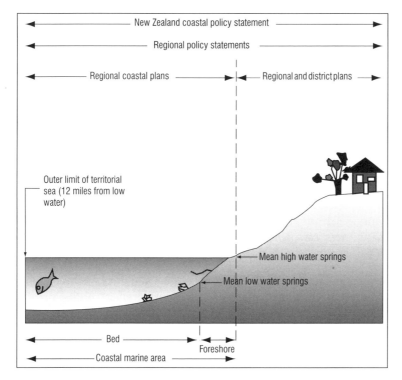

Hierarchy of policies and plans for guiding national, regional and local governments in managing the New Zealand coastal environment

Box 5.16

The Strymonikos, Greece European Union ICZM demonstration project

Thirty-five 'demonstration projects' were undertaken in Europe between 1996 and 1999, reflecting the ecological, economic and social diversity of the European coastal zone (European Union 2000). Projects were also selected on the basis of linkages to related European Union initiatives. Demonstration projects included examples from the Baltic Sea, North Sea, Atlantic coast and Mediterranean. The objectives of the demonstration projects were to:

- provide technical information about sustainable coastal zone management; and
- stimulate a broad debate among the various actors involved in the planning, management or use of European coastal zones.

One of the six demonstration projects in Greece was the Concerted Actions for the Management of the Strymonikos Coastal Zone (Koutrakis *et al.* 2003). The project area covered an area of 1,075 km^2 (land: 262, sea: 813) of the northern Greek mainland coast. Coastal management issues facing the small resident population of around 17,000 include the massive influx up to 150,000 tourists in the summer months, with associated pollution and environmental damage. Other management challenges are related to fisheries/aquaculture, forestry and mining.

Local authorities are aware of the need for strategic coastal planning but were constrained by the lack of baseline data, by complex and conflicting administrative arrangements, and by low levels of environmental awareness. The project aimed to demonstrate the benefits of integrated coastal management planning through a staged approach of:

- Project area profile – biological, physical, social, economic and administrative features; the planning and initiation of a monitoring programme for the zone's marine environment; the evaluation of environmental problems; the preliminary selection of domains of intervention; and management measures.
- A coordination scheme, involving bodies responsible for the project zone's management (now established), to set protection and management aims, decide on priority measures and coordinate their implementation.

The project also included environmental awareness activities, such as publication and distribution of awareness material and of conferences. An information centre on coastal zones (the first in Greece) is now established.

When the project started its anticipated results were:

- Better planning of management strategies and programmes of environmental protection.
- The identification of sustainable practices.

continued...

Box 5.16, continued

- Development of cooperative instruments and the promotion of social dialogue.
- Raised awareness of the values of the zone.
- Better regulation enforcement.
- Sharing of experience in Greece and in the context of the demonstration program.

At the conclusion of the project an action plan was produced that is an important local management tool. Many of the actions have been already implemented or are in the process of implementation. For example, sand extraction from the Strymon estuarine area has been halted. In addition, the study of the abiotic and biotic characteristics of the study area and the socio-economic study are completed and in use by ministerial and local authorities in the area. Moreover, the demonstration project helped to increase awareness on the values of the coastal zone generally and the Strymonikos area in particular. However, the project steering committee did not have the mandate (either administratively or legislatively) to implement other action plan recommendations. Nor were the institutional changes required for an ongoing plan implementation and coordinating process enacted.

The information centre continues its activities, targeted at school children throughout northern Greece and visitors to the area. At the end of the project the information centre was granted to the municipality of Agios Georgios. It is operated by municipality staff trained by the Fisheries Research Institute and the Greek Biotope Wetland Centre. Many of the organisations involved in coordination of the demonstration project are now involved the steering committee of the information centre.

As with all the demonstration projects their major impact on the coastal zone management in Europe was the impetus produced for the European Parliament and Council recommendation concerning the implementation of Integrated Coastal Zone Management in Europe (2002) outlined in Box 4.2. In addition, the Greek demonstration projects gained locally from the experience. For example, the prefecture of Kavala uses a GIS to control illegal housing while the region of Epirus has created the pathways to direct visitors in the inland areas.

On their completion, Greek demonstration projects developed a shared proposal to the Greek government. These proposals, among others, were used to formulate the national strategy for Greece on integrated coastal zone management currently being evaluated by the newly elected Greek government.

Box 5.17

Malta coastal area management plan

The small island state of Malta has been working under the umbrella of the Mediterranean Action Plan (MAP), part of regional seas program (Box 5.8), to develop a Coastal Area Management Plan (CAMP) (Vella 2002). The Malta CAMP is part of a series of thirteen plans (either completed, in preparation or under consideration) under the CAMP program that was initiated in 1989 throughout the southern Mediterranean (PAP/RAC 2001b).

The Malta CAMP sought to address strategic coastal land and marine use, tourism development, aquaculture, habitat management, integrated water resource management, and erosion/desertification control for the north-west coast of the main island of the Maltese archipelago. Malta planners followed the overall process for CAMP development (Pavasoviå 1999). The planning process included components for public consultation, sustainability analysis and a special analysis of tourism and health implications for coastal management. The participatory nature of the CAMP process encouraged cooperation between Maltese government agencies, the private sector (in particular hotel operators) and the non-governmental sector – a revolutionary approach to integrated planning and management in Malta.

A key outcome of the CAMP process was improved inter-agency communication and understanding of their respective roles and responsibilities.

In 2002 the Mediterranean Action Plan conducted a major review of the effectiveness the CAMP process, in particular their implementation effectiveness (PAP/RAC 2001a, 2001b) of CAMPs. The review found overriding support for the CAMP process while recognising the challenges of sustaining short-term project-based coastal management initiatives. As a result, the effectiveness of many CAMPs when measured by the number of planning recommendations actually implemented was relatively low. This was contrasted with the intangible benefits through plan development, including improved inter-agency coordination on coastal management issues and improved relationships between the governmental, private and non-governmental sectors, and heightened overall awareness in sustainable coastal management principles and practice. Indeed, an important outcome of the CAMP evaluation process was to focus on methods for sustaining long-term coastal management programs in the southern Mediterranean after the short-term CAMPs have been completed.

5.3.4 Local area integrated plans

Local integrated coastal plans will vary according to the particular issues addressed as well as the level of sophistication of the approach. Local plans tend to cover lengths of coast in the order of 10 to 100 km and generally involve only one local government. Planning at the local level is often a response to a particular set of issues requiring immediate attention or to facilitate current and future use of particular areas. Typical issues are dune stabilisation, demands for recreational facilities and access to coastal areas for development. In dealing with issues at the local scale, it is important to differentiate between issues which are best managed at the regional level and those which can be managed at the local level.

The contents of a plan are usually determined by the process of its production. If a public consultation process is used to identify the important issues, much of the plan's content is defined. On the other hand if local plans have been recommended by broader regional or whole-of-jurisdiction planning, then there may be issues and/or approaches already identified.

The goals and objectives in local coastal plans will be tangible, and action and development oriented especially on the foreshore. Often the coast is divided into areas and specific objectives are formulated for each area: one area may be developed for recreation and associated commercial facilities, another for recreation without commercial facilities. Because the goals and objectives are tangible, they are easier to implement and facilitate the identification of a set of criteria to evaluate the plan's performance. Information collected for planning is focused at the local scale. Requirements include information on biophysical features such as prevailing weather, seas and coastal conditions, and particulars of existing facilities and current and future access needs.

The number and scope of options available for policy and planning responses at the local level is often determined by community opinion. If sectors of the community are divided on the preferred outcomes of a plan a number of options with a diversity of actions and recommendations will be needed. If the issues are broad, then most options are 'best-fit outcomes' which attempt to reach a consensus amongst the community. Other factors such as funding, expertise and community opinions will influence the range of options (Chapter 4), which can range from administrative (e.g. changing town planning schemes) to engineering works (e.g. re-forming a dune).

Criteria should, if possible, be given against which the success of the plan can be measured. Given that the plans are intended to cover all relevant local coastal management issues, their recommendations may be expected to cover a wide range of subject areas. Recognising this, the major recommendations of each plan are often divided into groups such as environmental, planning, administrative and miscellaneous/sociocultural.

Three case studies are presented below to illustrate the range of approaches from the developing and developed worlds used in formulating a local coastal plan. The first case study analyses the Sri Lankan Special Area Management Plans (Box 5.18). The second study addresses the approach taken in Indonesia to develop local-level integrated coastal plans using Spermonde Archipelago as an example (Box 5.19). The final case study outlines the approach taken to local coastal planning on the Central Coast region of the Province of British Columbia, Canada (Box 5.20). The case studies demonstrate many, if not all, of the above features and analyse their strengths and weaknesses within their unique circumstances.

5.3.5 Site level integrated plans

Site planning is the art and science of arranging the uses of portions of land. Site planners designate these uses in detail by selecting and analysing sites, forming land use plans, organising vehicular and pedestrian circulation, developing visual form and materials concepts, readjusting the existing landforms by design grading, providing proper drainage, and finally developing the construction details necessary to carry out their projects.

(Rubenstein 1987: 3)

Box 5.18

Sri Lanka Special Area Management Plan

Hikkaduwa is a densely developed coastal tourist centre approximately 100 km south of Sri Lanka's capital Colombo and 150 km from the nation's international airport (White *et al.* 1997). Hikkaduwa is a popular international tourist destination, with over 300,000 tourist-guest-nights during 1992. However, tourism development is poorly planned and is causing significant impacts on the natural environment, including Sri Lanka's first marine sanctuary which abuts the town (White *et al.* 1994). Without the development of a planned approach to tourist management, and in light of other local problems, most notably coral mining, there is the potential for a gradual reduction in the natural coastal assets that draw international and local tourists to the area. This phenomenon is well documented from other areas in Asia (Chapter 4).

The SAM process was chosen for this site because of the broad number of local user groups required to take ownership of management issues in order to improve environmental quality and ensure a sustainable fishery and tourism industry. Additional emphasis was placed on the implementation of the plan being self-supporting locally, thereby reducing the need for continuing national or international financial contributions (the tourist industry at Hikkaduwa is mainly locally owned).

The processes for producing the SAM are shown in Box 3.14, which shows the focus on consensual planning using locally-based, full time facilitators to bring the plan together with participation from the broadest possible range of stakeholders. The strategies produced through the planning process and the expected results are shown in the figure below.

Each strategy is implemented through a set of defined actions coordinated through an implementation committee supported by national coastal resource management agencies.

Early lessons learned from the SAM process in Sri Lanka were summarised by (White *et al.* 1994) as:

- the SAM process must be open, participatory and work towards consensus;
- decisions must be clear and well documented;
- national government agencies must understand and accept the process;
- stakeholder groups must be equally represented in the management process;
- implementation results should be apparent within three years;
- monitoring and feedback results make the program tangible;
- in Sri Lanka, collaborative management is a more appropriate concept than community-based management for coastal resources; and
- community groups can make the difference between success or failure.

An initial external evaluation of the SAM was undertaken in 1996 and found that the process was successful against the evaluation criteria used, but concluded 'judgements about what determines success are necessarily provisional' (Lowry *et al.* 1997: 31).

continued...

Box 5.18, continued

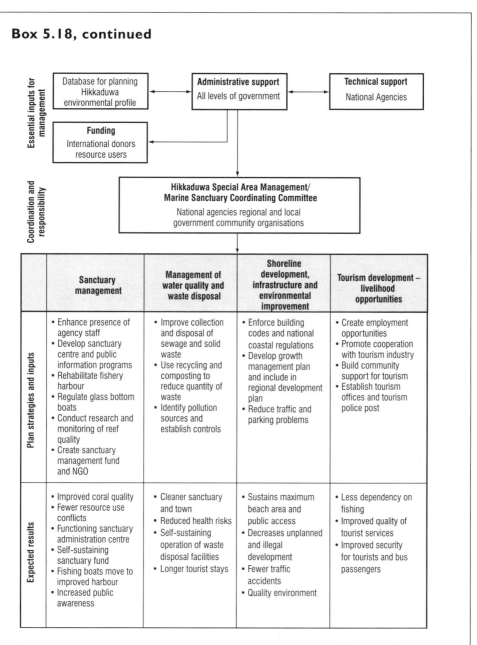

Essential inputs for management

| Database for planning Hikkaduwa environmental profile | Administrative support All levels of government | Technical support National Agencies |

| Funding International donors resource users |

Coordination and responsibility

Hikkaduwa Special Area Management/ Marine Sanctuary Coordinating Committee
National agencies regional and local government community organisations

	Sanctuary management	Management of water quality and waste disposal	Shoreline development, infrastructure and environmental improvement	Tourism development – livelihood opportunities
Plan strategies and inputs	• Enhance presence of agency staff • Develop sanctuary centre and public information programs • Rehabilitate fishery harbour • Regulate glass bottom boats • Conduct research and monitoring of reef quality • Create sanctuary management fund and NGO	• Improve collection and disposal of sewage and solid waste • Use recycling and composting to reduce quantity of waste • Identify pollution sources and establish controls	• Enforce building codes and national coastal regulations • Develop growth management plan and include in regional development plan • Reduce traffic and parking problems	• Create employment opportunities • Promote cooperation with tourism industry • Build community support for tourism • Establish tourism offices and tourism police post
Expected results	• Improved coral quality • Fewer resource use conflicts • Functioning sanctuary administration centre • Self-sustaining sanctuary fund • Fishing boats move to improved harbour • Increased public awareness	• Cleaner sanctuary and town • Reduced health risks • Self-sustaining operation of waste disposal facilities • Longer tourist stays	• Sustains maximum beach area and public access • Decreases unplanned and illegal development • Fewer traffic accidents • Quality environment	• Less dependency on fishing • Improved quality of tourist services • Improved security for tourists and bus passengers

Management inputs, strategies and expected results from the Hikkadawa Special Area Management Planning Process (Hikkaduwa Special Area Management and Marine Sanctuary Coordination Committee 1996)

continued…

Box 5.18, continued

The SAMs were developed with their long-term sustainability in mind. However, as United States international aid agency (USAID) funding for SAMs tailed off in the late 1990s the Sri Lankan government was unable to provide sufficient levels of funding. A recent evaluation of SAMs and their role in Sri Lankan coastal management concluded (Aeron-Thomas 2000):

Whilst this may be acceptable on a project basis, it is of limited benefit as a national policy when funding is uncertain and activities are geographically restricted, and has serious implications in terms of the sustainability of the process. Such a situation can create uncertainty, limit the feeling of ownership by government staff, and may lead to inconsistencies in implementation between sites. SAM may be an effective means to address critical local issues but it must be part of a wider programme to address national coastal resources management issues.

Similar problems were also evident at the Rekawa SAM in the late 1990s, overlain by allegations of local corruption and mis-management (Senaratna and Milner-Gullan 2002).

Despite these setbacks, the SAM model is widely viewed as the optimal approach to managing the intensive coastal management problems facing critical areas of the Sri Lankan coastline (IUCN 2002a). Indeed, significant new donor funding has been provided to Sri Lanka since the late 1990s to improve implementation at Hikkadawa and Rekawa and to significantly expand the SAM network (Asian Development Bank 1999). However, a key issue remains: ensuring that when the current injection of donor funds into Sri Lanka is completed around 2006, that sustainable methods are developed for the ongoing management of SAM implementation.

In addition, it is likely that the SAM process will be reviewed within the broader context of the management responses to the 26 December 2004 Indian Ocean Tsunami tragedy.

Box 5.19

Spermonde Archipelago, Indonesia local integrated coastal planning

The Indonesian national coastal and marine management project (MREP) (Box 5.11) identified the Spermonde Archipelago offshore of the province of Sulawesi Selatan as one of the Marine Coastal Management Areas (MCMA) within the Sulawesi Selatan Coastal Strategy (Box 5.13) in the late 1990s. Limited progress was made in developing a management plan until the Indonesian government included it in its COREMAP program in 2001.

continued...

Box 5.19, continued

Within the MCMA framework the focus for the Spermonde area is on more detailed planning through the formulation of a zoning plan. The zoning plan will support the provincial vision, goals and policies while addressing issues within the areas nominated for zoning. There are also provisions for more detailed local area planning. In the Spermonde Archipelago, Pulau Kapoposang and Papandangan are used as a case study for this level of planning. Pulau Kapoposang and Papandangan are located on the western boundary of the Archipelago and close to the Makassar Strait. Kapoposang was selected because it had an established community which expressed a desire to be involved in sustainable management of the island and adjacent reef, the potential for island- and reef-based tourism, and an active local reef fishery around the island and adjacent reefs. Tourism and improved marine resource use offer the most promise for economic development on the island. Papandangan was included because the adjacent reef has tourism potential if destructive fishing ceases. The community on this island uses Kapoposang's resources and is a source of problems such as destructive fishing. The potential for management for the area was recognised by the Governor of Sulawesi Selatan, who has recommended that the area be declared a marine tourism park.

The Spermonde is now included as a centre for the COREMAP component sponsored by the World Bank and should address the following major issues (Salam *et al.* 1996):

- Destructive fishing and collecting – cyanide and explosives are used in the commercial and subsistence fisheries. These methods are not sustainable and will ultimately lead to coral degradation.
- Low socioeconomic conditions – the residents of both islands are poorly educated, health and education facilities are limited and access to finance is difficult. Many residents are caught in a cycle of borrowing money at above market interest rates to finance the next season's fishing while committing their future catch to the lender.
- Over-exploitation – commercial and subsistence fisheries are heavily exploited and showing signs of overfishing; catch rates are declining despite increasing effort.
- Lack of information – information on the distribution and abundance of resources in the management area is limited and constrains the identification of high conservation areas.

The COREMAP program at the Spermonde will formulate a zoning scheme as the basis for the management plan. Primary objectives are the improvement of the socioeconomic conditions for island residents, protection of endangered species, rehabilitation of degraded areas, and the development of community-based resource management programs. General provisions in the management plan include restricting commercial fishing within the management area, prohibiting the mining of coral, substituting liquid gas for mangroves as a source of cooking fuel, and passive rehabilitation of habitats by prohibiting destructive

continued...

Box 5.19, continued

fishing practices. Access to freshwater is a major constraint to island development although all islands have a brackish lens. To maintain or improve this water source, mangroves are to be replanted and a ban imposed on further harvesting on mangroves.

The development of tourism in the area will be facilitated by improved tourist facilities and transport to the area. In addition, tourism operator awareness programs are proposed to ensure that operators are aware of the potential impacts their operations can have on coral reefs. Better education facilities and programs are proposed to improve the islands' social situation. A study of the mariculture potential of the area is also proposed. The problem of a lack of information will be addressed by encouraging researchers to include the area in their study, especially for the management of fisheries resources.

Five zones are proposed:

- conservation for protection, research and regeneration of resources, with access restricted to researchers only;
- traditional use for sustainable use of resources by traditional residents; commercial exploitation is banned;
- replenishment for protection for a specified time from exploitation to allow resources to recover for a maximum of five years;
- intensive use for protection of reef resources while allowing for general use; and
- buffer to provide a transition area between intensive and conservation zones.

Looking back at the provincial coastal policy developed through the MREP/MCRMP programs (Box 5.13), the Kapoposang study incorporates many elements of these policies.

Site-level integrated coastal management plans are detailed strategies for the use, protection, development and management of a small coastal area. Site plans are prepared:

- to provide a local context for detailed land and/or water- use decisions;
- as a condition of an approval for a planning application (e.g. subdivision, development, rezoning, or as a requirement of an Environmental Impact Assessment);
- to implement a local or regional coastal management plan;
- to review a previous plan; or
- to meet community demands.

Site planning focuses on localised issues and problems. Site plans can as a result be extremely varied, given the potential range of coastal management issues (Chapter 2). Typical among such issues are:

Box 5.20

Coastal planning – central coast Canada

Coastal planning in British Columbia, Canada has been initiated by federal, provincial, First Nations, local, and non-government agencies. As each of these groups has different jurisdiction over coastal resources, planning has traditionally been initiated at multiple scales with different objectives. While Fisheries and Oceans Canada (DFO) is the lead federal agency for ocean matters, it had done little in the way of coastal planning prior to enacting the Oceans Act in 1997. Since then DFO has initiated a number of coastal planning (which DFO terms 'Integrated Management') projects. Integrated Management (IM) is a means for the federal government to cooperate on issues under their jurisdiction, including fisheries management, habitat protection, endangered species, and navigation and transportation safety. In British Columbia, integrated management pilots are being integrated with local, provincial, and First Nations planning processes. Another federal agency, Parks Canada, has significant experience with management planning and currently is working on developing coastal and marine management plans in its National Marine Conservation Areas and national parks.

The province of British Columbia has been implementing coastal plans since the 1970s. The purpose of these plans was originally to address specific tenuring issues, but has recently been used to assist coastal communities which have experienced significant economic and population decline to identify opportunities to diversify and expand their economies. Recently, coastal planning has been used to resolve increasingly conflicting and competing interests on a limited land base. Where coastal plans are available, applicants wishing to obtain a tenure for a certain activity can refer to the plans to identify where their tenure application is most likely to be successful. Planning in British Columbia is hierarchical, where regional and sub-regional land use plans (1:250,000 to 1:500,000 scale) identify the broad issues in a particular coastal area (e.g. increasing tourism, need to protect certain resources) and identify which of these areas requires additional planning.

Local coastal plans are adaptive and flexible, where plans are anticipated to be reviewed in 5–10 year increments. Plans are non-binding on First Nations and may be changed subsequent to the signing of treaties. Recently, DFO has begun to participate in local coastal plans as part of their IM project. Coastal plans have been completed for most of British Columbia and have reduced the backlog of tenure applications province wide.

- identification of degraded or coastal resources requiring rehabilitation or protection;
- identification of environmental, landscape, recreation, and coastal resource values;
- threatened fragile coastal ecosystems, landforms or rare/threatened flora and fauna;
- natural hazard management (erosion, flooding, cliff-falls, potential sea-level rise);
- level of access to coastal areas and resources (multiple use paths – walking, cycling, skating – pedestrian ways, roads and car parks, harbours, moorings or boat ramps);
- demand for type, location and access to recreation facilities;
- need to upgrade existing recreation facilities;
- types and levels of environmental assessment; and
- clarification of ongoing operational coastal management responsibilities (who pays for what and where).

Site plans may be based on information contained in regional and local plans in the area and on consultations with government agencies and local authorities. Studies, however, may be needed to obtain details about specific sites conditions or issues.

Site-level coastal planning commonly uses approaches developed by landscape architects and planners (e.g. Rubenstein 1987; Thompson *et al.* 1997; Russ 2002), integrated with civil engineering skills (for earthworks, etc.).

The type of information needed in planning at this level will depend on the objectives for managing the site. If the main objectives for site planning are to stabilise coastal dunes, information collected should include those factors that will assist in identifying the causes of degradation. If the plan is to provide low-key recreational facilities, such as a car park, lookouts and picnic area, a different range of information has to be obtained, including geomorphology, soil type, runoff characteristics and vegetation types, and user data such as the type, level and timing of existing and future demand.

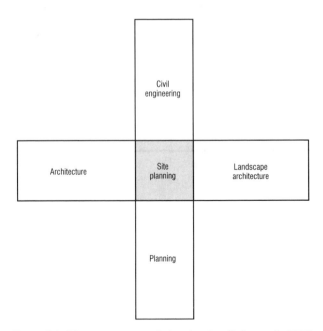

Figure 5.6 The components of site planning (Rubenstein 1987)

Site planning can often follow a relatively simple planning process, as conflicting uses or demands, which may require a consultative style of management planning, have been addressed by local or regional plans. Consequently, a linear step-by-step approach can be used which follows the general design process shown in Figure 5.7 which begins with ideas, develops and then refines those ideas in relation to site opportunities and constraints. This process can provide the framework for working design concepts to be developed which can evolve into sketch designs (Figure 5.7a and b) and which in turn help articulate the relationship between the various elements of a site (Rutledge 1971).

Site plans may divide the coast into sectors or precincts based on natural environmental features, beach/shore characteristics and current and proposed use or development. Within each sector, proposals for managing the foreshore are presented which may include provisions for access and parking, small structures such as toilets, shelters and other recreation facilities, conservation and rehabilitation works, landscaping, marine facilities, protection works and possible commercial uses.

Plans usually contain an implementation program, time frame and a budget as well as an outline of who is responsible for funding, the undertaking and staging of proposed

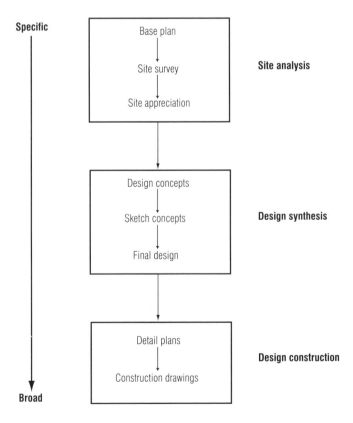

Figure 5.7a Generalised site planning design process

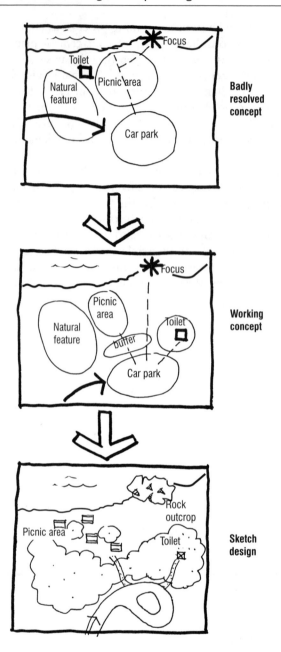

Figure 5.7b Concept evolution (adapted from Schmidt 1996 after Rutledge 1971)

works, and responsibilities for interim and long-term maintenance. A schedule of work and estimated costs may be required depending on the type and intensity of proposed works. Warnbro, Western Australia is a good example of site planning with a local planning framework. The works and outcomes are illustrated in Box 5.21.

Box 5.21

Dune rehabilitation planning, Warnbro, Western Australia

The coastal dunes at Warnbro in Perth, Western Australia were badly degraded due to uncontrolled access causing vegetation loss and blow-outs (see figure below). The area was used for many years by Perth residents for a range of destructive activities, including sandboarding, walking at random through the dunes, trail-bikes and four-wheel-drive vehicles.

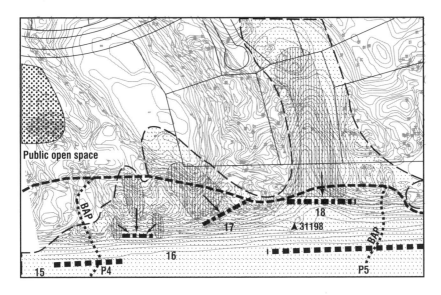

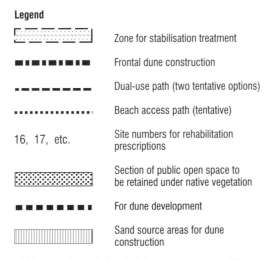

Legend

(dashed box)	Zone for stabilisation treatment
(dash-dot line)	Frontal dune construction
(dashed line)	Dual-use path (two tentative options)
(dotted line)	Beach access path (tentative)
16, 17, etc.	Site numbers for rehabilitation prescriptions
(dotted box)	Section of public open space to be retained under native vegetation
(square dashed line)	For dune development
(hatched box)	Sand source areas for dune construction

Warnboro Dune (Western Australia) rehabilitation sequence (Source: Quilty Environmental)

continued...

Box 5.21, continued

The urban expansion of Perth extended to the Warnbro area in the 1960s, but did not significantly impact on the dunes until the early 1990s. The broad-scale planning of this expansion, including location of major roadways, and the overall boundaries of the subdivision and extent of the foreshore reserve, is undertaken through regional-scale planning under the statutory Perth Metropolitan Region Scheme (Hedgcock and Yiftachel 1992).

Under Western Australia's planning legislation, conditions may be placed on developers to write and implement special area management plans as part of approval to subdivide land. It has been common practice in the state that management plans for public foreshores be undertaken as a condition of subdivision. These 'foreshore management plans' are site-level plans which specify the detail of environmental rehabilitation, managed access ways and the location of car parks and recreational facilities (Box 5.21). Subsequent to these works being undertaken the foreshore reserves themselves are transferred from private landowners to be managed by a government authority (state or local) as part of subdivision approval.

Planning processes were supported in the Warnbro area by the application by the state government for placement of a Soil Conservation Notice over the coastal land. The notice required that the land be stable before development could take place, resulting in development being delayed for several years until this could take place (S. Clegg – Ministry for Planning, personal communication, June 1997). These actions were supported by a local environmental community group the Warnbro Land Conservation District Committee.

Perth's dry and hot summers, linked with strong summer sea-breezes, require sensitive and well-planned dune rehabilitation. The foreshore management plan for the site detailed rehabilitation techniques, staging and costing (Quilty Environmental Consulting 1991). Dune rehabilitation techniques used were:

* earthworks;
* brush and mulch for temporary stabilisation;
* wind fences; and
* permanent revegetation (various species for the berm, frontal- and hind-dunes).

Individual site rehabilitation prescriptions were guided by a detailed site management plan, an extract from which is shown in Figure (a). The foreshore management plan also shows the potential location of access paths between the subdivision (and its car parks for visitors) and beach, together with a coast-parallel multiple-use (cycling/walking/rollerblading, etc.) path (see Figure (a)).

Rehabilitation of the Warnbro dunes was undertaken by specialists for the developer between 1991 and 1993, with follow-on maintenance for a further two years until the rehabilitation had successfully established a self-sustaining plant community. A total of US$600,000 was spent on the rehabilitation program.

continued...

Box 5.21, continued

(a) May 1993

(b) August 1993

(c) March 2004

continued...

Box 5.21, continued

The result was that fencing, beach access pathways and a north–south multiple use path allowed for an increasing intensity of public use without the damage that past uncontrolled use had caused.

The foreshore land was transferred from the developers to the local government (the City of Rockingham) in 1995. The final photograph shows the rehabilitation site in March 2004. The healthy condition of the rehabilitated dunes at the middle and right of the picture is contrasted with fire-damaged vegetation at the left of the photograph. This demonstrates the ongoing management challenges of foreshore management.

A program to monitor and review the plan's effectiveness is often developed to ensure that the works are meeting management objectives. Once the works are completed, they should be reviewed periodically to ensure that they are meeting their objectives. If rehabilitation works were undertaken, are they working? If recreation plans were implemented, are user expectations satisfied? Did unpredicted hazards arise, and are maintenance costs within the allocated budget? Specific aspects of the plan may need to be monitored (e.g. dune stabilisation) and changes made where appropriate.

5.4 Subject plans in coastal management

Subject plans are written to address one, or a limited number of subjects or issues. Subject plans can be developed at a range of spatial scales and can have different foci and statutory bases depending on the subject being addressed, the mode of implementation of the plan, and who is writing the plan. Subject plans can be written by individual government agencies, private companies or non-government organisations, depending on their involvement in particular coastal issues. The various coastal issues, opportunities and problems described in Chapter 2 can individually, or in combination, require the production of one or more subject plans (Table 5.5).

Subject planning for most resources differs from spatial planning in a number of areas. At the site specific level subject planning appears as spatial, but at higher scales planning is not spatially based. National programs are too broad to have a geographic focus, especially if there is a wide range of ecosystems in the country. Subject planning is often sector specific, while spatial planning tends to cut across sectors and integrates a number of sector management activities. Information needs are well defined, narrow and focused on that particular sector rather than a range of sectors. This sector-focused approach often limits the scope of community or stakeholder involvement to those who are directly involved in the particular sector. In sector planning, modelling is more prevalent and uses 'what if' scenarios extensively. There have been attempts to use models in spatial planning, but their use is not as well developed since the number of variables such as representing various interests groups is much larger and hence more complicated.

While there are considerable differences between the two forms of planning, similarities between spatial and integrated plans can be found. Both forms of planning

Table 5.5 Example subject plans used for coastal management

Subject plan groups	Example subject plans
Resource exploitation	Capture fisheries
	Aquaculture
	Oil and gas
Natural resource management	Key ecosystems (e.g. mangrove, coral reef, salt marsh)
	Water quality
Tourism and recreation	Tourism development
Infrastructure management	Coastal engineering (flood and coastal defence)
	Ports
	Recreational boating (marinas, moorings, boat ramps)
	Sewerage
	Solid waste disposal
	Offshore tourist pontoons

are issues driven, use similar planning principles and often have a legislative basis. The same approach to planning as described in the previous section is used, and community involvement is integral to either form, with the planning process being as important as the outcome

There are literally thousands of subject plans pertaining to the coast and its resources which could be used to illustrate subject planning; however, there are few examples of subject planning integrated with high order planning. For example, plans that focus on the management of aquaculture can focus on the overall development of the industry, such as in Shark Bay, Western Australia (Gasgoyne Development Commission 1994), or on particular species; for example the salmon aquaculture development approach in British Columbia (MAFF 2004).

Subject plans have a wide range of applications in a coastal planning framework (Table 5.5). They can be used when there is no apparent conflict between coastal uses, a circumstance which usually arises when such conflicts have been previously resolved, allowing single issue plans to be developed and implemented. Subject plans can follow from the outcomes of integrated coastal management plans, often forming the action statements of such plans, assisting in their implementation. This is especially common at the broader coastal planning levels, mainly from the international to regional scales. Sections of coast with one owner or manager can also lead to single issue plans being developed, again due to a lack of conflict.

Subject plans can also be used as the forerunners of integrated plans. In such circumstances there may not be a willingness to develop a fully integrated strategy without first undertaking some single-subject plans. However, this approach has the danger that a subject-by-subject planning approach further deepens divisions between sectors, possibly leading to a long-term increase in conflict between coastal uses.

However, in many coastal planning frameworks a temporal separation of integrated and subject plans does not occur, with integrated and subject plans being developed and implemented at the same time.

The development of tourism plans with integration between planning scales is illustrated in Box 5.22.

Box 5.22

Levels of tourism plans for coastal management

Regional

Major recreation opportunities may be identified, based on an assessment of current and future recreation demand and a general appraisal of site characteristics. Appropriate recreation sites are generally those which are likely to be heavily used, now and in the future; capable of sustaining that use for a range of popular coastal activities; appeal to several user groups; and are readily accessible, close to urban centres or on major travel routes. It is also important to ensure that a region provides a variety of sites as suggested by the Recreation Opportunity Spectrum (Box 4.12).

Local

As at the regional level, appropriate planning steps include assessment of recreation demand, application of regional-scale planning and consideration of site capacity and possible environmental and social impacts. At a local level, basic facilities (e.g. parking, boat launching ramps, commercial enterprises, ablution facilities) can be assigned to specific localities and their general standard determined. Strategies for altering sites can be devised; for example by managing access, adding more facilities, upgrading existing ones, or developing new sites.

The general level of access to various segments of the coast should be decided on, as should any requirements to exclude existing or potential recreation activities in order to reduce conflicts between user groups or to maintain a site's desired position on the 'recreation opportunity spectrum' (Box 4.12).

Site

Recreation planning for specific locations should ensure that facilities, pathways, signs, commercial enterprises, etc., meet the needs of particular user groups and are suitable for the recreation activities occurring at that site.

At specific recreational locations, users' recreation needs should be addressed, such as determining exact design of parking, ablution blocks, barbecues, tables, pathways and other facilities. Recreation needs are also relevant to landscape plans, which should provide shelter, space for children's play equipment and/or space for setting up recreational equipment such as sailboards.

5.5 Coastal management plan production processes

There are a number of ways to produce coastal management plans depending on the type of plan and the issues to be addressed by it. The processes used to produce a regional scale single-subject coastal plan are likely to be very different from a site-level integrated coastal plan. Of course, this is to be expected, given the diversity of coastal management plans. Nevertheless there are some generic steps which are shared by the great majority of coastal plans, be they integrated or subject (Figure 5.8).

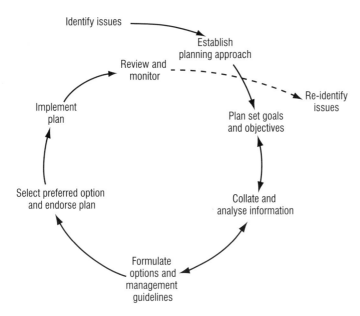

Figure 5.8 General steps in the formulation of a coastal management plan (Alder *et al.* 1997)

The general steps outlined in Figure 5.8 can, in some circumstances, be followed in a simple linear fashion; that is, the planning cycle is started at the top of the figure and followed to its conclusion in a step-wise manner. Circumstances where this is most commonly appropriate are when the need for consultation is limited, such as in the production of site-level plans, or for some subject plans.

There are frequently two common themes underlying the preparation of successful coastal management plans. First, there is the division of the planning process into the gaining of specialist and/or technical information about the coastal environment, including the pressures placed by people on that environment. Second, there is the identification and assessment of views of local people and users on the best use and management of the coast. This division between 'technical' and 'user/social' information is obviously blurred by the fact that users often have considerable technical knowledge. However, the division is useful in that the two categories of information are usually acquired in different ways. Information from users is gained though methods including the staging of public workshops, questionnaires and surveys, focused public forums, the production of a draft plan and the receipt of public submissions on the draft plan. In contrast, technical information is gained by using consultants, academics and other specialists.

The situation shown in Figure 5.8 becomes more complex when community consultation is involved. In these cases experience has shown that a linear step-by-step approach is generally less efficient than a more flexible community-driven approach. This is not to say that there are no well-defined processes for guiding, albeit gently, consensual styles of coastal management plans, as described below.

Consensual processes have evolved out of general everyday experience in undertaking planning exercises. Special planning studies which attempt to address complex issues and which involve a high degree of vested interest and public scrutiny have also contributed. Plan production processes for consensual-style coastal management plans are outlined in the next section.

5.5.1 Consensual-style coastal plan production processes

The consensual-style coastal plan is now the most commonly used plan production technique in integrated coastal planning exercises which attempt meaningful public consultation. There are three main reasons for this:

1   If the people, organisations and government agencies affected by a coastal plan are meaningfully included in its production, the plan's recommendations are likely to reflect their opinions, reducing conflict and making the plan easier to implement.
2   The community is more likely to participate in the implementation of a plan if it has been involved in its production, creating the potential for future cost-savings.
3   The involvement of relevant government agencies increases the likelihood that they will support the plan's implementation with either cash donations or the provision of staff time.

Coastal management plans produced through consensus-building require a special approach. Those responsible for the production of the plan usually need to spend more time 'planning the plan' than when more traditional techniques are used. This is because a detailed planning framework has to be constructed which is rigid enough to ensure that a plan is actually produced on time (and within budget), but flexible enough to allow the consensus building approach to work effectively. This delicate balance requires that plans:

• are issue driven;
• use steering committees with a wide range of representation;
• are based on a rigorous public participation program; and
• are focused on goals and objectives which can be implemented.

The rigour required to develop a fully consensual coastal planning program is shown from recent South African experience in Box 5.23.

Three major processes drive the consensual style of plan production: an administrative process, a public participation process, and the process of writing the plan itself (Figure 5.9). These three processes are discussed in separate sections below.

Unlike the linear step-by-step plan production process shown in Figure 5.8, the three processes shown in Figure 5.9 are generally run in parallel; however, despite this paralleling effect, the basic steps shown in Figure 5.8 still have to be carried out, the difference being that some are carried out at the same time or iteratively. As the planning process progresses through the various steps there are a number of activities and considerations to be undertaken, as detailed in Box 5.24.

Box 5.23

The challenge of transforming coastal management: the South African experience (Glavovic 2004)

South Africa's coastal policy (1999) was assessed as 'the world's first consensus based national policy for the sustainable development of a nation's coastal regions and natural resources' (Burbridge 2000, in Glavovic 2000). Prior to this, a bureaucratic and biophysical coastal management approach had prevailed, virtually ignoring issues relating to justice, democracy and poverty (Glavovic, in press a). The policy was founded on a process of extensive public participation and is strongly supported by diverse coastal stakeholders (CMPP 2000; Glavovic 2000). The policy introduced a new way of thinking about the coast and a new approach to coastal management: a collaborative and integrated approach that aims to improve the well-being of coastal communities and retain the diversity, health and productivity of coastal ecosystems (Glavovic, in press b).

The following enabling conditions were critical to the development of the new approach in South Africa (Glavovic 2004):

- A *'window of political opportunity'*: the political landscape changed dramatically in the 1990s with South Africa's transition to democracy creating a conducive environment for new ideas about deliberative and collaborative policy making. A cross-section of coastal stakeholders was brought together and there was enthusiastic support to initiate a new participatory coastal policy formulation process, which became known as the Coastal Management Policy Program (CMPP).
- *Commitment and activism by key role-players:* a small group of about half a dozen individuals was responsible for much of the hard work and creativity underpinning the CMPP. Drawn from government, the private sector and academia these individuals were in effect 'sustainability activists'. Their values-based approach and dedication infused the policy formulation process and gave rise to much of its success.
- *Access to enabling resources:* the policy itself was formulated during an intensive two-year period; it involved thousands of stakeholders from local coastal communities to the national level in an iterative process of policy development. The CMPP simply could not have been carried out without the financial support provided by the British Department for International Development.
- A *credible and innovative policy formulation process:* the support of key stakeholders, coupled with the commitment of the Minister of Environmental Affairs and Tourism and donor support, enabled the development of a unique policy formulation process. The CMPP sought to deepen participation and established a distinctive government–civil society partnership represented by the Policy Committee, the body responsible for making policy recommendations to the Minister of Environmental Affairs and Tourism. The Policy Committee was constituted as an independent 'not-for-profit' organisation. Five members represented central government and each of the four coastal provinces, and five members represented the business, labour, community

continued...

Box 5.23, continued

based organisations, environmental non-governmental organisations, and the sport and recreational sectors of civil society. Importantly, members had equal status and decisions were made by consensus. Creating and sustaining this government–civil society partnership throughout the CMPP was not without its challenges. But the credibility of the Policy Committee, coupled with the meaningful opportunities for public participation afforded by the CMPP, engendered public trust in the process. The policy recommendations, including principles, goals and objectives for coastal management, together with practical implementation guidelines and a plan of action, enjoyed overwhelming support from stakeholders around the coast. Cabinet approved these recommendations without alteration.

- *A Policy aligned with the dominant political agenda:* the policy calls for an unprecedented investment in coastal management to unlock the development potential of coastal resources, to reduce poverty, meet basic needs and sustain coastal livelihoods. The policy highlights the value of coastal ecosystems as a cornerstone for human development. It is a 'people-centred' policy but it emphasises the vital importance of maintaining the diversity, health and productivity of coastal ecosystems. It advocates a facilitatory style of integrated coastal management to ensure that the coast is managed as a complex, evolving system. In short, the policy is strongly aligned with the dominant political agenda in South Africa. As a result, once Cabinet had approved the policy, government committed significant resources to its implementation. Responsibility for implementing the policy now rests chiefly with the Ministry and Department of Environmental Affairs and Tourism.

Nearly four years have passed since the policy was approved and it is ten years since South Africa's first democratic elections. To what extent is the promise and potential of the coastal policy being realised? A full and independent evaluation of policy implementation efforts is required to properly answer this question; but preliminary indications reveal that much has been done to sharpen the focus on addressing coastal poverty and promoting sustainable coastal livelihoods (Glavovic 2004). A variety of institutional and legal developments has taken place, including the drafting of legislation to entrench the policy in law. A variety of awareness, education and training initiatives has been carried out, and various information and research-related activities are underway. In addition, a number of programs and projects have been set up to achieve the aims of the policy. Notwithstanding these well-intentioned efforts, there are still serious obstacles to be overcome. There are troubling reports that the draft coastal legislation diverges significantly from both the substance and 'spirit' of the policy. If the ideals and practical recommendations of the coastal policy are to be realised, the government–civil society partnership created through the CMPP needs to be remobilised. The substance and 'soul' of the policy, together with the lessons learned from the policy formulation experience, provide both a chart and navigational aids for continuing South Africa's coastal sustainability voyage through uncharted and often stormy waters (Glavovic, in press b).

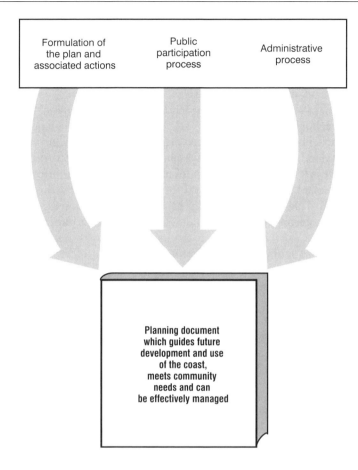

Figure 5.9 Typical plan production process for consensual-style coastal management plans (adapted from Alder *et al.* 1997)

Box 5.24

The general steps in the formulation of a consensual-style integrated coastal management plan

The general steps in the formulation of a coastal management plan are (see Figure 5.8):

- *Identify issues*: recognition that there are issues/problems which need to be addressed.
- *Establish the planning approach*: who will be responsible for the planning process? This is guided by decisions made in the administration process. The public participation program and the planning framework should complement each other so that the community is well represented and has the opportunity to effectively participate in the planning process.

continued...

Box 5.24, continued

- *Set broad goals and objectives*: stakeholders must participate in formulating goals and objectives. Regard should be given to how the assessment of the success of the plan in meeting its objectives will be evaluated. Tangible objectives such as maintaining ecosystems or reducing impacts to dune systems are easy to measure, but intangible objectives such as increasing stakeholders' enjoyment of a particular site are difficult to measure and monitor. As more information is obtained, goals and objectives need to be revised accordingly.

- *Collate and analyse information*: this includes all types of relevant information: biophysical, social and economic, and information obtained through public participation programs. There may be areas where information is lacking. If funding or staffing is available studies to obtain further information should be initiated.

- *Formulate options and management guidelines*: these will address issues identified through the consultative process and technical studies, and make provision for future use and development of the study area in consultation with stakeholders.

- *Select preferred options*: the planning team selects a final option after evaluating the range of options formulated. The option selected will depend on a number of factors such as available resources, community attitudes, existing plans and policies, and feasibility of implementation. The ease of implementing the selected option is critical to the long-term success of the planning project. The planning team also ensures that recommended management actions and guidelines are consistent, integrated and coordinated with other strategies and management plans; otherwise it makes implementation of the plan difficult. At this stage, the planning team determines the life of the plan so that the stakeholders and subsequent managers of the area know when to evaluate and review the effectiveness of the plan.

- *Endorse the plan*: the plan should be forwarded to relevant agencies for endorsement. The agencies which should endorse the plan will generally be made known during the plan production process.

- *Implement the plan*: a series of action plans (e.g. capital works program) with tangible benefits, is formulated so that implementation proceeds in an effective and efficient way. These action plans can be short- or long-term depending on the level of planning. At this stage, monitoring plans should be underway or initiated to enable managers to evaluate whether the plan is meeting its goals and objectives.

- *Review and monitor*: the life of a plan is usually between one and five years. New information, changes in government policy and direction, and changes in community values and attitudes will all influence the relevance of implemented plans.

It is important that monitoring be considered as an ongoing activity throughout the life of the plan to facilitate ongoing adaptation. The variables and criteria used in a monitoring program should be identified during the setting of goals and objectives.

An example of the processes used in developing a consensual-style integrated coastal management plan is shown by that used in the development of the Thames Estuary Management Plan (Box 5.25). The complexity of estuarine systems, coupled with often intense and competing demands, has been one of the most difficult aspects of coastal management planning (van Westen and Scheele 1996). Indeed, adaptive consensual-styles of management plans for large and complex estuarine systems appear to be an emerging norm, especially in the developed world (e.g. Government of Victoria 1990; Imperial *et al.* 1992; Imperial and Hennessey 1996; Inder 1997).

In contrast to the successful coastal planning processes outlined in Box 5.23 and Box 5.25, an example from Tonga in Box 5.26 demonstrates the problems that can arise when consensus breaks down.

(a) Administrative process

Planning does not happen spontaneously; there must be an organisational structure or force responsible for the plan's ultimate production. An administrative program is often used to accomplish this. Administrative programs are especially necessary for higher-order plans and serve a number of functions such as:

- establishing administrative and public consultative committees to assist in the planning process;
- coordinating and integrating existing plans and studies;
- involving relevant government agencies, industries, user groups and the general community;
- ensuring that the legislative requirements for planning are met;
- providing a mechanism for ensuring that the decision-making process is made by, and supported by, representatives from the community. This extends the decision-making process beyond the legislative or government process;
- providing funding and other resources for the planning process; and
- providing secretarial and administrative support.

The administrative process is generally initiated when the decision to formulate a plan is made. The next decision concerns the administrative structure to be used to guide the process. A commonly used structure of local steering committees and/or working groups, and their role in decision making in the plan's production, is shown in Box 5.27.

Membership of local plan steering committees is usually kept under ten if possible, simply for administrative efficiency; however, efficiency needs to be balanced against demands to be as inclusive as possible. In some cases there may be a need for large steering committees. These can reduce administrative time and effort in the long term by being able to reach agreement among the major stakeholders relatively quickly through steering committee meetings. Often steering committees are made up of elected government representatives. These representatives assist in making sure that the drafted plan is politically acceptable, does not conflict with government policies, and where possible complements government initiatives.

Steering committee members evaluate their need for working groups to support them in undertaking the study. Working groups are usually formed to investigate specific

Box 5.25

**Thames Estuary Management Plan production process
(Kennedy 1996)**

The Thames Estuary Management Plan (Box 2.1) was produced using a clearly
defined consensual process. This method was adopted in order to encourage
understanding between user groups and to promote a sense of ownership among
stakeholders. The process used to produce the plan is summarised in the table
below.

Thames Estuary Management Plan – production process

Phases	Advantages/considerations/issues
Scoping (Product: issues report)	• Identification of 'real issues' • Widespread agreement that there are things that can be resolved via coastal management planning • Rejection of non-issues • Establish partnerships • Building trust • Allay fears of unwanted compromise • Never present a 'fait accompli'
Planning (Product: business plan)	• The process and its design • Administrative arrangements • Estimate cost and time • Allow contingency for problem areas • Differences in organisational culture • Range of political, ethical and aesthetic perspectives • Define project limitations (what can be resolved via a mechanism that is already in place) • Risk assessment • Relationship with other plans
Management structures	• Allocate time for this realistically • Make room for minority views • Ensure effective lines of communication – up, down and across
Information gathering (Product: topic papers)	• Try to get agreed technical information • Information and draft policies provided by the practitioners themselves • Joint fact finding • Agreeing what is technically viable and then deciding what to do within the context of the wider economic, environmental and social framework
Presenting the plan (Product: Thames Estuary Management Plan)	• Making a strategic plan operational • Building trust • Allowing time for discussion • Respecting differences • Understanding group dynamics

Box 5.26

Four years in the life of Fanga'uta Lagoon, Tonga (Kaly and Morrison 2004)

The Fanga'uta Lagoon system lies at the heart of Nuku'alofa, the capital of Tonga, on the island of Tongatapu. It is an enclosed tropical Pacific Island lagoon covering an area of about 27 km², about one tenth of the total area of the island. The ecological boundaries of the lagoon system do not end at its lagoon shores, but extend onto the land, and out into sea to the north and east of Nuku'alofa. The lagoon is the heart of the city and functions as a recreational area, fishing ground, rubbish dump, pig rearing area, and medicinal plant source, while supporting other ecosystems around it and absorbing run-off laden with sediments and pollution from the land.

Year 1: 1993 – the year that Fanga'uta Lagoon flipped

Aerial photographs taken in 1992 show Fanga'uta Lagoon with relatively clear waters. In these photos it is possible to see areas of seagrasses, coral rubble and patch reefs to depths of at least 2 m. There is information that by this time the lagoon was occasionally turning green, that turbidity was increasing, and that fish catches were declining. These were seen as gradual changes, not noticed by many, or of particular concern to more than a few fishermen.

Then some time in 1993 the lagoon changed for good. It lost its clear waters, fish kills started to occur and foam was often seen forming on its narrow muddy beaches. The many species of seagrasses in the lagoon became covered in algae and more and more mangrove areas were being cut and damaged. Many of the beaches were by that time converted to seawalls and sewage was a common component of storm water entering through drains.

Year 2: 1998 – monitoring and community consultation for management begins

In 1998 a project commenced to help protect the natural resources of Tongatapu for future generations to use. The Tonga environmental management and planning Project (TEMPP), funded by AusAID, worked with Department of Environment (DoE) to look at the health of the Fanga'uta lagoon system. Under the project, a range of studies was undertaken to determine the condition of the lagoon ecosystem and the values to surrounding communities. The DoE, in collaboration with other ministries, non-government organisations (NGOs) and the community, reviewed these studies and developed strategies for protecting the lagoon's values while allowing for its many important uses.

Year 3: 2000 – establishment of a Fanga'uta Lagoon management plan

The strategies developed under the TEMPP project were brought together to create an Environmental Management Plan (EMP) for the lagoon ecosystem. A draft EMP was circulated to communities, government ministries and NGOs; their views, concerns and suggestions were gathered during a series of meetings

continued...

Box 5.26, continued

around Tongatapu. This culminated in the production of a multi-user EMP, designed to help implement strategies that were expected to result in the sustainable use of the lagoon.

Year 4: 2004 – is management working?

The Fanga'uta Lagoon Environmental Management Plan, despite its careful design and consultations, has not been a sustained success. There are many economic, social, cultural, governance and environmental reasons why the EMP could not be fully implemented. The most important of these are that:

1   The plan was not designed using a sufficiently bottom-up approach. Although communities were consulted, they did not seek the plan or take responsibility for their part in it.
2   Neither the government nor communities around the lagoon really believed it was necessary to limit their activities. Although things were changing for the worse, a strong confidence in nature and God to always provide made action seem unnecessary.
3   Neither the government nor villages could see an economic way of managing the lagoon. For example, the cost of putting in a large sewage system is so high it is not presently possible, so sewage continues to leak into the lagoon from hundreds of septic tanks and the hospital. Land is so strictly controlled, where else would pigpens be put?
4   Communities were mostly concerned with stopping others from neighbouring villages from fishing in their traditional areas. This was being used to develop no-take areas in return for rights to control the fishing areas in front of the village. There is no clear legal or other mechanism for doing this or enforcing it.
5   The plan was developed in the last few months of a 3–4 year development project with many other components. This means there was no follow through with the plan. A fully functional co-management plan would have required at least several years of focused effort to work.

issues or aspects of the planning process and to provide advice to the steering committee, and commonly consist of interest groups and technical people from government. Attendance at working group meetings may change according to the issues discussed.

(b) Public participation

A public participation program ensures that the local community and user groups have the opportunity to fully participate in the plan production process (Figure 5.10). Ideally they should be a part of the entire planning process, but this is difficult to attain for a number of reasons including the funding and resources needed for such a high level of

Box 5.27

Typical membership and decision-making roles of local integrated coastal plan steering committees

Steering committee membership is typically made up of:

- chairperson – someone who has standing within the community, should be aware of coastal management issues and be dedicated to the task;
- local elected representatives and/or elders;
- members of key community groups – e.g. coastal ratepayers, progress associations;
- members of key community groups – e.g. recreation fishers, surfers, retirement clubs;
- representatives of government agencies; and
- other members as required.

  Local plan steering committee usually makes decisions on:

- terms and definitions associated with the plan; these should be specified with mutual agreement reached on the intentions of terms;
- the spatial boundaries, scope (degree of local/regional content and nature of the plan need to be specified early in the process);
- what planning techniques will be used – could include various coastal management and planning techniques (see Chapter 4);
- what reporting procedures are required– clarification on the powers of the steering committee and those who report to it; and
- the resources funds and staff needed to support the committee's operation.

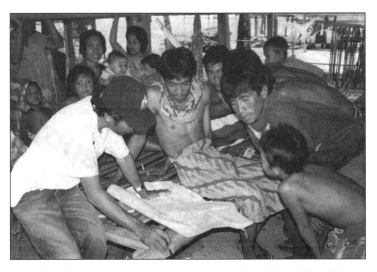

Figure 5.10 Stakeholder meeting for Take Bone Rate Marine Plan, Indonesia

involvement. Where involvement is possible, the public should be given several opportunities to be involved in all aspects of the plan's formulation. The initial stages of the public participation process should ensure that the community is aware of and understands the processes. A number of useful principles used to guide public participation processes for large infrastructure developments in Western Australia provide a valuable introduction to this section (Box 5.28).

Public participation in integrated coastal planning often involves many perspectives. People living in the study area will be the most affected by planning (Figure 5.11). They should have an opportunity to have a say in the region's future and feel confident that decision makers are aware of their views, and have considered them, before plans are finalised. The public can also often identify values that need to be managed and priority issues which need to be addressed in planning. And public involvement is essential if a plan is to be supported and easily implemented.

As mentioned throughout this chapter, planning varies with scale. This concept also applies in public participation. Public participation at the national level is very different from that at the site level. At the national level it is both logistically and practically impossible to directly involve all individuals with interests in planning. To ensure that individuals do have the opportunity to influence the planning process, representatives are usually included on the planning team, and where possible public meetings and

---

Box 5.28

Guiding principles for public participation for large infrastructure projects in Western Australia (Department of Resources Development 1994)

The following principles are essential in the design and implementation of a public participation program.

* Public participation is an integral part of, and complementary to, planning and decision-making processes.
* Public participation programs should occur throughout the life of a proposal.
* A public participation program should recognise the diversity of values and opinions that exist within and between communities.
* A public participation programme must be designed to deal with controversy.
* Specialised public participation techniques are required for contentious or complex issues.
* The timing of the public participation program is crucial to its success.
* The information content of the public participation program must be comprehensive, balanced and accurate.
* A public participation program must be custom designed.
* Public participation should always be a two-way process between the proponent and community.

A public participation program requires adequate amounts of time, money and skilled staff.

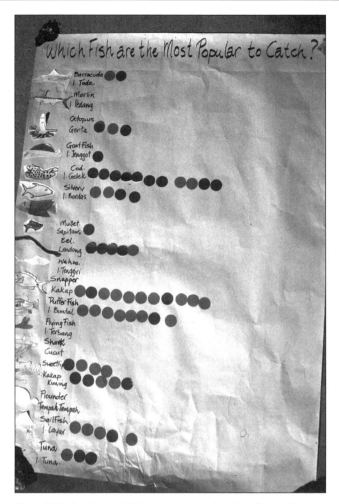

Figure 5.11 Involving children in coastal planning, Cocos Island

written submissions are used. At the site level it is possible to have individuals directly involved in planning either through membership of the planning team or through extensive liaison and ongoing workshops within the area.

Public participation is a continuing process. The community should be involved from the time the decision to formulate a plan is made through to its implementation and review. Ways of achieving this can include combinations of:

• initial advice of the intent to undertake a planning process with periodic briefings on the progress of the project;
• representation on steering and/or work committees;
• community workshops using facilitators who help participants to identify issues and values and management options;
• media campaigns which use press releases, newspaper, radio and television to encourage community involvement;

- surveys;
- meetings with specific interest groups;
- public submissions throughout the planning process.

The techniques to apply will depend on funding, staffing, social and political acceptability, and the complexity of the issues and the community.

Stakeholders within the study area should ideally have the opportunity to participate in the planning process, and where possible other agencies and interested parties from outside the study area should be consulted. The community should be consulted on all aspects of the planning process and invited to assist in the collection and collation of information. Community members can also be involved in setting goals and objectives, selecting preferred options, and determining implementation actions for the plan. The level of participation will depend on the issues being addressed, the planning approach and the resources available.

The planning process aims to produce a coastal management plan which has broad community ownership, especially of its recommended actions. Local needs and the issues covered by the plan will determine how this aim will be achieved. The factor which perhaps more than any other dictates the style and extent of participation is the type and intensity of use of the coast under study. If a section of coast is used by millions of people each year there is little point trying to involve everyone in the production of the plan, but rather to focus on representatives of key user groups. In contrast, a plan for a small section of a more isolated or less intensively used part of the coast could involve the majority of its users. Indeed, depending on the local perceptions of the impact of planning decisions, public participation can be quite extensive and active at the local level compared to national programs.

Central to a participatory style of coastal management plans is a series of community workshops or meetings (Table 5.6). A simple coastal management plan may have only one phase of public workshops in order to establish coastal resident and user perceptions of the coast under study, and the relationship of this region to the broader district. These workshops can then be used as input into the structure as well as the content of the plan. Issues raised at the workshops may change the opinions of the steering committee about the priority issues in the area.

It is important to recognise that there are basically two types of audiences involved in public participation: individuals and groups; and that the same techniques are not necessarily applicable to both groups. Table 5.6 outlines what techniques are the most appropriate for these audiences at various planning scales.

For example, at the site plan level of coastal management planning common methods for public participation include:

- notification of proposed works;
- displays;
- explanatory brochures or leaflets and websites;
- call for formal submissions or comments;
- series of formal and informal meetings and workshops with key individuals, the community or groups with a particular interest in the study; and
- advertisements in the media.

Table 5.6 Matrix of community participation techniques in the planning process

	Regional scale plan	Local scale management plan	Specific development proposal	Discussion/ background paper	Policy statement	Technical paper
Group techniques						
Community consultative committee	R	O	O	NA	O	NA
Technical consultative committee	R	O	O	O	O	O
Workshop	R	R	NA	NA	O	NA
Seminar	O	O	NA	NA	O	NA
Search conferences	R	O	NA	NA	O	NA
Public meetings	R	R	O	NA	O	NA
Public forum	O	NA	O	NA	O	NA
Small group meetings	O	O	O	NA	NA	NA
Design-in	O	NA	O	NA	NA	NA
Individual techniques						
Public submissions	O	O	R	NA	O	NA
Individual discussion	O	O	R	NA	O	NA
Contact with affected residents	R	R	R	NA	O	NA
Project team contact	R	R	R	O	O	O
Site office	O	O	O	NA	NA	O
Participant observation	NA	NA	O	NA	NA	NA
Surveys and questionnaires	O	O	O	O	O	O
Opinion polls	NA	NA	R	NA	NA	NA
Telephone hotline	O	O	O	NA	NA	NA

Key: R = recommended; O = optional; NA = considered not applicable

Source: Department of Planning and Urban Development (1993).

With more elaborate plans, or those which address issues of conflict and/or controversy, there may be the need for further public workshops. These can be held at key times in the development of the plan to:

- provide an opportunity to review outcomes to date together with possible actions to address issues raised; and
- review and evaluate the plan's recommendations and to seek input to the implementation of key recommendations.

Efforts should be made to ensure the widest representation at workshops of coastal residents and users. Advertisements or features in the local press are the usual means of achieving this. In addition, invitations should be sent to key bodies such as ratepayers' organisations, progress associations, local non-government organisations and user groups. During workshops the names and addresses of attendees should be recorded in order to ensure that they receive invitations to subsequent workshops and are sent copies of draft and final plans. Depending on the social and cultural context of the area, other innovative techniques such as compensating fishers who participate instead of fishing, or involving religious leaders to facilitate participation, need to be considered.

Effective community workshops need careful planning. Thought should be given to the structure and content of the workshop, as well as the best location and time. A range of techniques can be used to ensure that the maximum benefit is gained from the workshops, including new analytical techniques such as saliency testing (City of Mandurah *et al.* 1996).

If required, a number of smaller forums or one-on-one interviews may be held with key users and non-users. These can include high school students, commercial/business groups, residents' associations and coastal-based sporting groups. Consultations should also be held with key government agencies if they are not represented on the steering committee. Industry association representatives or representatives from specific companies may also be interviewed; this is essential if they well be affected by the plan.

Findings of the consultations, public workshops and other meetings should be summarised into a written report and be made available to the public. The report should highlight the issues raised and the management options available to address the issue.

(c) Producing the plan

The responsibility for producing the plan – the written document or map – is generally undertaken by the steering committee with the assistance of a technical and professional staff. At lower planning levels the steering committee may be actively involved, but at higher levels such as the national level they rarely do the groundwork of data collections, mapping, etc. These tasks are most often undertaken by technical and professional staff.

Irrespective of who produces the plan, the production of the actual document generally follows (Figure 5.7) with the technical and professional staff or working groups undertaking a range of activities such as revising goals and objectives, data analysis, mapping, producing various reports and drafting the plan text. The team assists in setting and revising goals and objectives based on advice and direction from the steering committee and analysis of information. Regard should be given to including measures for evaluating the success of the plan in meeting these objectives. Tangible objectives such as maintaining ecosystems or reducing impacts to dune systems are easy to measure, but intangible objectives such as increasing stakeholders' enjoyment of a particular site are difficult to measure and monitor. As more information is obtained, goals and objectives need to be revised accordingly.

In producing the plan, information is collected, collated and analysed. All types of information – biophysical, social and economic in a range of formats – may be required, such as statistics, digital, maps and geographic information systems and information obtained through public participation programs. There may be areas where information is lacking. If funding or staffing is available, developers or managers should initiate those studies to obtain the required information. If the issues are significant or the study area is diverse, background papers on particular issues or subjects should be prepared so that stakeholders or participants are better placed to make informed decisions.

Once the information is analysed, including the community's views, the planning team prepares a report of the various options for managing the issues and meeting objectives. This report should be released for public comment after endorsement at a steering committee meeting. Comment may be sought through additional workshops, requests for written comments or both. The report should summarise the major issues raised, and present the steering committee's opinion on management actions required. There are generally two styles for such reports, each with their advantages and disadvantages:

- an issues and options paper which is followed by a draft coastal management plan; or
- a draft coastal management plan.

Draft plans, which summarise findings and make a series of specific recommendations for action, have traditionally been the preferred choice. Comment is sought on each recommendation. An issues and options paper, as the name implies, gives a range of options, and comment is sought on the preferred options.

The main advantage of releasing a draft plan is that the opinion of a steering committee is clearly stated. Another advantage is that if public comment on the draft plan is favourable, there can simply be final endorsement of the draft plan, which then becomes the final plan.

The main disadvantage of releasing a draft plan, especially if it is professionally printed and designed, is that can look too final and people can feel that, whatever the merit of their comments, it will not be changed. The danger is that a sense of ownership, or commitment to the implementation of the plan, is not engendered in the plan's stakeholders.

Whatever choice is made about the form of the report, effort is generally made to ensure that:

- the response period is long enough for all who wish to review and provide comment (this is usually two to three months);
- distribution is wide – to relevant state government departments, politicians, libraries of educational institutions, study workshop participants, local public libraries, local community groups and clubs;

- copies are made available free of cost, in order to encourage the widest opportunity for comment;
- local newspapers are used to advise the public that the document is available for comment; and that
- websites are used to make documents available and provide a forum for feedback and discussion (subject to the technology being available and a sufficient numbers of consultees having Internet access).

Once the responses from stakeholders and interested parties are collated and analysed, and any additional information is analysed, the planning team selects a final option. The option selected will depend on a number of factors such as available resources, community attitudes, existing plans and policies, and feasibility of implementation. Ease of implementation of the selected option is critical to the long-term success of the planning project. Relevant to this is ensuring that recommended management actions and guidelines are consistent, integrated and coordinated with other strategies and management plans (e.g. regional transport strategy or local tourism strategy). The plan should be forwarded to relevant agencies for endorsement.

5.6 The implementation of coastal management plans

> In most nations it is generally much more difficult to secure commitment to management than for the creation or establishment of management plans.
>
> (Kenchington 1990: 218)

> Two conclusions are constantly reinforced whenever public program implementation is subjected to scrutiny:
>
> 1 No one is clearly in charge of implementation; and
> 2 Domestic programs virtually never achieve all that is expected of them.
>
> (Ripley and Franklin 1986: 2)

It is worth concluding this chapter with a brief section on the implementation of coastal management plans. The ever-present danger is that the production of the plan is viewed as the end of the planning process, instead of the beginning of the real actions for its implementation. After all, most coastal management plans were started in the first place to help solve problems, not to just sit on the shelf.

There are many reasons why more plans grace bookshelves than become the dog-eared guide for the people on the ground. The implementation of plans can be difficult to achieve. Perhaps the most important are the emotions felt by those who are actually undertaking the planning exercise. Finishing and publishing a plan can be an exciting and satisfying experience, one giving a real sense of achievement. There can be, quite naturally, a lull in the motivation of these individuals after the plan is finished. Picking themselves up and implementing the plan can be tough going.

A related phenomenon is the division of responsibilities which can occur when a specialist planning group that has produced the plan is distinct from those who are charged with its implementation. The result of planners not involving the staff who will implement the plan is that implementation staff lack the background knowledge and rationale, and that sense of ownership that ensures effective implementation. In turn, staff charged with implementing the plan may be unsure about the plan's objectives. This is especially true if the implementation staff were peripheral to the planning process. These apprehensions and mistrusts can be managed by involving implementing agencies and staff throughout the planning process, and through a well designed implementation program which makes a conscious effort to ensure all involved understand why the plan was developed and what it is aiming to achieve.

Implementing plans formulated at strategic and operational levels involves translating objectives into actions. At the strategic level, implementation can involve several diverse activities: the drafting of relevant legislation; the formulation of policy for a wide range of issues; the establishment and resourcing of programs to identify and declare potential protected areas; and the development of other national programs. An important failure of implementation at this level is the setting of unrealistic expectations for what the plan can actually assist in achieving (OECD 1993).

At the operational level, the translation of management objectives and guidelines into on-the-ground and on-the-water management is known as operational or day-to-day management. It is these activities, such as enforcement, surveillance and EIA, which provide most of the active management of current and future uses of an area as well as meeting the wider community's perception of coastal management. Many stakeholders perceive this form of management as the 'doing' part of planning. These practical actions, together with communication and education, research, monitoring and site planning are all considered by managers to be necessary for sound implementation.

The problem of allocating responsibilities for plan implementation increases as the complexity and/or geographic coverage of the plan increases. A wide-ranging plan will inevitably require the involvement of a wider field of stakeholders.

Implementation can be considered as having three major components:

- managing the resource and resource users;
- ensuring that stakeholder expectations are met; and
- meeting statutory requirements in a cost effective manner.

An additional component that could be added to the above list is evaluation of the implementation measures and ensuring that the results of this evaluation are fed back into the planning cycle.

Figure 5.12 is a simple representation of the interactions of the three major components in the implementation of management plans.

The three components of users, resources and statutory requirements shown in Figure 5.12 interact, the areas of intersection representing management activities. Effective implementation is achieved by developing a balance between the three components. The area of intersection of all three components represents optimal implementation, where effective and efficient management is achieved. Here the resources are effectively managed, while users' expectations are satisfied and statutory requirements met. Implementation therefore tries to minimise regulation and resource costs while maximising community support and participation.

Managers have at their disposal a variety of tools for plan implementation, including planning and programming, staff training, education, stakeholder involvement,

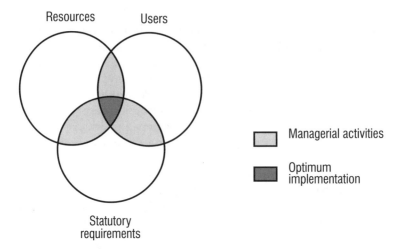

Figure 5.12 Interaction of the major components in implementing management plans

surveillance and enforcement, environmental impact assessment, research, and project monitoring and evaluation. Many of these tools are described in Chapter 4 and illustrated in a general model for implementing plans in Box 5.29.

Box 5.29

A general model of plan implementation

Many of the tools described in this chapter can be used to implement a plan. A general model for implementing plans is illustrated in the figure below. This model is based on a generalised regional- or local-level integrated plans, and assumes that levels of resources are available for implementation as identified in the plan.

Once a plan is endorsed, several ongoing activities take place to implement the plan. Some take place simultaneously while others take place in a sequence. Implementation usually commences with the programming of implementation requirements. This stage focuses on the establishment and management of day-to-day activities and projects. The requirements of this stage depend on the focus and scale of the plan – site-level plans will usually have focused on the direction of tangible management activities, and hence implementation programming will be minimal.

Staff who will be involved in implementation activities should have ideally been involved in producing the plan. This ensures that all staff are aware of the reasons behind the various planning outcomes and subsequent management activities used in their day-to-day responsibilities. However, different staff have varying levels of involvement in the range of implementation activities. They should be given training and job preparation. This assists in team building, and ensures that all staff understand the reason for the programs and the basis of their personal work plan.

continued...

Box 5.29, continued

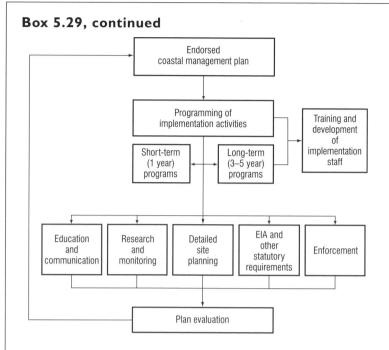

Generalised flow diagram of plan implementation

Training and staff development are important components of the implemetation process and they serve two main functions. First, they create and build an effective management team. Usually, no one person has all the skills and expertise required in the operational management of an area. By building a team and focusing on a common goal of plan implementation, one person's deficiencies may be compensated by another's strengths. Team building ensures that everyone knows their role in management of an area. Second, it ensures that staff have the skills and expertise to perform their expected tasks and to contribute to the overall management of the area.

An implementation program will use a number of activities to meet management objectives, including education, research, surveillance and enforcement. Education should be an ongoing program of activities to raise awareness of issues, alter user behaviour and facilitate involvement in management throughout the life of the management plan. The focus of education will change as the level of awareness of stakeholders improves, and issues are managed. Each program will need to be developed for the area's needs and the available resources.

Research programs, within a plan implementation framework, should be designed to fill in the information gaps identified in the planning process and provide input into any ongoing adaptive capacities built into planning processes. Research activities outside of the implementation program (e.g. university programs) should be coordinated to ensure that research is relevant to management, and that staff and financial resources dedicated to plan implementation are maximised, while being scientifically rigorous.   continued...

Box 5.29, continued

Surveillance and enforcement requirements for implementing a plan are established in relation to the management objectives and priorities, the needs of various monitoring programs, and available resources. Surveillance is a multi-functional monitoring tool in day-to-day management (see Chapter 4). It not only detects and deters infringers, but it can gather information for monitoring, plan adaptation and research purposes. The objectives of an enforcement program usually include improving user compliance of rules and regulations – to help 'do the right thing'. Similarly, education and communication programs seek the same objective, and therefore when enforcement programs are formulated managers should ensure that they are complementary to education programs to maximise their effectiveness.

Box 5.30

Moving from planning to implementation for the Thames Estuary Management Plan (Kennedy 1996)

A strategy for implementation of the Thames Estuary Management Plan (English Nature 1996) was launched in July 1996 in parallel to public consultation on the estuary management plan itself. The objectives of the implementation strategy were to:

- define the key elements and priorities for implementation;
- outline the initial work programme and time-scale for implementation;
- suggest appropriate management and administrative structures; and
- identify the financial and other key resources required to support implementation.

It took three months to agree on the institutional arrangements for the future of the project and, as a result, there was a time lag in moving forward the recommendations of the plan into action on the ground. This resulted in a review of the Thames Estuary Management Plan in 1999 and, following further consultation, its development into Management Guidance for the Thames Estuary. Several steps were taken to ensure successful implementation of the revised plan:

- Production of a three tier document – full strategy; condensed principles for action; rationalised action plan in manageable elements.
- Formation of the Thames Estuary Partnership as a company limited by guarantee and a charity.
- Formation of issue-based action groups to provide neutral forums for discussion and project implementation.
- Interpretation of the main action plan to specific, operational sub-action plans.
- Initial support from a few core funders, later diversified to ensure sustainable funds.

continued...

Box 5.30, continued

- Increased credibility from successful flagship projects.
- Continuous widening, strengthening and nurturing of the partnership.

   Implementation of the management guidance for the Thames Estuary follows a 3–4 year cycle closely related to the original phases of its development. The cycle evolves as a more formalised structure emerges, reflecting a small business with a coastal management function. Phases of the second cycle of plan implementation in the Thames Estuary are shown in the table below.

Thames Estuary 3–4 year implementation cycle

Phases	Advantages/considerations/issues
Scoping	• Identification of current issues • Rejection of non-issues • Exploration of wider partnership needs • Consideration of original action plan and other sub-action plans • Consideration of political agenda • Relation to other management plans • Increase partner interest
Planning	• The process and its design • Administrative arrangements • Estimate cost and time • Development and adoption of management guidance for the Thames Estuary • Interpretation of management guidance to 3-year operational action plan • Prioritisation of actions • Develop funding strategy • Production of 3-year business plan
Changes to management structures	• Review management structures in light of current business plan • Consider representation • Formalise, strengthen and nurture partnership • Ensure effective lines of communication – up, down and across • Refresh and maintain relevance as a business
Implementation	• Develop subject • Specific, operational sub-action plans • Support action groups as forums for discussion and project implementation • Secure project-based funding • Implement key flagship projects • Communicate widely to raise awareness • Increase creditability • Link with other agendas, i.e. government
Review	• Review management guidance for the Thames Estuary action plan • Review operational sub-action plans • Monitor and evaluate progress • Reassess partnership's needs

continued...

Box 5.30, continued

Funding required for the successful implementation of Management Guidance for the Thames Estuary follows a cycle related to the different phases of implementation, as shown in the figure below.

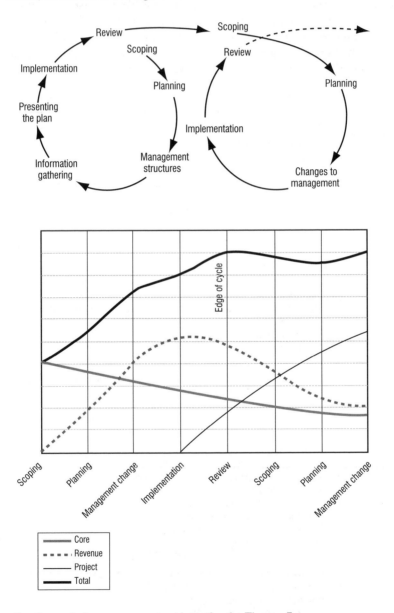

Funding cycle for management guidance for the Thames Estuary

continued...

Box 5.30, continued

Core funding begins high to initiate implementation of the plan and decreases as the partnership becomes more defined.

Other sources of revenue funding increase from the planning stage onwards as confidence in the partnership increases.

Project-based funding begins during the implementation phase; reflects the need to deliver projects to prove the worth of the Partnership and supplements reduced core funds.

As the partnership moves into the second cycle of implementation, funds diversify, aiding financial stability. There will always be a critical period at the end of every cycle during the transition between different types of funding.

An innovative approach to the problem of bridging the often large chasm between the finalisation of a plan and its implementation is shown by the Thames Estuary Management Plan (Box 5.30).

Two important tools in assessing the effectiveness of a plan or program are program monitoring and evaluation, which are discussed in the next two sections.

The political aspect of coastal management implementation should also be borne in mind. If a coastal management planning exercise has attained a high political profile, then the political interest in the plan usually peaks at the time the plan is released. Politicians are often given the opportunity (and some demand) to publicly release the plan. This gives the politician an opportunity to show he or she is doing something positive, innovative and forward-looking for the coast. Giving a politician a stake in the success of the project can help to maintain political interest until tangible results of the plan's implementation can be seen. This can be very important, as the first stage of the plan's implementation is often the most fragile, especially if funds are required.

Tangible outcomes should be included early in the plan's implementation to maintain the interest and support of politicians and stakeholders. More politically astute coastal managers have been known to deliberately stage the implementation of plans to give some early 'wins' to ensure the ongoing interest of politicians and other senior stakeholders.

Implementation of plans and strategies may also be distributed across a range of programs or initiatives rather than using a specific implementation program. This approach requires fewer resources but it does not ensure implementation will be coordinated or complete. It is an approach used to implement components of Indonesia's coastal management plans (Box 5.31).

Finally, the costs of plan implementation can often be realised only after the plan has been completed. Clearly, this can lead to major implementation problems. Explicitly including costings for implementation within the planning process is emerging as one mechanism to counter the 'this costs too much to implement' argument. In the case of the Thames Estuary (Box 5.30), a plan implementation 'prospectus' was developed in order to clarify the requirements and costs for implementation.

Box 5.31

Implementing coastal management planning in Indonesia

The policies, zoning plans and subject plans developed as part of Indonesia's national Marine Resources Evaluation Program now called Marine and Coastal Resource Management Program (Box 5.9) are implemented through a number of government initiatives.

One of these initiatives is COREMAP, a national program focused on coral reef rehabilitation and management, which commenced in 1998. The second phase of COREMAP program commenced in 2004 and has four major components:

- public awareness and participation;
- locally based management of priority coral reef sites;
- institutional strengthening at all levels; and
- collaborative planning. (Putra and Mulyana 2003)

These components of COREMAP are focused on implementing management objectives at varying government levels, as well as within industry and the community. Although this program is focused on coral reefs, other coastal environments such as mangroves and seagrasses are considered since they play an important role in maintaining coral reefs.

Whereas MCRMP is primarily a resource information management and planning program, COREMAP is a program which can assist in implementing MCRMP outcomes. Sulawesi Selatan province is one of the first provinces to participate in COREMAP (Box 5.13). Province-wide COREMAP initiatives such as multi-agency enforcement patrols to address the problem of blast fishing and cyanide fishing will contribute towards the provincial policy of sustainable resource use. In addition, a number of initiatives at the community level are being implemented. Awareness programs for cyanide fishers and developing alternative income generating activities for fishers are two examples which will assist in implementing a number of provincial MREP policies such as 'raising public awareness of the value of resources and processes so as to encourage responsible resource use' and 'all coral reefs in Sulawesi Selatan waters will be protected from unsustainable exploitation and damage due to human activity'.

The COREMAP program also proposes the formulation of community management groups to provide a bottom-up approach to management. These groups can identify issues and problems which may be widespread and better managed at the regional or provincial level as part of MCRMP. The MCRMP and COREMAP programs are a good example of how top–down and bottom–up approaches to coastal management can be undertaken simultaneously and merged at the provincial planning level.

5.7 Chapter summary

This chapter has focused on taking the theory and concepts of coastal management and planning discussed in Chapter 3 and the tools for management described in Chapter 4, and translating them into a structured planning framework. We have shown the power of planning as a management tool. A hierarchical geographic planning framework was selected as the basis for structuring this chapter for a number of reasons, most notably its wide ranging application and generic nature. This approach to planning is not necessarily the answer for all circumstances. Consequently, we discussed alternative frameworks – strategic, operational, subject, integrated, statutory and non-statutory.

Much of this chapter has focused on how to develop integrated coastal plans from the whole-of-government level to site planning, how to integrate these plans vertically or horizontally and the advantages and disadvantages of each.

In the last section of the chapter we discussed the aspect most important to ensuring that plans leave the shelf and become part of the ongoing management of coastal areas: their implementation and evaluation.

Finally, we again emphasise the importance of the emerging evaluation activities that are beginning to pervade coastal planning practice round the world. The net result of these evaluations is to provide invaluable information for the design of new coastal planning initiatives, or for the improvement of existing programs. The lessons learned from such evaluative activity are directly applicable to coastal planning practice within the jurisdiction that is subject to the evaluation. Also, the combined influence of such reflective activity is a significant contributor to global best-practice. For example, the lessons learned from the Philippines (Box 5.32) are important not only in that nation but internationally, also serving to move both the theory and practice of coastal planning and management forward. This forward-looking view is presented in the next, and final, chapter.

Box 5.32

Lessons learned from Philippines coastal planning projects

Coastal management has been practised in the Philippines for almost three decades to try to stem the increasing tide of destruction to coastal habitats and the decline of fisheries. Nevertheless, coastal resources continue to decline and deteriorate at alarming rates. While numerous experiments in coastal management have been conducted around the world, few have been evaluated with sufficient scientific rigour in order to distinguish successful from unsuccessful outcomes (White 2004). In many cases, the lessons emerging from these projects are neither evaluated nor learned. There is currently a great deal of emphasis in the Philippines to provide substance upon which to evaluate the successes and failures of coastal management in the country (White 2004).

A key lesson generated by the history of coastal management projects in the Philippines since 1974 (summarised in Courtney and White 2000; Courtney *et al.* 2000) is that it is extremely difficult to plan and implement successful integrated coastal management programs without a multi-sector approach. Such programs

continued...

Box 5.32, continued

must have sufficient support from the national and local government and its partners and a strong level of acceptance among the resource dependent communities. Successful coastal management programs are still relatively localised where the geographic scope is small and the number of stakeholders limited. Nevertheless, this is changing rapidly as more multi-municipal or city and bay-wide coastal plans are being developed and implemented. A few lessons pointing out new directions are (White 2004):

- Baseline information is a prerequisite for coastal management planning and comparative analyses of 'with' and 'without' project scenarios for present and future learning.
- Coastal plans that build on good information (environmental profiles and baseline data) that evolves with the planning process are more likely to succeed.
- Quality technical expertise is a key determinant of success so that planning and planning processes do not head down dead-end paths.
- Participation at all levels that engenders learning and some basic level of awareness is a prerequisite to the implementation of coastal plans.
- The sustainability of interventions cannot be determined without sufficient time for field testing whereby participants begin to see results and take responsibility for their actions.
- An integrated planning process is essential to bring together the divergent efforts of various government, non-government and other organisations involved in management.
- Real and practical results at the field level such as improved income from fish catch, other resource use or alternatives such as tourism are a critical sustaining force at the community level.
- Even community-based management which appears relatively successful and autonomous requires continuing support and mentoring from local and/or national government, NGOs and the private sector.
- Political leadership is always required to start and sustain successful coastal management and planning programs through provision of resources, legal mandate, moral support and overall direction for local resource stakeholders.

Chapter 6

Conclusions and future directions

In this second edition we have endeavoured to provide a continued foundation for coastal planning and management by mixing theory with examples of best-practice from around the world. What has this approach told us about the current status of coastal planning and management, and about how theory and practice have evolved in recent years, and what pointers has it given us possible directions for the future?

The main theme of the book has been that the coast, with its intensity of land and water use, is a place where the issues of economic development and environmental management, and their interactions with social and cultural values, are brought into sharp relief. If there are problems with any of these issues, in any area of a coastal nation, the symptoms are likely to show up at the coast first.

Several other themes and principles emerge from the book. They are necessarily broad in scope, given the enormity of the issues and challenges facing coastal managers, but we summarise them in order to stimulate further discussion and research.

THE CENTRAL ROLE OF SUSTAINABLE DEVELOPMENT PRINCIPLES AND APPROACHES

Coastal programs are now firmly based on principles of sustainable development, the precautionary principle, and inter-generational equity. The continued challenge for coastal planners and managers is to transform sustainable development principles into tangible management outcomes. We hope that the tools and techniques described in this book go some way towards meeting this challenge.

THE INSEPARABLE NATURE OF COASTAL PLANNING AND MANAGEMENT

Coastal planning and management activities are generally so strongly linked that in successful coastal programs they are almost indistinguishable. The interweaving of planning and management to create a single coastal program can help to break down institutional boundaries or possible professional rivalries between planners and managers, and is to be encouraged.

THE INCREASING EMPHASIS ON CONSENSUAL STYLES OF COASTAL PLANNING AND MANAGEMENT

Consensual planning processes have become the most widely used approach for integrated coastal plans at the regional and local levels. Increased community empowerment and the problems caused by more directive planning styles have lead to different community-based, collaborative and co-management methods of coastal management planning. Innovative consensus building tools have to be used to ensure that consensus does not equate to 'lowest common denominator', resulting in bland outcomes. This is especially so where conflict arises, often in the case of siting hazardous and/or polluting industries. Early indications of the use of consensual styles of planning in these cases suggest that they can be successful when adequate resources are allocated to them, although they are yet to be used in extreme cases of conflict.

COASTAL MANAGEMENT IS A SHARED CONCERN

Responsibility for sustainably managing the coast is shared by all levels of government, from international to local, along with coastal users, residents, private companies and advocacy groups. Governments are increasingly realising the long-term benefits of engaging all stakeholders on the coast in coastal program development. This partnership approach has rapidly evolved from just a 'good idea' into a cornerstone of many coastal initiatives around the world.

GOOD COASTAL MANAGEMENT IS FOUNDED ON AN APPRECIATION OF LOCAL CULTURAL FACTORS

Western, and especially rational, approaches to coastal planning and management, while successful in many countries, especially those with European land-tenure systems, require modification if they are to be successfully integrated into other local cultural settings. Traditional and local knowledge about coastal resources and their management can be invaluable in formulating management prescriptions. The bringing together of western and traditional management tools and techniques has shown success in many developing countries. Indeed, there appears to be now a genuine two-way flow of knowledge and experience in coastal management between developed and developing countries.

THE CROSSING OVER OF PLANNING AND MANAGEMENT TECHNIQUES

Coastal programs have become a melting pot for various planning and management techniques which have crossed over from other disciplines. Land-use planning techniques, such as separating conflicting uses through zoning, blend with economic analysis and risk management, co-management and a host of other approaches to help address coastal issues. Coastal planners and managers are increasingly being encouraged to add to – and occasionally stir – this melting pot to find innovative ways of addressing coastal problems and opportunities.

DESIGNING A MIXED COASTAL PLANNING SYSTEM CAN BE SUCCESSFUL

Issues requiring coastal management and planning cut across jurisdictions, occur at widely different scales, and involve a diversity of stakeholders. No single plan can be expected to cope with all coastal issues, but management practices and plans can be substantially improved by mixing integrated coastal plans at different scales, orientations and statutory bases. Cascading planning systems designed to link broad strategic plans to detailed local planning initiatives are an example of such integration.

THE IMPORTANCE OF MEASURING SUCCESS

A plethora of coastal plans exists around the world, addressing vastly different issues, often in very different ways. But a common feature of most of these plans is the continued absence of quantitative evidence of their effectiveness – this despite the often considerable resources used in their formulation and implementation. Politicians, government departments, donors and the public are now expecting coastal programs to provide clear demonstrations of success. Performance measures, evaluation criteria and success indicators have become an increasing part of the coastal manager's lexicon. Yet the science and economics of measuring just how successful coastal programs are is only now beginning in earnest. Coastal program managers are increasingly required to include monitoring and evaluation measures into program design at the outset – a difficult task without a set of commonly accepted coastal management performance measures. Some long-standing programs are in the process of being questioned and critically reviewed on the basis that their performance has been unable to be effectively evaluated in the past.

6.1 Facing the future

Chapter 2 could invoke contrasting responses in the reader: pessimism at its rather depressing list of often chronic problems, painting a not too bright future for the coast; or excitement and optimism about the challenges that these problems present. A realistic coastal planner/manager is one who would absorb a little of both and plan to pragmatically tackle the major challenges facing the world's coast, while being creative and flexible in dealing with the inherent limitations of the workings of government and private sector bureaucracies. To this planner/manager we offer our six virtues of coastal planning: to seek, to understand, to develop, to link, to bring into mainstream, to sustain. And the challenges that go with them.

To seek
- the money and willingness to implement plans;
- true economic values of coastal resources and implementing management responses which reflect those values;
- an appropriate balance between traditional and local management practices and knowledge, and typical western approaches to coastal management;
- the mechanisms which allow developing countries to sustainably exploit coastal resources and avoid many of the mistakes of developed countries and for developed countries to more effectively learn from their own mistakes;
- optimal solutions to resource sharing on the coast, especially emerging industries and uses, such as recreational and tourism demands;
- workable strategies for ensuring equitable access to coastal resources for all sections of society;
- approaches that are able to adapt quickly to changing pressures, community and political expectations, and the increasing pace of change in the global economy;

- locally sustainable and tailored economic and social growth for the coastlines of developing countries.

To understand
- the values and expectations of all stakeholders in coastal management;
- the role of traditional and local user knowledge;
- the ecosystems (including human ecosystems) on which management decisions are based; and
- deal with uncertainties in decision making;
- the social and biophysical inter-relationships between catchments and coasts and oceans;
- the capacities required for coastal management, including training, monitoring and scientific studies.

To develop
- and maintain appropriate stewardship of coastal resources tailored to social and cultural settings of coastal nations;
- meaningful indicators for the evaluation of coastal initiatives.

To link
- coastal initiatives at all scales – from international to local;
- managers to other managers to further develop theoretical and practical management and planning approaches;
- integrated and subject plans.

To bring into mainstream
- monitoring and evaluation at all stages of coastal management;
- sustainable economic and social development.

To sustain
- community expectations after plans are completed;
- the momentum going from Agenda 21, and related international initiatives;
- the energy of local coastal managers.

6.2 Afterword

The enormous problems facing the world's coastlines are unlikely to diminish with time. Population increase, technological change, economic growth and ever more waste generation make it likely that the problems will become even more acute. The key questions are thus not if, or when, these pressures will occur, but whether the coast can be managed to sustainably absorb them, and whether those management actions can be assessed as successful and replicated elsewhere. And in this of course lies the fascination of being involved with the management of the coast – the huge challenge it presents to forge creative, innovative and long-term solutions to apparently intractable management problems.

We do not pretend with this book to have offered all the solutions, but rather to have provided a wide selection of methods and models to guide the search for environmentally, culturally and economically appropriate planning and management outcomes. We will judge our success by the extent to which we have stimulated the searchers and helped them to meet their challenges with a critical and optimistic outlook.

Appendix

Some definitions of the coastal zone for planning and management

The following are definitions of the coastal zone used to define areas within which coastal management policies apply. Chapter 1 describes the advantages and disadvantages of each type of definition.

Fixed distance definitions

Table A.1 Example fixed definition boundaries of the coastal zone

Country	Inland boundary	Ocean boundary
New South Wales, Australia	1 km from open coast HWM 1 km around all bays, estuaries, coastal rivers to limit of mangroves	3 nautical miles seaward of the mainland and offshore islands
Brazil	2 km from MHW	12 km from MHW
Costa Rica	200 m from MHW	MLW
China	10 km from MHW	15 m isobath (depth)
Spain	100 m from highest storm or tide line	12 nautical miles (limit of territorial sea)
Sri Lanka	300 m from MHW	2 km from MLW

Source: Sorensen and McCreary (1990); NSW Government (1997).

Sri Lanka Coast Conservation Act (1990):

> The area lying within a limit of three hundred metres landward of the Mean High Water Line and a limit of two kilometres seaward of the Mean Low Water Line and in the case of rivers, streams, lagoons, or any other body of water connected to the sea either permanently or periodically, the landward boundary shall extend to a limit of two kilometres measured perpendicular to the straight line base line drawn between the natural entrance points (defined by the mean low water line) thereof and shall include waters of such rivers, streams and lagoons or any other body of water so connected to the sea.

Dubai Decree 22 (2001):

> The coastal zone of Dubai extends from a landwards boundary 1 km inland of the mean high water mark (MHWM) to a seawards limit 10 nautical miles offshore and the entire extent of the Dubai Creek as well as Al Mamzar lagoon and any other excavated water bodies connected to the Arabian Gulf.

Variable distance definitions

The South Pacific Regional Environment Programme (SPREP 1993):

> The coastal zone is a region of indeterminate and variable width. It extends from and includes, the wholly marine (i.e. the seabed, the overlying waters and their resources) to the wholly terrestrial (i.e. beyond the limits of marine incursion and the reach of salt spray). Linking these two environments is the tidal area which forms a transition between land and the sea.

Definition according to use

United States Federal Coastal Zone Management Act (1990) Section 304 (note that each coastal state must interpret the federal definition through the production of maps and charts):

> The term 'coastal zone' means the coastal waters (including the lands therein and thereunder) and the adjacent shorelands (including the lands therein and thereunder), strongly influenced by each other and proximity to the shorelines of the several coastal states, and includes islands, transitional and intertidal areas, salt marshes, wetlands and beaches. The zone extends, in Great Lakes waters, to the international boundary between the United States and Canada and, in other areas, seaward to the outer limit of State title and ownership ... (continues with list of Acts) ... The zone extends inland from the shorelines only to the extent necessary to control shorelands, the uses of which have a direct and significant impact on the coastal waters.

Delaware Coastal Zone Act (1971):

> "The coastal zone" is defined as all that area of the State, whether land, water or subaqueous land between the territorial limits of Delaware in the Delaware River, Delaware Bay and Atlantic Ocean, and a line formed by certain Delaware highways and roads as follows:

> Beginning at the Delaware–Pennsylvania line at a place where said line intersects U.S. Route 13; thence southward along the said U.S. Route 13 until it intersects the right-of-way of U.S. Route I-495; thence along the said I-495 right-of-way until the said I-495 right-of-way intersects Delaware Route 9 south of Wilmington ... thence along Maintenance Road 395 to the Maryland state line.

Australian Commonwealth Coastal Policy (1995):

> For the purpose and actions of the Commonwealth, the boundaries of the coastal zone are considered to extend as far inland and as far seaward as necessary to achieve the Coastal Policy objectives, with a primary focus on the land–sea interface.

United Kingdom Government Environment Committee Report on Coastal Zone Protection and Planning (1992):

> We conclude that definitions of the coastal zone may vary from area to area and from issue to issue, and that a pragmatic approach must therefore be taken at the appropriate national, regional or local level.

World Bank Environment Department (1993):

> For practical planning purposes, the coastal zone is the *special area* (original emphasis), endowed with special characteristics, of which the boundaries are often determined by the special problems to be tackled.

OECD Environment Directorate (1992, 1993):

> What constitutes the coastal zone depends on the purpose at hand. From both the administrative and scientific viewpoints, the extent of the zone will vary depending on the nature of the problem. Accordingly, the boundaries of the coastal zone should extend as far inland and as far seaward as necessary to achieve the objectives of management.

Australian Commonwealth (1980) Inquiry recommendation:

> Any definition of the coastal zone should be flexible, and should depend on the issue being confronted.

New South Wales Government Draft Revised Coastal Policy (1994) – Option 5 (not selected – see Table A.1 above):

> … an issues based definition where the boundaries of the coastal zone extend as far inland and as far seaward as necessary to achieve the policy's objectives, with a focus on the land sea interface …

Helsinki Commission – Baltic Marine Environment Protection Commission Recommendation for Implementation of Integrated Marine and Coastal Management of Human Activities in the Baltic Sea Area (2003):

> The coastal area(s) (same as coastal zone) is defined as the zone following the Baltic Sea coastline, extending 3 km landwards from the mainland coast to the adjacent marine offshore areas. The offshore areas extend from the outer border of the coastal areas as far offshore as it in each case is relevant for the sustainability

of the marine and coastal biodiversity and geodiversity, in particular these areas are used or intended to be used in a way that conflict or may conflict with the aims of Article 3 of the Helsinki Convention. These zones thus cover the Baltic Sea waters, the underlying seabed and coastal terrestrial areas including the biota as well as abiotic resources.

Notes

2 Coastal management issues

1 Examples of texts, conference and workshop proceedings which outline coastal problems around the world include (only the most recent published references of conference series are shown):

Asia-Pacific: recently initiated Coastal Zone Asia Pacific series CZAP (2002) 'Improving the State of the Coastal Areas', in *Coastal Zone Asia Pacific Conference*, Bangkok, Thailand, 11–17 May, Thailand Coastal Development Institute.

Africa: (Burns *et al.* 2000; Coastal Development Authority 2000; Tanzanian Coastal Management Partnership 2001).

Australasia: Proceedings of the Coast to Coast (Australia) conference series Hamilton, B. and Kay, R.C. (1998) 'Proceedings of the Coast to Coast "1998"', in *Coast to Coast '98*, Perth: Ministry for Planning, Government of Western Australia, Victorian Coastal Council (2000) 'Beyond the Beach', in *Coast to Coast*, Melbourne, Victoria, 6–9 March, CRC for Coastal Zone Estuary and Waterway Management (2002) 'Source to Sea', in *Coast to Coast*, Tweed Heads, NSW, 4–8 November: CRC for Coastal Zone Estuary and Waterway Management, and the New Zealand Coastal Society conferences. Textbook by Harvey, N. and Caton, B. (2003) *Coastal Management in Australia*, Oxford: Oxford University Press, focusing on Australia.

Europe (including Eastern Europe and Scandinavia): Eurocoast and European Union for Coastal Conservation Eurocoast Federation (2000) 'Littoral', in *Responsible Coastal Zone Management – The Challenge of the 21st Century*, Dubrovnik, Croatia: Eurocoast Federation (2002) 'Littoral', in *The Changing Coast*, Porto, Portugal.

North America: proceedings of the US Coastal Zone and US Coastal Society conferences NOAA Coastal Services Center (2003) 'Coastal Management Through Time', in *Coastal Zone*, Baltimore, Maryland: NOAA. The Coastal Society (2002) 'Converging Currents: Science, Culture and Policy at the Coast', in *Proceedings of the Eighteenth International Conference*, Galveston Island, Texas; 19–22 May. Coastal Zone Canada Coastal Zone Canada Association (2002) 'Managing Shared Waters', in *Coastal Zone Canada 2002*, Hamilton, Ontario: Coastal Zone Canada Association, conference series. The textbook of Beatley, T., Brower, D.J. and Schwab, A.K. (2002) *An Introduction to Coastal Zone Management*, Washington, DC: Island Press. There are also many individual articles in the journals *Coastal Management* and *Ocean and Coastal Management*.

In addition, there are conferences on various coastal management problems on a sector-by-sector or subject-by-subject basis.

4 Major coastal management and planning techniques

1 The site dedicated solely to this purpose can be found at http://www. coastalmanagement. com and is managed by the principal author.
2 'Environmental Impact Analysis' is another term for Environmental Impact Assessment. Environmental Impact Statements (EIS) are generally the written reports required of an EIA process.

References

Adger, N. (2000) 'Environmental and Ecological Economics', in *Environmental Science for Environmental Management* (T. O'Riordan, ed.), Harlow: Prentice Hall.

Adler, E. (2004) 'Regional Seas Programme – recent developments', personal communication.

Aeron-Thomas, M. (2000) 'Integrated Coastal Zone Management in Sri Lanka, Improving Policy-Livelihood Relationships in South Asia', *Policy Review Paper 4*, York: Stockholm Environment Institute, Water and Development Group.

Agardy, M. (1990) 'Integrating Tourism in Multiple Use Planning for Coastal and Marine Protected Areas', in *Proceedings of 1990 Congress on Coastal and Marine Tourism: A Symposium and Workshop on Balancing Conservation and Economic Development*, Honolulu, Hawaii, 25–31 May.

Agardy, T. and Alder, J. (in press) 'Coastal Ecosystems and Coastal Communities', in *Millennium Ecosystem Assessment: Condition and Trends Working Group Report* (R. Hassan and N. Ash, eds), Washington, DC: Island Press.

Alcala, A. and Russ, G. (1990) 'A Direct Test of the Effects of Protective Management on Abundance and Yield of Tropical Marine Resources', *J. Cons. Int. Explor. Mer.*, 46: 40–7.

Alcock, D. (1991) 'Education and Extension: Management's Best Strategy', *Australian Parks and Recreation*, 27(1): 15–17.

Alder, J. (1993) 'Permits, an Evolving Tool in the Day-to-Day Management of the Cairns Section of the Great Barrier Reef Marine Park', *Coastal Management*, 21(1): 25–36.

Alder, J. (1994) 'Have Six Years of Public Education Changed Community Awareness of and Attitudes Towards Marine Park Management?', *Proceedings of the Seventh International Coral Reef Symposium*, Guam.

Alder, J. (1996) 'Education or Enforcement – the Better Deal?', in *Great Barrier Reef Science, Use and Management*, Townsville, QLD: James Cook University.

Alder, J. and Ward, T. (2001) 'Australia's Oceans Policy: Sink or Swim?', *Journal of Environment and Development*, 10(3): 266–89.

Alder, J., Kay, R.C., Clegg, S., Carman Brown, A. and Macklin, K. (1997) 'The West Coast: A Manual for Coastal Planning in Western Australia' (Draft), Perth: Unpublished draft for the Western Australian Ministry for Planning.

Alder, J., Sloan, N.A. and Uktolseya, H. (1994) 'A Comparison of Management Planning and Implementation in Three Indonesian Marine Protected Areas', *Ocean and Coastal Management*, 24: 179–98.

Alder, J.A., Sloan, N. and Uktolseya, H. (1995) 'Advances in Marine Protected Area Management in Indonesia: 1988–1993', *Ocean and Coastal Management*, 25(1): 63–75.

Alexander, E.R. (1986) *Approaches to Planning: Introducing Current Planning Theories, Concepts and Issues*, New York: Gordon and Breach.

Alix, J.C. (1989) 'Community-Based Resources Management: The Experience of the Central Visayas Regional Project-I', in *Coastal Area Management in Southeast Asia: Policies, Management Strategies and Case Studies* (T.-E. Chua and D. Pauly, eds), Manila: ICLARM.

Allen, A. and You, N. (2002) *Sustainable Urbanisation: Bridging the Green and Brown Agendas*, London: Development Planning Unit, University of London.

AMCORD (1995) *A National Resource Document for Residential Development*, Canberra: Australian Government Printing Service.

Anon (1997) 'Gill-Netting Banned to Protect Endangered Dugong', *Courier Mail*, 31 March 1997.

Anthias (1994) 'The Ras Mohammed National Park Newsletter', *Egyptian Environmental Affairs Agency*.

Archer, J. and Knecht, R.W. (1987) 'The U.S. National Coastal Zone Management Program – Problems and Opportunities in the Next Phase', *Ocean and Coastal Management*, 15: 103–20.

Archer, M.S. (1998) *Critical Realism: Essential Readings*, London: Routledge.

Armstrong, M. (1986) *A Handbook of Management Techniques*, London: Kogan Page.

Arnstein, S. (1969) 'A Ladder of Citizen Participation', *American Institute of Planners*, 35: 216–24.

Ashby, W.R. (1969) 'Self-Regulation and Requisite Variety', in *Systems Thinking* (F.E. Emery, ed.), Harmondsworth: Penguin Books.

Asian Development Bank (1991) *Environmental Evaluation of Coastal Zone Projects: Methods and Approaches*, Manila: Asian Development Bank.

Asian Development Bank (1999) *Report and Recommendation of the President to the Board of Directors on a Proposed Loan to the Democratic Socialist Republic of Sri Lanka for the Coastal Resource Management Project*, Rrp:Sri 31287, November.

Atapattu, S. (2005) Personal communication. Reef impacts at Hikkaduwa, Sri Lanka.

Atkins Project Team (2003) *ICZM in the UK: A Stocktake. Final Scoping Report*, London: Department for Environment Food and Rural Affairs.

Atkins Project Team (2004a) *ICZM in the UK: A Stocktake. April 2004 Newsletter*, London: Department for Environment Food and Rural Affairs.

Atkins Project Team (2004b) *ICZM in the UK: A Stocktake. Final Report*, London: Department for Environment Food and Rural Affairs.

AusAid (2005) 'N.E Indian Ocean Earthquake/Tsunami Affected Area'. http://www.ausaid.gov.au/ (accessed 13 February 2005).

Australian Bureau of Statistics (2003) 'Australian national accounts, state accounts', ABS Cat 5220.0, 12 November.

Australian Bureau of Statistics (2004) 'Australian demographic statistics September quarter 2003', ABS Cat 3101.0, 18 March.

Baines, G. (1985) 'Coastal Area Conservation in Tropical Islands', *IUCN Bulletin*, 16(7–9): 79–80.

Balmford, A., Bruner, A., Cooper, P., Costanza, R., Farber, S., Green, R.E., Jenkins, M., Jefferiss, P., Jessamy, V., Madden, J., Munro, K., Myers, N., Naeem, S., Paavola, J., Rayment, M., Rosendo, S., Roughgarden, J., Trumper, K. and Turner, R.K. (2002) 'Economic Reasons for Conserving Wild Nature', *Science*, 297(3) (9 August): 950–53.

Bangda (1996) *Technical Manual for Preparation of a Provincial Coastal and Marine Management Strategy* (English version), Jakarta: Government of Indonesia.

Bason, D. (1990) *Bom!*, Jakarta: Departemen Kehutanan, The Nature Conservancy, Yayasan Pusaka Alam Nusantar.

Bateman, I. (1995) 'Environmental and Economic Appraisal', in *Environmental Science for Environmental Management* (T. O'Riordan, ed.), Harlow: Longman.

BBC (2005). 'Help Rebuild Tsunami Village'. http://www.bbc.co.uk/birmingham/content/articles/2005/ 01/05/michelle_mills_tsunami_feature.shtml (accessed 6 February 2005).

Beatley, T., Brower, D.J. and Schwab, A.K. (2002) *An Introduction to Coastal Zone Management*, Washington, DC: Island Press.

Beer, S. (1981) *Brain of the Firm*, Chichester: John Wiley.

Belfiore, S. (2001) 'Integrated Coastal Management in the European Union: Prospects for a Common Strategy', in *Sustainable Coastal Management: A Transatlantic and Euro-Mediterranean Perspective* (B. Cicin-Sain, I. Pavlin and S. Belfiore, eds), Dordecht: Island Press.

Belfiore, S., Balgos, M., McLean, B., Galofre, J., Blaydes, M. and Tesch, D. (2003) *A Reference Guide on the Use of Indicators for Integrated Coastal Management*, ICAM Dossier 1, IOC Manuals and Guides No. 45, Paris: Intergovernmental Oceanographic Commission of UNESCO.

Bell, S. and Morse, S. (1999) *Sustainability Indicators: Measuring the Immeasurable*, London: Earthscan.

Bell, S. and Morse, S. (2003) *Measuring Sustainability: Learning by Doing*, London: Earthscan.

Bergin, T. (1993) 'Marine and Estuarine Protected Areas (MEPAs): Where Did Australia Get It Wrong?', in *Proceedings of the Fourth Fenner Conference on the Environment*, 9–11 October 1991, Sydney, Australia: IUCN.

Berkes, F. (ed.) (1989) *Common Property Resources: Ecology and Community-Based Sustainable Development*, London: Belhaven Press.

Bertalanffy, L.V. (1969) *General System Theory: Foundations, Development, Applications*, New York: George Briziller.

Beukenkamp, P., Gunther, P., Klein, R., Misdorp, R., Sadacharan, D. and de Vrees, L. (1993) 'World Coast '93 Proceedings', in *World Coast '93*, Noordwijk: Coastal Zone Management Centre, The Netherlands National Institute for Coastal and Marine Management.

Bhaskar, R. (1998) *The Possibility of Naturalism: A Philosophical Critique of the Contemporary Human Sciences*, London: Routledge.

BIEAP (2001) *Consolidated Environmental Management Plan for Burrard Inlet*, Vancouver: Burrard Inlet Environmental Action Program.

Blair, S. and Truscott, M. (1989) 'Cultural Landscapes – Their Scope and Their Recognition', *Historic Environment*, 7(2): 40–4.

Blowers, A. (ed) (1993) *Planning for a Sustainable Environment*, London: Earthscan/Town and Country Planning Association.

Boelaert-Suominen, S. and Cullinan, C. (1994) *Legal and Institutional Aspects of Integrated Coastal Area Management in National Legislation*, Rome: Development Law Service Legal Office, Food and Agriculture of the United Nations.

Born, S.M. and Miller, A.H. (1988) 'Assessing Networked Coastal Zone Management Programs', *Coastal Management*, 16: 229–43.

Bower, B.T. (1992) *Producing Information for Integrated Coastal Management Decisions: An Annotated Seminar Outline*, Washington, DC: National Oceanic and Atmospheric Administration (NOAA).

British Columbia Forest Service (2000) *Adaptive Management Initiatives in the BC Forest Service*, http://www.for.gov.bc.ca/hfp/amhome/amhome.htm (accessed 1 May 2004).

Brown, A.L. and McDonald, G.T. (1995) 'From Environmental Impact Assessment to Environmental Design and Planning', *Australian Journal of Environmental Management*, 2: 65–77.

Brown, T.C. (1984) 'The Concept of Value in Resource Allocation', *Land Economics*, 60: 231–46.

Brown, V.A. (1995) *Turning the Tide: Integrated Local Area Management for Australia's Coastal Zone*, Canberra: Commonwealth Department of the Environment, Sports and Territories.

Brush, R.O. (1976) 'Perceived Quality of Scenic and Recreational Environments: Some Methodological Issues', in *Perceiving Environmental Quality: Research Applications* (K.H. Craik and E.H. Zube, eds), New York: Plenum Press.

Bryner, G.C. (1987) *Bureaucratic Discretion: Law and Policy in Federal Regulatory Agencies*, New York: Pergamon.

Buckingham-Hatfield, S. and Evans, B. (eds) (1996a) *Environmental Planning and Sustainability*, Chichester: Wiley.

Buckingham-Hatfield, S. and Evans, B. (1996b) 'Achieving Sustainability through Environmental Planning', in *Environmental Planning and Sustainability* (S. Buckingham-Hatfield and B. Evans, eds), Chichester: Wiley.

Burbridge, P.R. (2000) 'Observations about the CMPP', personal communication.

Burbridge, P.R. and Humphrey, S. (2003) 'Introduction to Special Issue on the European Demonstration Programme on Integrated Coastal Zone Management', *Coastal Management*, 31(2): 121–6.

Burby, R.J., Cigler, B.A., French, S.F., Kaiser, E.J., Kartez, J., Roenigk, D., Weist, D. and Whittington, D. (1991) *Sharing Environmental Risks: How to Control Governments' Losses from Natural Disasters*, Boulder, CO: Westview Press.

Burke, L., Selig, L. and Spalding, M. (2002) *Reefs at Risk in Southeast Asia*, Washington, DC: World Resources Institute.

Burkett, E. (2003) 'Mighty Wind', *New York Times*, 15/06/2004: 48.

Butler, R.W. (1980) 'The Concept of a Tourist Area Cycle of Evolution: Implications for Management of Resources', *Canadian Geographer*, 24(1): 5–12.

Byron Shire Council (2000) *Draft Byron Coastal Values Study: Background Information for the Byron Coastline Management Study and Plan*, July 2000, Mullumbimby, NSW: Byron Shire Council and Natural Heritage Trust.

Callan, S. and Thomas, J.M. (1996) *Environmental Economics and Management: Theory, Policy, and Applications*, Chicago, IL: Irwin.

Callon, M. (1987) 'Society in the Making: The Study of Technology as a Tool for Sociological Analysis', in *The Social Construction of Technological Systems: New Directions in the Sociology and History of Technology* (W.E. Bijker, T.P. Hughes and T.J. Pinch, eds), Cambridge, MA: MIT Press, 83–103.

CALM (1991) *South Coast Region: Regional Management Plan: 1992–2002*, Perth, Australia: West Australian Department of Conservation and Land Management.

Campbell, S. and Fainstein, S. (eds) (1996) *Readings in Planning Theory*, Cambridge, MA: Blackwell.

Campbell, S. and Fainstein, S. (eds) (2002) *Readings in Planning Theory*, 2nd edn, Cambridge, MA: Blackwell.

Castledine, G. and Herrick, R. (1995) 'Subdivision: Assessment of Landscaped [sic] Value and Public Interest', *Australian Environmental Law News*, 2: 40–4.

CBD (2004) *Integrated Marine and Coastal Area Management (IMCAM) Approaches for Implementing the Convention on Biological Diversity*, Technical Series, 14, January: Convention on Biological Diversity, AIDEnvironment, Netherlands National Institute for Coastal and Marine Management/ RIKZ Coastal Zone Management Centre.

CBD Secretariat (2004) 'CBD Approach and Ecosystem based Management', February, personal communication.

CEET (2004) 'Cooperative Institute for Coastal and Estuarine Environmental Technology', http:// ciceet.unh.edu/ (accessed 25 April 2004).

Center for Urban and Regional Studies (1991) *Evaluation of the National Coastal Zone Management Program*, Washington, DC: National Coastal Resources Research and Development Institute, University of North Carolina Center for Urban and Regional Studies of the Department of City and Regional Planning.

Centre for Water Policy and Development (2001) *Integrated Coastal Zone Management in Bangladesh: A Policy Review*, Livelihood-Policy Relationships in South Asia Working Paper 1, Leeds: Centre for Water Policy and Development, University of Leeds.

Cesar, H. (1996) *The Economic Value of Indonesian Coral Reefs*, Washington, DC: Environment Department, World Bank.

Cesar, H., Lundin, C., Bettencourt, S. and Dixon, J. (1997) 'Indonesian Coral Reefs – An Economic Analysis of a Precious but Threatened Resource', *Ambio*, 26(1): 345–58.

Chapman, D.M. (1999) *Natural Hazards*, South Melbourne: Oxford University Press.

Checkland, P. and Holwell, S. (1998) *Information, Systems and Information Systems: Making Sense of the Field*, Chichester and New York: Wiley.

Checkland, P. and Scholes, J. (1990) *Soft Systems Methodology in Action*, Chichester: Wiley.

Chon, K.S. (2000) *Tourism in Southeast Asia: A New Direction*, New York: Haworth Hospitality Press.

Christensen, C.R. (1982) *Business Policies and Case Studies*, Homewood, IL: R.D. Irwin.

Christensen, N.L., Bartuska, A.M., Brown, J.H., Carpenter, S., D'Antonio, C., Francis, R., Franklin, J.F., MacMahon, J.A., Noss, R.F., Parsons, D.J., Peterson, C.H., Turner, M.G. and Woodmansee, R.G. (1996) 'The Report of the Ecological Society of America Committee on the Scientific Basis for Ecosystem Management', *Ecological Applications*, 6(3): 665–91.

Christensen, V. and Walters, C.J. (2004) 'Ecopath with Ecosim: methods, capabilities and limitations', *Ecological Modelling*, 172: 109–39.

Christie, P. (in press) 'Is Integrated Coastal Management Sustainable', *Ocean and Coastal Management*.

Christie, P. and White, A.T. (1997) 'Trends in Development of Coastal Area Management in Tropical Countries: From Central to Community Orientation', *Coastal Management*, 25: 155–81.

Chua, T.-E. and Pauly, D. (eds) (1989) *Coastal Area Management in Southeast Asia: Policies, Management Strategies and Case Studies*, Manila: ICLARM.

Chuenpagdee, R. and Alder, J. (2002) 'A Review of Coastal Projects in Asia Pacific', in *Coastal Zone Asia Pacific Conference (CZAP)*, Bangkok, Thailand, 11–17 May.

Cicin-Sain, B. (1993) 'Sustainable Development and Integrated Coastal Zone Management', *Ocean and Coastal Management*, 21: 11–44.

Cicin-Sain, B. and Knecht, R.W. (1998) *Integrated Coastal and Ocean Management: Concepts and Practices*, Washington, DC: Island Press.

CIESIN (2004) Gridded Population of the World (GPW), Version 3, Palisades, NY: Center for International Earth Science Information Network (CIESIN), Columbia University, International Food Policy Research Institute (IFPRI), and World Resources Institute (WRI).

City of Mandurah, Peel Development Commission and Western Australian Planning Commission (1996) *Mandurah Coastal Strategy*, Perth: City of Mandurah, Peel Development Commission and Western Australian Planning Commission.

Clark, J.R. (1996) *Coastal Zone Management Handbook*, Boca Raton, FL: CRC Press.

Clark, R.N. and Stankey, G.H. (1979) *The Recreation Opportunity Spectrum: A Framework for Planning, Management and Research* (General Technical Report PNW-98), Portland, OR: USDA, Forest Service.

Clark, T.W. (2002) *The Policy Process: A Practical Guide for Natural Resource Professionals*, New Haven, CT: Yale University Press.

Clarke, A.C. (2005). *Once and Future Tsunamis*. Clarke Foundation (2004). http://www.clarkefoundation.org/.

Clarke, B. (2004) 'More than a Sum of its Parts: a Reflection of Australia's Coastcare Program under NHT I, 1995–2002', in *Coast to Coast 2004: Hobart +10*, Hobart, Tasmania, 19–23 April.

Cleary, J. (1997) 'An "All of Government" Approach to Landscape Management in Western Australia', unpublished manuscript, Perth: West Australian Government Department of Conservation and Land Management.

Cleary, J. (2004) Unpublished material supplied by the author to the second edition of this book.

Clemett, A. (2002) *Coastal Zone Management in Sri Lanka: A Policy Process Analysis*, Improving Policy–Livelihood Relationships in South Asia Briefing Note 7, York: Stockholm Environment Institute, Water and Development Group.

Cloke, P., Crang, P. and Goodwin, M. (eds) (1999) *Introducing Human Geographies*, London: Arnold; New York: Oxford University Press.

CMPP (2000) *White Paper for Sustainable Coastal Development in South Africa*, Cape Town: Department of Environmental Affairs and Tourism, Coastal Management Policy Program.

Coast Conservation Department (1990) *Coastal Zone Management Plan*, Colombo: Coast Conservation Department.

Coast Conservation Department (1996) *Revised Coastal Zone Management Plan Sri Lanka, 1996–2000*, Colombo: Coast Conservation Department.

Coastal America Partnership (2001) *A Decade of Commitment to Protecting, Preserving and Restoring America's Coastal Heritage*, Washington, DC: Coastal America Partnership.

Coastal Committee of New South Wales (1994) *Draft Revised Coastal Policy for New South Wales*, Sydney: Government of New South Wales.

Coastal Society, The (2002) 'Converging Currents: Science, Culture and Policy at the Coast', in *Proceedings of the Eighteenth International Conference*, Galveston Island, Texas, 19–22 May.

Coastal Zone Canada Association (2002) 'Managing Shared Waters', in *Coastal Zone Canada 2002*, Hamilton: Coastal Zone Canada Association.

Coastalmanagement.com (2005) 'Coastal Zone Management Reponses to the 26 December 2004 Tsunami'. http://www.coastalmanagmeent.com/tsunmai.html (accessed 12 February 2005).

Coastwest/Coastcare (2000) *A Report to the Community 1995–2000*, Perth: Western Australian Ministry for Planning.

Coccossis, H. and Nijkamp, P. (eds) (1995) *Sustainable Tourism Development*, Aldershot: Avebury.

Coccossis, H., Mexa, A., Collovini, A., Parpairis, A. and Konstandoglou, M. (2001) *Defining, Measuring and Evaluating Carrying Capacity In European Tourism Destinations*, December, Athens: Environmental Planning Laboratory of the University of the Aegean, Greece.

Cohen, J.E., Small, C., Mellinger, J., Gallup, A. and Sachs, J. (1997) 'Estimates of Coastal Populations', *Science*, 278(5341): 1209–13.

Colebatch, H.K. (1993) 'Policy-making and Volatility: What is the Problem?', in *Policy-making in Volatile Times* (A. Hede and S. Prasser, eds), Sydney: Hale and Iremonger.

Coles, T.F. and Tarling, J.P. (1991) *Environmental Assessment: Experience to Date*, Lincoln: Institute of Environmental Assessment.

Commonwealth House of Representatives Standing Committee on Environment and Conservation (1980) *Australian Coastal Zone Management*, Canberra: Australian Government Publishing Service.

Commonwealth of Australia (1995) *Living on the Coast: The Commonwealth Coastal Policy*, Canberra: Australian Government Printing Service.

Commonwealth of Australia (1998) *Australia's Oceans Policy*, Canberra: Commonwealth of Australia.

Considine, M. (1994) *Public Policy: A Critical Approach*, South Melbourne: Macmillan Education Australia.

Cornforth, R. (1992) 'Bridging the Gap: The Distance Between Customary and Parliamentary Law Making in the Management of Natural Resources in Western Samoa', in *SPREP/UNEP Workshop on Strengthening Environmental Legislation in the South Pacific*, 23–27 November 1992, Apia, Western Samoa.

Cornforth, R. (2004) *Environmental Governance in Samoa*, 10 April 2004, personal communication.

Costannza, R.E. (1992) *Ecological Economics*, New York: Columbia University Press.

Council and the European Parliament (2000) *Integrated Coastal Zone Management: A Strategy for Europe*, COM/00/547 2000, Brussels: European Parliament.

Couper, A. (ed.) (1983) *The Times Atlas of the Oceans*, London: Times Books.

Court, J., Wright, C. and Guthrie, A. (1994) *Assessment of Cumulative Impacts and Strategic Assessment in Environmental Impact Assessment*, Canberra: Commonwealth Environmental Protection Agency.

Courtney, C.A., Deguit, E.T. and Yambao, A.C. (2001) *Monitoring and Evaluation: A Key to Sustainability of Coastal Resource Management Programs*, September, Cebu City, Philippines: Coastal Resource Management Project.

Courtney, C.A. and White, A.T. (2000) 'Integrated Coastal Management in the Philippines: Testing New Paradigms', *Coastal Management*, 28(1): 39–53.

Courtney, C.A., White, A.T. and Anglo, E. (2000) *Philippine Case Study: Managing Coastal Resources – Drawing Lessons and Directions from the Philippine Experience*, Sustainable Development Report, Manila: Asian Development Bank.

Crawford, B., Cobb, J.S. and Friedman, A. (1993) 'Building Capacity for Integrated Coastal Management in Developing Countries', *Ocean and Coastal Management*, 21: 311–37.

Crawley, B. (2003) 'Samoa Coastal Management Project Case Study: Resilience-Coastal Infrastructure and Communities Resilient to Natural Hazards', in *NAPA Samoa Workshop*, Apia, Samoa, 5–7 March.

CRC for Coastal Zone Estuary and Waterway Management (2002) 'Source to Sea', in *Coast to Coast*, Tweed Heads, NSW, 4–8 November: CRC for Coastal Zone Estuary and Waterway Management.

CRMP/DENR (2003) *Monitoring and Evaluating City/Municipal Plans and Programs for Coastal Resource Management*, Cebu City: Coastal Resources Management Project of the Deprtment of Environment and Natural Resources, supported by the United States Agency for International Development.

Crocombe, R. (ed.) (1995) *Customary Land Tenure and Sustainable Development: Complementarity or Conflict?*, Noumea and Suva: South Pacific Commission and University of the South Pacific.

Cronbach, L.J. (1992) *Designing Evaluations of Educational and Social Programs*, San Francisco, CA: Jossey Bass.

Crow, A. (2004) *XTM 1.0 Specification and CZAP Database Topic Map Schema*, Technical Working Paper 1, Perth: OneCoast.

Cullingworth, J.B. (1993) *The Political Culture of Planning: American Land Use Planning in Comparative Perspective*, New York: Routledge.

Cullingworth, J.B. and Caves, R.W. (2003) *Planning in the USA: Policies, Issues and Processes*, New York: Routledge.

Culliton, T.J. (1998) *Population: Distribution, Density and Growth*, Silver Spring, MD: National Oceanic and Atmospheric Administration (NOAA).

CZAP (2002) 'Improving the State of the Coastal Areas', in *Coastal Zone Asia Pacific Conference*, Bangkok, Thailand, 11–17 May: Thailand Coastal Development Institute.

Daft, R.L. (2000) *Organizational Theory and Design*, Cincinnati, OH: Southwestern College Publishing.

Daniel, T.C. (1976) 'Criteria for Development and Application of Perceived Environmental Indices', in *Perceiving Environmental Quality: Research Applications* (K.H. Craik and E.H. Zube, eds), New York: Plenum Press.

Davis, G., Wanna, J., Warhurst, J. and Weller, P. (1993) *Public Policy in Australia*, Sydney: Allen and Unwin.

Dawes, R.M. (1988) *Rational Choice in an Uncertain World*, Fort Worth, TX: Harcourt Brace Jovanovich College Publishers.

de Wilde, K. and Islam, M.R. (2002) 'The Dimension of Co-Governance in Coastal Development: A Case Study From Bangladesh', in *Coastal Zone Asia Pacific Conference*, Bangkok, Thailand, 11–17 May.

DEH (2004) *Framework for a Cooperative Approach to Integrated Coastal Zone Management*, April 2004, Canberra: Government of Australia Department of Environment and Heritage.

Delaware State Government (1999) *Delaware Coastal Management Program Federal Consistency Certification for Federal Permit and License Applicants*, Dover, DE: Department of Natural Resources and Environmental Control.

DENR (2003) *Monitoring and Evaluating Municipality/City Plans and Programs for Coastal Resources Management*, Cebu City: Coastal Resource Management Project. Department of Environment and Natural Resources.

Department of Conservation (1994a) *New Zealand Coastal Policy Statement*, Wellington: Department of Conservation.

Department of Conservation (1994b) *Te Kupu Kaupapahere Takutai Mo Aotearoa*, Wellington: Department of Conservation.

Department of Conservation and Land Management (1983) *Why Have Marine Parks and Marine Nature Reserves?*, Perth: Department of Conservation and Land Management.

Department of Fisheries and Oceans (1999) *Canada's Ocean Industries: Contribution to the Economy 1988–1996, Ocean Tourism Industry*, Department of Fisheries and Oceans, http://www.dfo-mpo.gc.ca/communic/statistics/Oceans/rascal/part_034.html (accessed 2004).

Department of Planning and Urban Development (1993) *Public Participation in the Planning Process*, Perth: Western Australian Government Department of Planning and Urban Development.

Department of Resources Development (1994) *Working with Communities: A Guide for Proponents*, Perth: Western Australian Government Department of Resources Development.

DKP (2003) *Perkembangan Pembangunan Sektor Kelautan dan Perikanan*, Jakarta: Departmen Kelautan dand Perikanan (DKP).

Dolak, N. and Ostrom, E. (eds) (2003) *The Commons in the New Millennium: Challenges and Adaptation*, Boston, MA: MIT Press.

Dolan, R. and Goodell, H.G. (1986) 'Sinking Cities', *American Scientist*, 74: 38–47.

Donaldson, B., Eliot, I. and Kay, R.C. (1995) *Final Report of the Review of Coastal Management in Western Australia: A Report to the Minister for Planning*, Perth: Coastal Management Review Committee.

Doxey, G.V. (1975) 'A Causation Theory Visitor-Resident Irritations: Methodology and Research Inferences', in *Proceedings of the Travel Research Association Sixth Annual Conference*, San Diego, California.

Dresner, S. (2002) *The Principles of Sustainability*, London: Earthscan.

Drijver, C. and Sajise, P. (1993) 'Community-Based Resource Management and Environmental Action Research', in *Proceedings of the Experts' Workshop on Community Based Resource Management: Perspectives, Experiences and Policy Issues*, Los Banos: Environmental and Resource Management Project and UPLB.

Dutton, I. and Saenger, P. (1994) 'Expanding the Horizon(s) of Marine Conservation: The Challenge of Integrated Coastal Management', in *Marine Protected Areas and Biosphere Reserves: 'Towards a New Paradigm'*, Canberra: Australian Nature Conservation Agency.

Dutton, I., Boyd, W.E., Luckie, K., Knox, S. and Derrett, R. (1995) 'Measuring Coastal Landscape and Lifestyle Values: An Interpretive Approach', *Australian Journal of Environmental Management*, (December): 245–56.

Eagleton, T. (1990) *The Significance of Theory*, Cambridge, MA: Basil Blackwell.

Edgren, G. (1993) 'Expected Economic and Demographic Developments in Coastal Zones World Wide', in *World Coast '93*, Noordwijk: Coastal Zone Management Centre, The Netherlands National Institute for Coastal and Marine Management.

Edwards, S.F. (1987) *An Introduction to Coastal Zone Economics: Concepts, Methods and Case Studies*, New York: Taylor & Francis.

Ehler, C. (1995) 'Integrated Coastal Ocean Space Management: Challenges for the Next Decade', in *Coastal Ocean Space Utilization III* (N.D. Croce, S. Connell and R. Abel, eds), London: E & F Spon.

Ehler, C. and Basta, D. (1993) 'Integrated Management of Coastal Areas and Marine Sanctuaries', *Oceanus*, 36(3): 6–14.

Elkington, J. (1997) *Cannibals with Forks: The Triple Bottom Line of 21st Century Business*, Oxford: Capstone.

English Nature (1996) *Thames Estuary Mangement Plan: Strategy for Implementation*, London: English Nature.

Environment Australia (2003) *Bilateral Agreement between the Commonwealth of Australia and the New South Wales to deliver the Natural Heritage Trust*, August, Canberra: Natural Heritage Trust.

Environment Committee (1992) *Coastal Zone Protection and Planning: Volume 2, Minutes of Evidence and Appendices*, 17–II, 12 March: House of Commons Environment Select Committee.

ETSO (2003) *Report on Renewable Energy Sources (RES)*, Brussels: European Transmission System Operators.

Eurocoast Federation (2000) 'Littoral', in *Responsible Coastal Zone Management – The Challenge of the 21st Century*, Dubrovnik: Eurocoast Federation.

Eurocoast Federation (2002) 'Littoral', in *The Changing Coast*, Porto: Eurocoast Federation.

European Parliament and Council (2000) *Recommendation Concerning the Implementation of Integrated Coastal Zone Management in Europe* (September), Brussels: European Parliament and Council.

European Union (2000) *ICZM Demonstration Projects*, http://europa.eu.int/comm/environment/iczm/projects.htm (accessed 20 May 2004).

European Union (2002) *European Parliament and of the Council concerning the implementation of Integrated Coastal Zone Management in Europe*, 2002/413/EC.

European Union (unknown) *Glossary of European Union Terminology*, http://europa.eu.int/scadplus/leg/en/cig/g4000s.htm (accessed 12 April 2004).

Fabbri, P. (ed) (1990) *Recreational Uses of Coastal Areas : A Research Project of the Commission on the Coastal Environment, International Geographical Union*, Dordrecht: Kluwer Academic.

Fabos, J.G. and McGregor, A. (1979) *Assessment of Visual/Aesthetic Landscape Qualities*, Melbourne: Centre For Environmental Studies, University of Melbourne.

Faludi, A. (ed) (1973) *A Reader in Planning Theory*, Oxford: Pergamon.

Feeny, D., Berkes, F., McKay, B. and Acheson, J. (1990) 'The Tragedy of the Commons: Twenty-two Years Later', *Human Ecology*, 18(1): 1–19.

Fekete, J. (1988) 'Introductory Notes for a Postmodern Value Agenda', in *Life After Postmodernism: Essays on Value and Culture* (J. Fekete, ed.), London: Macmillan.

Feldman, M. (1989) *Order without Design: Information Production and Policy Making*, Stanford, CA: Stanford University Press.

Ferguson, S.D. (1999) *Communication Planning: An Integrated Approach*, Thousand Oaks, CA: Sage.

Ferrer, E., M. (1992) *Learning and Working Together: Towards a Community – Based Coastal Resources Management*, Quezon City: College of Social Work and Community Development, University of Philippines.

Field, B.C. (1994) *Environmental Economics*, New York: McGraw-Hill.

First Pacific Island Regional Ocean Forum (2004) 'Vision and Principles', in *First Pacific Island Regional Ocean Forum*, Fiji: South Pacific Commission.

Fisk, G.W. (1996) Unpublished material supplied by the author to the first edition of this book.

Florida Department of Environmental Protection (2004) *Coastal Management Program*, Florida Department of Environmental Protection, http://www.dep.state.fl.us/cmp/ (accessed 29 April 2004).

Folmer, H., Gabel, H.L. and Opschoor, H. (eds) (1995) *Principles of Environmental and Resource Economics*, Cheltenham: Edward Elgar.

Food and Agriculture Organisation (2002) *State of the World Fisheries and Aquaculture (SOFIA)*, Rome: Food and Agriculture Organisation.

Food and Agriculture Organisation (2003) *Code of Conduct for Responsible Fisheries*, Rome: Food and Agriculture Organisation of the United Nations.

Food and Agriculture Organisation (2005) 'Tsunami Reconstruction: Impacts of the Tsunami on Fisheries, Aquaculture and Coastal Livelihoods – 1 Sri Lanka (As of 3 February 2005)' http://www.fao.org/tsunami/fisheries/srilanka.htm (accessed 10 February 2005).

Frassetto, R. (ed) (1989) *Impact of Sea-Level Rise on Cities and Regions. Proceedings of the First International Meeting 'Cities on Water'*, 11–13 December 1989, Venice, Italy: Centro Internazionale Citta d'Acqua.

Gasch, R. and Twele, J. (2002) *Wind Power Plants: Fundamentals, Design, Construction and Operation*, London: James and James Science Publishers.

Gascoyne Development Commission (1994) *Gascoyne Aquaculture Development Plan*, Perth: Western Australian Government Department of Fisheries and Gascoyne Development Commission.

Gaughan, D.J., Leary, N., Mitchell, R.W. and Blight, I.W. (2004) 'A Sudden Collapse in Distribution of Pacific Sardine (*Sardinops sagax*) off Southwestern Australia Enables an Objective Re-assessment of Biomass Estimates', *Fishery Bulletin*, 102(4): 617–33.

Gavaghan, H. (1990) 'The Dangers Faced By Ships in Port', *New Scientist*, September: 19.

General Accounting Office (1992) *Coastal Barriers: Development Occurring Despite Prohibition against Federal Assistance*, RCED-92-115, July: United States General Accounting Office.

Gerrard, S. (1994) 'Managing Risks in the Coastal Zone: Assessment, Perception and Communication', in *Coast to Coast '94*, Hobart, Tasmania.

Gerrard, S. (1995) 'Environmental Risk Management', in *Environmental Science for Environmental Management* (T. O'Riordan, ed.), Harlow: Longman.

Gerrard, S. (1996) 'Shout, Speak, Whisper or Listen? Evaluating Risk Communication in Europe', in *International Workshop on Perception, Communication and the Social Representation of Environmental Risks*, 24–27 October, Bremen, Germany.

Gerrard, S. (2000) 'Environmental Risk Management', in *Environmental Science for Environmental Management* (T. O'Riordan, ed.), Harlow: Prentice Hall.

GESAMP (1996) *The Contributions of Science to Integrated Coastal Management*, Some: FAO.

GESAMP (1999) *Report of the 28th Session of Group of Experts on the Scientific Aspects of Marine Environmental Protection: GESAMP Reports and Studies No. 66*, Geneva, 20–24 April 1998, Geneva, Switzerland: World Meteorological Organization.

Gilchrest, W. (2001) *Coastal Resources Conservation Act of 2001* (introduced in the House), 20 December, Washington, DC: 107th Congress, 1st Session, HR3577.

Gilpin, A. (1995) *Environmental Impact Assessment: Cutting Edge for the Twenty-First Century*, Cambridge: Cambridge University Press.

Gilpin, A. (2000) *Environmental Economics: A Critical Overview*, Chichester: Wiley.

Gipe, P. (1993) 'The wind industry's experience with aesthetic criticism', *Leonardo*, 26: 243–8.

GIWA (2004) Summary of Environmental Impacts on Large Marine Ecosystems (Draft), Global International Waters Assessment.

Glasson, J., Therivel, R. and Andrew, C. (1994) *Introduction to Environmental Impact Assessment*, London: UCL Press.

Glavovic, B. (2000a) *The South African Coastal Policy Formulation Experience: The Process, Perceptions and Lessons Learned*, Cape Town: Department of Environmental Affairs and Tourism, Coastal Management Policy Programme, Common Ground Consulting.

Glavovic, B. (2000b) 'A New Policy for South Africa', *Coastal Management*, 28: 261–71.

Glavovic, B. (2004) 'The Challenge of Transforming Coastal Management: The South African Experience', unpublished material supplied by the author to the second edition of this book.

Glavovic, B. (in press a) 'ICM as a Transformational Practice of Consensus Building: A South African Perspective,' *Journal of Coastal Research*, special issue 39.

Glavovic, B. (in press b) 'Lessons Learned from South Africa's Coastal Policy Experience', *Journal of Coastal Research*, Special Issue 39.

Global Vision (1996) *Preparation Document for the Awareness and Participation Component of COREMAP: Draft No. 2*, Silver Springs, MD: Global Vision Inc.

Godschalk, D.R., Brower, D.J. and Beatley, T. (1989) *Catastrophic Coastal Storms: Hazard Mitigation and Development Management*, Durham, NC: Duke University Press.

Godschalk, D.R. (1992) 'Implementing Coastal Zone Management: 1972–1990', *Coastal Management*, 20: 93–116.

Goldberg, E.B. (1994) *Coastal Zone Space – Prelude to Conflict?*, Paris: UNESCO.

Goldin, I. and Winters, L.A. (eds) (1995) *The Economics of Sustainable Development*, Cambridge: Cambridge University Press.

Goodhead, T. and Johnson, D. (eds) (1996) *Coastal Recreation Management: The Sustainable Development of Maritime Leisure*, London: E & F Spon.

Government of Australia (1975) *Great Barrier Reef Marine Park Act 1975*, Canberra: Government of Australia.

Government of Dubai (2001) *Conservation of Coastal Zone of Dubai*, Dubai: United Arab Emirates, Dubai Municipality.

Government of Victoria (1990) *Making the Most of the Bay: A Plan for the Protection and Development of Port Phillip and Corio Bays*, Melbourne: Government of Victoria.

Gray, W.M. (1968) 'Global Review of the Origin of Tropical Disturbances and Storms', *Monthly Weather Review*, 96: 669–700.

Great Barrier Reef Marine Park Authority (2003a) *Great Barrier Reef Marine Park Zoning Plan*, Townsville, QLD: Great Barrier Reef Marine Park Authority.

Great Barrier Reef Marine Park Authority (2003b) *Remote Natural Areas and No Structure Sub-zones in the Great Barrier Reef Marine Park*, RAP Information Sheet, Townsvile: Great Barrier Reef Marine Park Authority.

Green, M. and Paine, J. (1997) 'State of the World's Protected Areas at the end of the Twentieth Century', in *IUCN World Commission on Protected Areas Symposium: Protected Areas in the 21st Century: From Islands to Networks*, Albany, Australia, 24–29 November.

Grigalunas, T.A. and Congar, R. (1995) *Environmental Economics for Integrated Coastal Area Management: Valuation Methods and Policy Instruments*, Nairobi: United Nations Environment Programme.

Grumbine, R.E. (1994) 'What is ecosystem management?' *Conservation Biology*, 8: 27–39.

Gubbay, S. (1989) *Coastal and Sea Use Management: A Review of Approaches and Techniques*, Ross-on Wye, UK: Marine Conservation Society.

Gubbay, S. (1994) 'Local Authorities and Integrated Coastal Zone Management Plans', *Marine Update: Newsletter of the World Wide Fund for Nature* (June): 1–4.

Gubbay, S. (ed.) (1995) *Marine Protected Areas*, London: Chapman & Hall.

Gubbay, S. (2002) *Just Coasting: An Assessment of the Commitment of the Devolved Administrations and the English Regions to Integrated Coastal Management*: Report to the Wildlife Trusts and WWF, London: WWF; http://www.wwf.org.uk/researcher/issues/livingseas/.

Guerrier, Y., Alexander, N., Chase, J. and O'Brien, M. (eds) (1995) *Values and the Environment: A Social Science Perspective*, Chichester and New York: J. Wiley.

Guiney, J.L. and Lawrence, M.B. (2000) *Hurrican Mitch 22 October – 5 November*, Miami, FL: National Hurricane Center; http://www.nhc.noa.gov/1998mitch.html (accessed 16 March 2004).

Haar, C.M. (1977) *Land-Use Planning: A Casebook on the Use, Misuse and Re-use of Urban Land*, Boston, MA: Little, Brown.

Haeruman, H. (1986) 'The Exploitation of Marine Resources and Environmental Problems in Indonesia', in *Symposium on The Sustainable Development of Coastal and Marine Resources in Eastern Indonesia*, Jakarta: KLH/EMDI/UNDP.

Hale, L.Z. (1996) 'Involving Communities in Coastal Management', in *Proceedings of the Coast to Coast '96 Conference*, Adelaide, 16–19 April: University of Adelaide.

Hale, L.Z., Amaral, M., Issa, A.S. and Mwandotto, B.A.J. (2000) 'Catalyzing Coastal Management in Kenya and Zanzibar: Building Capacity and Comittment', *Coastal Management*, 28(1): 75–85.

Hall, D., Hebbert, M. and Lusser, H. (1993) 'The Planning Background', in *Planning for a Sustainable Environment* (A. Blowers, ed.), London: Earthscan/Town and Country Planning Association.

Hall, M.C. (2001) 'Trends in Ocean and Coastal Tourism: The End of the last Frontier?' *Ocean and Coastal Management*, 44(9): 601–18.

Hamilton, B. and Kay, R.C. (1998) 'Proceedings of the Coast to Coast 1998', in *Coast to Coast '98*, Perth: Ministry for Planning, Government of Western Australia.

Hanley, N., Shogren, J.F. and White, B. (2001) *Introduction to Environmental Economics*, New York: Oxford University Press.

Harger, J. (1986) 'Community Structure as a Response to Natural and Man-made Environmental Variables in the Pulau Seribu Island Chain', in *MAB-COMAR Regional Workshop on Coral Reef Ecosystems: Their Management Practices and Research/Training Needs*, Jakarta: UNESCO, MAB-COMAR and LIPI.

Harmon, J.E. and Anderson, S.J. (2003) *The Design and Implementation of Geographic Information Systems*, New York: Wiley.

Hart, G.M. (1978) *Values Clarification for Counselors: How Counselors, Social Workers, Psychologists, and Other Human Service Workers Can Use Available Techniques*, Springfield, IL: Charles Thomas.

Harvey, N. and Caton, B. (2003) *Coastal Management in Australia*, Oxford: Oxford University Press.

Haub, C. (1996) 'Future Global Population Growth', in *Population Growth and Environmental Issues* (S. Ramphal and S.W. Sinding, eds), Westport, CT: Praeger.

Hawkins, J.P. and Roberts, C.M. (1993) 'The Growth of Coastal Tourism in the Red Sea: Present and Possible Future Effects on Coral Reefs', in *Proceedings of the Colloquium on Global Aspects of Coral Reefs: Health, Hazards and History*, Miami, FL: University of Miami.

Hay, J.E., Chou, L.M., Sharp, B. and Thom, N.G. (eds) (1994) *Environmental and Related Issues in the Asia-Pacific Region: Implications for Tertiary-Level Environmental Training*, Bangkok: United Nations Environment Programme (UNEP), Regional Office for Asia and the Pacific (ROAP), Network for Environmental Training at Tertiary Level in Asia and the Pacific (NETTLAP).

Heady, F. (1996) *Public Administration: A Comparative Perpective*, New York: Marcel Dekker.

Health and Safety Executive, UK (1988) *The Tolerability of Risk at Nuclear Power Stations*, London: HMSO.

Hedgcock, D. and Yiftachel, O. (eds) (1992) *Urban and Regional Planning in Western Australia*, Perth: Paradigm Press.

Heinz Center (2002) *Human Links to Coastal Disasters*, Washington, DC: H. John Heinz III Center for Science, Economics and the Environment.

Heinz Center (2003) *The Coastal Zone Management Act: Developing A Framework For Identifying Performance Indicators*, Washington, DC: H. John Heinz III Center for Science, Economics and the Environment.

Helsinki Commission – Baltic Marine Environment Protection Commission (2003) *Implementation of Integrated Marine and Coastal Management of Human Activities in the Baltic Sea Area*, Recommendation 24/10.

Henocque, Y. and Denis, J. (2001) *A Methodological Guide: Steps and Tools Towards Integrated Coastal Area Management*, Paris: UNESCO.

Herman, J.L., Morris, L.L. and Fitz-Gibbon, C.T. (1987) *Evaluator's Handbook*, Newbury Park, CA: Sage.

Hershey, T. and Wilson, J. (1991) *Dynamite Fishing*, Port Moresby: Melanesian Environment Foundation and Papua New Guinea Integrated Human Development Trust.

Hershman, M.J., Good, J.W., Bernd-Cohen, T., Lee, V. and Pogue, P. (1999) 'The Effectiveness of Coastal Zone Management in the United States', *Coastal Management*, 27(2–3): 113–38.

Hewitt, K. (1997) *Regions of Risk: A Geographical Introduction to Disasters*, Harlow: Longman.

Hikkaduwa Special Area Management and Marine Sanctuary Coordination Committee (1996) *Special Area Management Plan for Hikkaduwa Marine Sanctuary and Surrounding Area*, Colombo: Coastal Resources Management Project, Coast Conservation Department, National Aquatic Resources Agency.

Hildebrand, L.P. and Norrena, E.J. (1992) 'Approaches and Progress Toward Effective Integrated Coastal Zone Management', *Marine Pollution Bulletin*, 25(1–4): 94–7.

Hildebrand, L.P. and Sorensen, J. (2001) 'Draining and Swamp and Beating Away the Alligators: Baseline 2000', *Intercoast*, Spring: 20–1.

Hinrichsen, D. (1990) *Our Common Seas: Coasts in Crisis*, London: Earthscan.

Hinrichsen, D. (1994) 'Coasts Under Pressure', *People and the Planet*, 3(1): 6–9.

Hinrichsen, D. (1998) *Coastal Waters of the World*, Washington, DC: Island Press.

Holling, C.S. (1978) *Adaptive Environmental Assessment and Management*, New York: John Wiley and Sons.

Hossain, H., Dodge, C.P. and Abed, F.H. (eds) (1992) *From Crisis to Development: Coping with Disasters in Bangladesh*, Dhaka: University Press.

Houghton, J.T. and Ding, Y. (eds) (2001) *Climate Change: The Scientific Basis*, Cambridge: Cambridge University Press.

House of Commons Environment Select Committee (1992) *Coastal Zone Protection and Planning*, 17–I, 12 March: House of Commons Environment Select Committee: Volume I, Report, Together with the Proceedings of the Committee Relating to the Report.

House, P.W. and Shull, R.D. (1988) *Rush to Policy: Using Analytic Techniques in Public Sector Decision Making*, New Brunswick, NJ: Transaction Books.

Huckle, J. and Martin, A. (2001) *Environments in a Changing World*, Harlow: Pearson Education.

Hurst, D.K. (1995) *Crisis and Renewal: Meeting the Challenge of Organizational Change Managing Innovation and Change*, Boston, MA: Harvard Business School Press.

Hussey, D.E. (1991) *Introducing Corporate Planning: Guide to Strategic Management*, Oxford: Pergamon.

Huttche, C., White, A.T. and Flores, M.M. (2002) *Sustainable Coastal Tourism Handbook for the Philippines*, Cebu City: Coastal Resource Management Project of the Department of Environment and Natural Resources.

IMO (2003) *Marine Environment Protection Committee (MEPC) – 50th session: 1 and 4 December 2003*, 1/12/2003, London: International Maritime Organization.

Imperial, M.T., Robadue Jnr, D. and Hennessey, T.M. (1992) 'An Evolutionary Perspective on the Development and Assessment of the National Estuary Program', *Coastal Management*, 20(4): 311–42.

Imperial, M.T. and Hennessey, T.M. (1996) 'An Ecosystem-Based Approach to Managing Estuaries: An Assessment of the National Estuary Program', *Coastal Management*, 24(2): 115–40.

Inder, A. (1997) 'Partnership in Planning and Management of the Solent', in *Partnership in Coastal Zone Management: Proceedings of the Eurocoast '96 Conference* (J. Taussik and J. Mitchell, eds), Cardigan: Samara Publishing.

Innes, J.E. (1996) 'Planning Through Consensus: A New Vision to the Comprehensive Planning Ideal', *Journal of the American Planning Association*, 62(4): 460–72.

Institute of Environmental Assessment (UK) and The Landscape Institute (UK) (1995) *Guidelines for Landscape and Visual Impact*, London: E & FN Spon.

IPCC (1990) *Strategies for Adaption to Sea Level Rise*, November: IPCC Coastal Zone Management Subgroup Ministry of Transport and Public Works, The Netherlands.

IPCC (1992) *Global Climate Change and the Challenge of the Rising Sea*, May: IPCC Coastal Zone Management Subgroup Ministry of Transport and Public Works, The Netherlands.

IPCC (ed.) (2001) *Climate Change: Working Group II: Impacts, Adaptation and Vulnerability*, Cambridge: Cambridge University Press.

Iqbal, M.S. (1992) *Assessment of the Implementation of the East African Action Plan and the Effectiveness of its Legal Instruments*, Nairobi: UNEP.

ITSU (International Tsunami Warning Center) (2005). http://www.prh.noaa.gov/itic/ (accessed 27 December 2004).

IUCN (2002a) *Regional Technical Assistance for Coastal and Marine Resources Management and Poverty Reduction in South Asia: Situation Analysis Report – Sri Lanka Component*, ADB RETA 5974, Male: Asian Development Bank.

IUCN (2002b) *Community-Based Marine Protected Areas in Samoa*, International Union for Conservation of Nature and Natural Resource, http://www.iucn.org/pareport/species_ samoa.htm (accessed 15 March 2004).

IUCN (2003) *Regional Technical Assistance for Coastal and Marine Resources Management and Poverty Reduction in South Asia: Integrated Coastal Zone Management Strategy – Sri Lanka Component*, ADB RETA 5974, Male: Asian Development Bank.

IWICM (1996) 'Enhaning the Success of Integrated Coastal Management: Good Practices in the Forumulation, Design and Implementation of Integrated Coastal Management Initiatives', in *International Workshop on Integrated Coastal Management in Tropical Developing Countries: Lessons Learned from Successes and Failures*, Xiamen, China, 24–28 May: GEF/UNDP/IMO Regional Programme for the Prevention and Management of Marine Pollution in the East Asian Seas and Coastal Management Centre, Quezon City, Philippines.

Jacobs, M. (1991) *The Green Economy: Environment, Sustainable Development and the Politics of the Future*, London: Pluto Press.

Japanese Research Group on the 26 December 2004 Earthquake Tsunami Disaster of Indian Ocean. http://www.drs.dpri.kyoto-u.ac.jp/sumatra/index-e.html (accessed 11 February 2005).

Jaques, E. (1996) *Requiste Organization: A Total System for Effective Managerial Organization and Managerial Leadership for the 21st Century*, Arlington, VA: Cason Hall and Co.

Jentoft, S. (1989) 'Fisheries Co-Management: Delegating Government Responsibility to Fisherman's Organizations', *Marine Policy*, 13(2): 137–54.

Jinendradasa, S.S. and Ekaratne, S.U.K (2000) 'Post-bleaching Changes in Coral Settlement at the Hikkaduwa Nature Reserve in Sri Lanka', *Proceedings of the 9th International Coral Reef Symposium*, Bali, Indonesia, 23–27 October, volume 1, pp. 417–20.

Johannes, R. (1984) 'Traditional Conservation Methods and Protected Areas in Oceania', in *Proceedings on the World Congress on National Parks*, Washington, DC: Smithsonian Institution.

Johannes, R. (ed.) (1989) *Traditional Ecological Knowledge: A Collection of Essays*, Gland, Switzerland and Cambridge: Internatonal Union for the Conservation of Nature.

Johannes, R.E., Freeman, M.M.R. and Hamilton, R.J. (2000) 'Ignore Fishers' Knowledge and Miss the Boat', *Fish and Fisheries*, 1: 257–71.

Johnson, D. (1974) *The Alps at the Crossroads*, Melbourne: Victorian National Parks Association.

Johnston, C. (1989) 'Whose Views Count?: Achieving Community Support for Landscape Conservation', *Historic Environment*, 7(2): 33–7.

Johnston, R.J., Grigalunas, T.A., Opaluch, J.J., Mazzotta, M. and Diamantedes, J. (2002) 'Valuing Estuarine Resource Services Using Economic and Ecological Models: The Peconic Estuary System Study', *Coastal Management*, 30(1): 47–65.

Jones, V. and Westmacott, S. (1993) *Management Arrangements for the Development and Implementation of Coastal Zone Management Programmes*, Noordwijk: Coastal Zone Management Centre The Netherlands, National Institute for Coastal and Marine Management.

Jubenville, A., Twight, B.W. and Becker, R.H. (1987) *Outdoor Recreation Management: Theory and Application*, State College, PA: Venture.

Kahawita, B.S. (1993) 'Coastal Zone Management in Sri Lanka', in *World Coast '93*, Noordwijk: Coastal Zone Management Centre The Netherlands, National Institute for Coastal and Marine Management.

Kaluwin, C. (1996) 'ICM Takes Different Approach in Pacific Islands', *Intercoast Network*, 27(Summer): 4, 10.

Kaly, U. and Morrison, J. (2004) Unpublished material supplied by the author to the second edition of this book.

Kausher, A., Kay, R.C., Asaduzzaman, M. and Paul, S. (1994) *Climate Change and Sea-Level Rise: The Case of the Bangladesh Coast*, Dhaka: Bangladesh Unnayan Parishad, Dhaka; Centre for Environmental and Resource Studies, New Zealand; Climatic Research Unit, England.

Kausher, A., Kay, R.C., Asaduzzaman, M. and Paul, S. (1996) 'Climate Change and Sea-Level Rise: The Case of the Bangladesh Coast', in *The Implications of Climate Change and Sea-Level Change for Bangladesh* (R.A. Warrick and Q.K. Ahmad, eds), Dordrecht: Kluwer Academic.

Kay, R. (1999) 'Coastal Information Access through the Internet in Australia: A Status Report', in *InfoCoast 1999: 1st European Symposium on Knowledge Management and Information for the Coastal Zone*, Noordwijkerhout: Coastlink UK, Coastal and Marine Observatory at Dover, UK and EUCC-UK, Brampton, UK.

Kay, R.C. (2000a) 'Is There a Theory of Coastal Zone Management?', in *Coasts at the Millennium: The Coastal Society 17th International Conference*, Portland, OR, 9–12 July: The Coastal Society.

Kay, R.C. (2000b) 'Challenging Assumptions in Integrated Coastal Planning', in *Integrated Management of the Jervis Bay Region*, Jervis Bay: NSW National Parks Service.

Kay, R.C. and Christie, P. (2001) 'Coastal Management and the Internet: A Status Report', *Coastal Management*, 29(3): 157–82.

Kay, R.C. and Crow, A. (2002) 'OneCoast Launch', *Tiempo: Global Warming and the Third World*, December (42): 26–7.

Kay, R.C. and McKellar, R. (2000) 'What Matters for your Coast? Moving Forward in Coastal Management by Looking Inwards', in *11th New South Wales Coastal Conference*, Yamba, NSW: New South Wales National Parks Service.

Kay, R.C., Alder, J., Houghton, P. and Brown, D. (2003) 'Management Cybernetics: A New Institutional Framework for Coastal Management', *Coastal Management*, 31(3): 213–27.

Kay, R.C., Carman-Brown, A. and King, G. (1995) 'Western Australian Experiences in Preparing and Implementing Coastal Management Plans: Some Implications for Shoreline Management Planning in England and Wales', in *Proceedings of the Ministry of Agriculture, Fisheries and Food Annual Conference of River and Coastal Engineers*, Keele, UK, 6–7 July.

Kay, R.C., Cole, R.G., Elisara-Laulu, F.M. and Yamada, K. (1993) *Assessment of Coastal Vulnerability and Resilience to Sea-level Rise and Climate Change. Case Study: 'Upolu Island, Western Samoa. Phase I: Concepts and Approach*, Apia: South Pacific Regional Environment Programme (SPREP).

Kay, R.C., Eliot, I., Caton, B., Morvell, G. and Waterman, P. (1996a) 'A Review of the Inter-governmental Panel on Climate Change's "Common Methodology for Assessing the Vulnerability of Coastal Areas to Sea-Level Rise" ', *Coastal Management*, 24(1): 165–88.

Kay, R.C., Ericksen, N.J., Foster, G.A., Gillgren, D.J., Healy, T.R., Sheffield, A.T. and Warrick, R.A. (1994) *Assessment of Coastal Hazards and their Management for Selected Parts of the Coastal Zone Administered by the Tauranga District Council*, Hamilton: University of Waikato Centre for Environmental and Resource Studies.

Kay, R.C., Kirkland, A. and Stewart, I. (1996b) 'Planning for Future Climate Change and Sea-Level Rise Induced Coastal Change in Australia and New Zealand', in *Greenhouse: Coping With Climate Change*, Wellington: CSIRO Publishing.

Kay, R.C., Panizza, V., Eliot, I.E. and Donaldson, B. (1997) 'Reforming Coastal Management in Western Australia', *Ocean and Coastal Management*, 35(1): 1–29.

Keeley, D. (1994) 'Balancing Coastal Resource Use: In Search of Sustainability', in *Coast to Coast '94*, Hobart: Tamanian Department of Environment and Land Management.

Kelleher, G. (1993) 'Progress Towards a Global System of Marine Protected Areas', in *Proceedings of the Fourth Fenner Conference on the Environment*, Sydney: IUCN.

Kelly, P.M., Granich, S.L.V. and Secrett, C.M. (1994) 'Global Warming: Responding to an Uncertain Future', *Asia Pacific Journal on Environment and Development*, 1(1): 28–45.

Kenchington, R.A. (1993) 'Tourism in Coastal and Marine Environments – A Recreational Perspective', *Ocean and Coastal Management*, 19(1): 1–16.

Kenchington, R.A. (1990) *Managing Marine Environments*, New York: Taylor & Francis.

Kenchington, R.A. (1992) 'Tourism Development in the Great Barrier Reef Marine Park', *Ocean and Shoreline Management*, 15: 57–78.

Kenchington, R.A. and Crawford, D. (1993) 'On the Meaning of Integration of Coastal Zone Management', *Ocean and Coastal Management*, 21(1–3): 109–27.

Kennedy, K. (1996) Unpublished material supplied by the author to the first edition of this book.

Ketchum, B.H. (ed) (1972) *The Water's Edge: Critical Problems of the Coastal Zone*, Cambridge, MA: MIT Press.

King, G. and Bridge, L. (1994) *Directory of Coastal Planning and Management Initiatives in England*, Maidstone: National Coasts and Estuaries Advisory Group.

King, G. (1996) Unpublished material supplied by the author to the first edition of this book.

King, J.A., Morris, L.L. and Fitz-Gibbon, C.T. (1987) *How to Assess Program Implementation*, Newbury Park, CA: Sage.

Kirkby, J., O'Keefe, P. and Timberlake, L. (eds) (1991) *The Earthscan Reader in Sustainable Development*, London: Earthscan.

Knecht, R., Cicin-Sain, B. and Fisk, G.W. (1996) 'Perceptions on the Performance of State Coastal Zone Management Programs in the United States', *Coastal Management*, 24(2): 141–64.

Kolstad, C.D. (2000) *Environmental Economics*, New York: Oxford University Pres.

Koutrakis, E., Lazaridou, T. and Argyropoulou, M.D. (2003) 'Promoting Integrated Management in the Strymonikos Coastal Zone (Greece): A Step-by-Step Process', *Coastal Management*, 31(2): 195–200.

Kraus, R.G. and Curtis, J.E. (1986) *Creative Management Recreation, Parks and Lesiure Services*, St Louis, MO: Times Mirror/Mosby.

Lange, E. (1994) 'Integration of Computerised Visual Simulation and Visual Assessment in Environmental Planning', *Landscape and Urban Planning*, 30: 99–112.

Lasswell, H.D. and McDougal, M.S. (1992) *Jurisprudence for a Free Society: Studies in Law, Science and Policy*, New Haven, CT: New Haven Press.

Latin, H.A. (1993) 'Reef Conservation: Disciplinary Conflicts Between Scientists and Policymakers', in *Proceedings of the Seventh International Coral Reef Symposium*, Guam: University of Guam Marine Laboratory, Mangilao, Guam.

Laurie, I.C. (1975) 'Aesthetic Factors in Visual Evaluation', in *Landscape Assessment: Values, Perceptions and Resources* (E.H. Zube, R.O. Brush and J.G. Fabos, eds), Stroudsberg, PA: Dowden, Hutchinson and Ross.

Lee, K.N. (1993) *Compass and Gyroscope: Integrating Science and Politics for the Environment*, Washington, DC: Island Press.

Leeworthy, V.R. and Bowker, J.M. (1997) *Nonmarket Economic Users Values of the Florida Keys/Key West*, Silver Spring, MD: National Oceanic and Atmospheric Administration.

Leung, H.L. (1989) *Land Use Planning Made Plain*, Kingston, ON: Ronald P. Fryre and Co.

Lieber, S.R. and Fesenmaier, D.R. (eds) (1983) *Recreation Planning and Management*, London: E & F N Spon.

Lim, C.P., Matsuda, Y. and Shigemi, Y. (1995) 'Co-management in Marine Fisheries: The Japanese Experience', *Coastal Management*, 23(3): 195–222.

Lipton, D.W. and Wellman, K.F. (1995) *Economic Valuation of Natural Resources*, Washington, DC: US National Oceanic and Atmospheric Administration, Coastal Ocean Program.

Lloyd's Register (2001) *A History of the Lloyd's Register of Ships*, London: Lloyd's Register.

Lobkowicz, N. (1967) *Theory and Practice: History of a Concept from Aristotle to Marx*, Notre Dame, IN: University of Notre Dame Press.

Local Government Engineering Department (LGED) (1992) *Guidelines on Environmental Issues Related to Physical Planning*, Dhaka: LGED, Rural Development and Cooperatives, Government of Bangladesh.

Lowenthal, D. (1978) *Finding Valued Landscapes*, Toronto: Institute of Environmental Studies, University of Toronto.

Lowry, K. (2003) 'The Landscape of ICM Learning Activities', *Coastal Management*, 30(4): 299–324.

Lowry, K. and Wickramaratne, H.J.M. (1987) 'Coastal Area Management in Sri Lanka', in *Ocean Yearbook 7*: 263–93.

Lowry, K., Olsen, S. and Tobey, J. (1999) 'Donor Evaluations of ICM Activities: What Can be Learned from Them?', *Ocean and Coastal Management*, 42(9): 767–89.

Lowry, K., Pallewatte, N. and Dainis, A.P. (1997) 'Assessing Sri Lanka's Special Area Management Projects', *Intercoast*, Fall: 16–17, 24, 31.

Mace, P.M. (1996) 'Developing and Sustaining World Fisheries Resources: The State of Science and Management', in *2nd World Fisheries Congress*, Brisbane: CSIRO Publishing.

MacEwen, A. and MacEwen, R. (1982) *National Parks: Conservation or Cosmetics?*, Sydney: George Allen and Unwin.

MAFF (2004) *Finfish Aquaculture*, Ministry of Fisheries, Food and Agriculture, Government of British Columbia, http://www.agf.gov.bc.ca/fisheries/finfish_main.htm (accessed 31 May 2004).

Magee, B. (1998) *The Story of Philosophy*, London: Dorling Kindersley.

Malafant, K. and Radke, S. (1995) 'The Terabyte Problem in Environmental Databases', in *Proceedings of the PACON Conference*, Townsville, QLD: James Cook University.

Malta Observatory (2003) *Sustainability Indicators*, Islands and Small States Institute, http://www.um.edu.mt/intoff/si-mo/firstpg.html (accessed 10 May 2004).

Marsden, D. and Oakley, P. (1990) *Evaluating Social Development Projects*, Oxford: Oxfam.

Mathieson, A. and Wall, G. (1982) *Tourism: Economic and Social Impacts*, New York: Longman.

Mau, R. (2003) *Green Island Recreation Area and Green Island National Park Management Plans*, Cairns: Queensland Environmental Protection Agency.

Maunter, T. (1996) *Penguin Dictionary of Philosophy*, London: Penguin Books.

Maypa, A. (2003) 'Contrasting Experiences from the Philippines: Apo and Sumilon Islands', in *The Fishery Effects of Marine Reserves and Fishery Closures* (F.R.A.C.M.R. Gell, ed.), Washington, DC: World Wildlife Fund, 36–41.

McCay, B. and Acheson, J. (1987) *The Question of the Commons: The Culture and Ecology of Communal Resources*, Tucson, AR: The University of Arizona Press.

McCloskey, M. (1979) 'Litigation and Landscape Esthetics [sic]', in *Our National Landscape, A Conference on Applied Techniques for Analysis and Management of the Visual Resource*, Berkeley, CA: PSW Forest and Range Experiment Station, USDA Forest Service.

McCool, S.F., Stankey, G.H. and Clark, R.N. (1985) *Proceedings – Symposium on Recreational Choice Behavior*, Missoula, MT: United States Forest Service.

McGlashan, D.J. and Barker, N. (in press) 'The partnership approach to Integrated Coastal Management', in *The British Sea: Towards a Sustainable Future* (H. Smith and J. Potts, eds), London: Routledge.

McHarg, I. (1969) *Design With Nature*, Garden City, NY: Natural History Press.

McKellar, R. and Kay, R.C. (2001) 'Values Based Coastal Planning', in *1st Western Australian Coastal Conference*, Esperance, WA: WA Conference Coordinating Committee.

McLain, R.J. and Lee, R.G. (1996) 'Adaptive Management: Promises and Pitfalls', *Environmental Management*, 20(4): 437–48.

McLean, R.F. and Tsyban, A. (2001) 'Coastal Zones and Marine Ecosystems', in *Climate Change 2001: Impacts, Adaptation and Vulnerability. Contribution of Working Group II to the Third Assessment Report of the IPCC* (J. McCarthy, O. Canziani, N. Leary, D. Dokken and K.S. White, eds), Cambridge: Cambridge University Press.

Meinig, D.W. (ed) (1979) *The Interpretation of Ordinary Landscapes, Geographical Essays*, New York: Oxford University Press.

Mercer, D. (1995) *A Question of Balance: Natural Resources Conflicts and Issues in Australia*, Sydney: The Federation Press.

Metoc PLC (2000) *An Assessment of the Environmental Effects of Offshore Wind Farms*, ETSU W/35/00534/REP, Harwell: Energy Technology Support Unit.

Milazzo, M.J. (1997) *Reexamining Subsidies in World Fisheries*, Washington, DC: World Bank.

Mileti, D.S. (1999) *Disasters by Design: A Reassessment of Natural Hazards in the United States*, Washingthon, DC: Joseph Henry Press.

Miller, M.L. (1993) 'The Rise of Coastal and Marine Tourism', *Coastal and Ocean Management*, 20(3): 181–99.

Milne, N., Christie, P., Oram, R., Eisma, R.L. and White, A.T. (2003) *Integrated Coastal Management Process Sustainability Reference Book*, Cebu City: University of Washington, School of Marine Affairs, Silliman University and the Coastal Resources Project of the Philippine Department of Natural Resources.

Ministry for Population and Environment (1992) *Strategy on Coral Reef Ecosystem Conservation and Management*, Jakarta: Government of Indonesia.

Ministry for the Environment (1988) 'Impact Assessment in Resource Management', in *Working Paper 20*, Wellington: New Zealand Ministry for the Environment.

Ministry of Home Affairs (1996) *Technical Manual for Preparation of a Provincial Coastal and Marine Management Strategy (English Version)*, Jakarta: Government of Indonesia.

Mintzberg, H. (1994) *The Rise and Fall of Strategic Planning*, New York: Prentice Hall.

Miossec, J.K. (1976) *Eléments pour une Theorie de l'Espace Touristique*, Aix-en-Provence: Les Cahiers due Tourisme C-36 CHET.

Mitchell, J.K. (1982) 'Coastal Zone Management: A Comparative Analysis of National Programs', in *Ocean Yearbook 3* (E.M. Borgese and N. Ginsburg, eds), Chicago: University of Chicago Press.

Moffat, D., Ngoile, M.N., Linden, O. and Francis, J. (1998) 'The Reality of the Stomach: Coastal Management at the Local Level in Eastern Africa', *Ambio*, 27(8): 590–98.

Moll, H. and Suharsono (1986) 'Distribution, diversity and abundance of coral reefs in Jakarta Bay and Kepulauan Seribu', in B.E. Brown (ed.) *Human Induced Damage to Coral Reefs: Results of a Regional UNESCO (COMAR) Workshop with Advanced Training*, UNESCO Reports in Marine Science, 40. Diponegoro University, Jepara and National Institute of Oceanology, Jakarta, Indonesia.

Mowforth, M. and Munt, I. (2003) *Tourism and Sustainability*, London: Routledge.

Mukhi, S., Hampton, D. and Barnwell, N. (1988) *Australian Management*, Sydney: McGraw-Hill.

Mumby, P.J., Edwards, A.J., Arias-Gonzalez, E.E., Lindeman, K.C., Blackwell, P.G., Gall, A., Gorczynska, M.I., Harborne, A.R., Pescod, C.L., Renken, H., Wabnitz, C.C.C. and Llewellyn, G. (2004) 'Mangroves enhance the biomass of coral reef fish communities in the Caribbean', *Nature*, 427(6974): 533–6.

Myers, N. and Kent, T. (1998) *Perverse Subsidies: Tax $s Undercutting our Economies and Environments Alike*, Winnipeg: International Institute for Sustainable Development.

Myers, N. and Kent, T. (2001) *Perverse Subsidies*, Washington, DC: Island Press.

National Atlas of the United States, 5 March 2003. http://nationalatlas.gov (accessed 30 April 2004).

National Oceanic and Atmospheric Administration (NOAA) (1996) *Biennial Report to the Congress on the Administration of the Coastal Zone Management 1994–1995*, Washington, DC: National Oceanic and Atmospheric Administration.

National Oceanic and Atmospheric Administration (NOAA) Coastal Services Center (2003) 'Coastal Management Through Time', in *Coastal Zone*, Baltimore, MD: NOAA.

National Research Council (1993) *Toward A Coordinated Spatial Data Infrastructure for the Nation*, Washington, DC: National Academy Press.

National Research Council (1995) *Science, Policy and the Coast: Improving Decision Making*, Washington DC: National Academy Press.

National Tsunami Hazard Mitigation Program (USA) (2001) *Designing for Tsunamis: Seven Principles for Planning and Designing for Tsunami Hazards*. http://www.pmel.noaa.gov/tsunami-hazard/Designing_for_Tsunamis.pdf.

Nauta, D. (1972) *The Meaning of Information*, Paris: Mouton.

New South Wales Government Department of Public Works (1990) *Coastal Management Manual*, June, Sydney: NSW Public Works Department.

New South Wales Government (1997) *NSW Coastal Policy: A Sustainable Future for the New South Wales Coast*, Sydney: New South Wales Government.

Newman, M.C., Roberts, M.H. and Hale, R.C. (2002) *Coastal and Estuarine Risk Assessment*, Boca Raton, FL: Lewis Publishers.

O'Riordan, T. (1981) *Environmentalism*, London: Pion.

O'Riordan, T. (1995) *Environmental Science for Environmental Management*, Harlow: Longman.

O'Riordan, T. and Cameron, J. (eds) (1994) *Interpreting the Precautionary Principle*, London: Earthscan.

O'Riordan, T. and Sewell, W.R.D. (1981) *Project Appraisal and Policy Review*, Chichester: Wiley.

O'Riordan, T. and Vellinga, P. (1993) 'Integrated Coastal Zone Management: The Next Steps', in *World Coast '93*, Noordwijk: Coastal Zone Management Centre The Netherlands, National Institute for Coastal and Marine Management.

O'Riordan, T. and Voisey, H. (eds) (1998) *The Transition to Sustainability: The Politics of Agenda 21 in Europe*, London: Earthscan.

O'Riordan, T., Kemp, R. and Perdue, M. (1987) *Sizewell B: The Anatomy of An Enquiry*, London: Macmillan.

OCHA (2005) 'ReliefWeb'. http://www.reliefweb.int (accessed 1 February 2005).

OECD (1989) *Environmental Policy: How to Apply Economic Instruments*, Paris: Organisation for Economic Cooperation and Development.

OECD (1992) *Recommendation of the Council on Integrated Coastal Zone Management*, Paris: OECD.

OECD (1993) *Coastal Zone Management: Integrated Policies*, Paris: OECD.

OECD (1998) 'Towards Sustainable Development: Indicators to Measure Progress', in *Proccedings of the Rome Conference*, Rome, 8–9 October, Paris: OECD.

Olsen, S. (1995) 'The Skills, Knowledge, and Attitudes of an Ideal Coastal Manager', in *Educating Coastal Managers*, Proceedings of the Rhode Island Workshop, March 4–10: University of Rhode Island, Coastal Resources Center.

Olsen, S., Sadacharan, D., Samarakoon, J.I., White, A.T., Wickremeratne, H.J.M. and Wijeratne, M.S. (1992) *Coastal 2000: Recommendations for a Resource Management Strategy for Sri Lanka's Coastal Regions, Volumes I and II*, Colombo: Coast Conservation Department, Coastal Resources Management Project, Sri Lanka and Coastal Resources Center, University of Rhode Island.

Olsen, S., Tobey, J., Robadue, D. and Ochoa, E. (1996) *Coastal Management in Latin America and the Caribbean: Lessons Learned and Opportunities for the Inter-American Development Bank*, Narragansett, RI: University of Rhode Island.

Olsen, S., Tobey, J. and Kerr, M. (1997) 'A Common Framework for Learning from ICM Experience', *Ocean and Coastal Management*, 37(2): 155–74.

Olsen, S., Tobey, J., Robadue, D. and Ochoa, E. (2002) *A World of Learning in Coastal Management: A Portfolio of Coastal Resources Management Program Experiences and Products*, Narragansett, RI: University of Rhode Island.

Olsen, S. (2003a) 'Assessing Progress Toward the Goals of Coastal Management', *Coastal Management*, 30(4): 325–45.

Olsen, S. (2003b) 'Frameworks and Indicators for Assessing Progress in Integrated Coastal Management Initiatives', *Ocean and Coastal Management*, 46(3–4): 347–61.

Owen, J.M. and Rogers, P.J. (1999) *Program Evaluation: Forms and Approaches*, Melbourne: Allen and Unwin.

Owens, D.W. (1992) 'National Goals, State Flexibility, and Accountibility in Coastal Zone Management', *Coastal Management*, 20: 143–65.

PAP/RAC (2001a) *Compendium of PAP Technical Reports and Studies*, Split, Croatia: Priority Actions Programme Regional Activity Centre (PAP/RAC) of the Mediterranean Action Plan.

PAP/RAC (2001b) 'Coastal Area Management Programmes: Improving the Implementation', in *MAP/PAP/METAP Workshop*, Malta, 17–19 January, Split, Croatia: Priority Actions Programme Regional Activity Centre (PAP/RAC) of the Mediterranean Action Plan.

PAP/RAC (2001c) *MAP Coastal Area Management Programme: Strategic Framework for the Future*, Split, Croatia: Priority Actions Programme Regional Activity Centre (PAP/RAC) of the Mediterranean Action Plan.

Paris, C. (ed.) (1982) *Critical Readings in Planning Theory*, Oxford: Pergamon.

Parsons, W. (1995) *Public Policy: An Introduction to the Theory and Practice of Policy Analysis*, Lyme, CN: Edward Elgar.

Patlis, J.M., Dahuri, R., Knight, M. and Tulungen, J. (2001) 'Integrated Coastal Management in a Decentralized Indonesia: How it Can Work', *Pesisir dan Laut*, 4(1): 1–16.

Patmore, J.A. (1983) *Recreation and Resources*, London: Blackwell.

Pauly, D. and Chuenpagdee, R. (2003) 'Development of Fisheries in the Gulf of Thailand Large Marine Ecosystem: Analysis of an Unplanned Experiment', in *Large Marine Ecosystems of the World: Change and Sustainability* (G. Hempel and K. Sherma, eds), Amsterdam: Elsevier Science.

Pavasoviå, A. (1999) *Formulation and Implementation of CAMP Projects: Operational Manual*, Split, Croatia: Priority Actions Programme Regional Activity Centre (PAP/RAC) of the Mediterranean Action Plan.

PCE (1999) *Setting Course for a Sustainable Future; The Management of New Zealand's Marine Environment*, Wellington: Parliamentary Commissioner for the Environment.

Pearce, D.G., Markandya, A. and Barbier, E. (eds) (1989) *Blueprint for a Green Economy*, London: Earthscan.

Pearce, D.G. (1993) 'The Conditions for Sustainable Development', in *Blueprint 3: Measuring Sustainable Development* (D. Pearce, R.K. Turner, R. Duborg and G. Atkinson, eds), London: Earthscan.

Pearce, D.G. (1987) *Tourist Today: A Geographical Analysis*, New York: Longman.

Pearce, D.G. (1989) *Tourist Development*, New York: Longman.

Pernetta, J.C. and Elder, J.L. (1993) *Cross-Sectoral, Integrated Coastal Area Planning (CICAP): Guidelines and Principles for Coastal Area Development*, Gland: IUCN.

Petts, J. (1999) *Handbook of Environmental Impact Assessment: Environmental Impact Assessment in Practices: Impacts and Limitations*, Oxford: Blackwell Scientific.

Platt, R.H. (1991) *Land Use Control: Geography, Law and Public Policy*, Englewood Cliffs, NJ: Prentice Hall.

Platt, R.H., Miller, H.C., Beatley, T., Melville, J. and Mathenia, B.G. (1992) *Coastal Erosion: Has Retreat Sounded?*, Boulder, CO: Institute of Behavioural Science, University of Colorado.

Pomeroy, R.S., Pollnac, R.B., Predo, C.D. and Katon, B.M. (1997) 'Impact Evaluation of Community-Based Coastal Resource Management Projects in the Philippines', *NAGA The ICLARM Quarterly*, 19(4): 9–12.

Presidential/Congressional Commission on Risk Assessment and Risk Management (1997) *Framework for Environmental Health Risk Management*, 2 volumes, Washington, DC: National Academy of Sciences.

Prosser, G. (1986) 'An Introduction to a Framework for Natural Area Planning', *Australian Parks and Recreation*, 22(2): 3–10.

Puget Sound Water Quality Action Team (2000) *Puget Sound Water Quality Management Plan*, adopted 14 December, Olympia, WA: Office of the Governor.

Putra, S. and Mulyana, Y. (2003) 'Linking Coral Reef Conservation into Integrated Coastal Management as part of Indonesia's Sea Large Marine Ecosystem: An Experience of Coral Reef Rehabilitation and Management Program (COREMAP III)', in *5th World Parks Congress*, Durban: IUCN.

Queensland Government (1995) *Coastal Protection and Management Act 1995*, Brisbane: Government of Queensland.

Quilty Environmental Consulting (1991) *Warnbro Dunes Foreshore Stabilisation and Management Plan*, Perth: Quilty Environmental Consulting for Australian Housing and Land.

Ralston, B. (2004) *GIS and Public Data*, Florence, KY: OnWord Press.

Redclift, M. (1987) *Sustainable Development: Exploring the Contradictions*, London: Routledge.

Redclift, M.R. (1995) 'Values and Global Environmental Change', in *Values and the Environment: A Social Science Perspective* (Y. Guerrier, ed.), Chichester and New York: J. Wiley.

Reid, D. (ed) (1995) *Sustainable Development: An Introductory Guide*, London: Earthscan.

Ren, M.-E. (1992) 'Human Impact on Coastal Landform and Sedimentation – The Yellow River Example', *GeoJournal*, 28(4): 443–8.

Rennie, H. (2004) Unpublished material supplied by the author to the second edition of this book.

Rennie, H.G. (1993) 'The Coastal Environment', in *Environmental Planning in New Zealand* (P.A. Memom and H.C. Perkins, eds), Palmerston North: Dunmore Press.

Republic of Indonesia (1990) *Conservation of Living Resources and their Ecosystems. Act No.5 of 1990*, Jakarta: Ministry of Forestry.

Resource Assessment Commission Coastal Zone Inquiry (1993) *Values and Attributes Concerning the Coastal Zone*, Canberra: Commonwealth of Australia.

Ribe, R.G. (1989) 'The Aesthetics of Forestry: What has Empirical Preference Research Taught Us?' *Environmental Management*, 13(1): 55–74.

Ripley, R.B. and Franklin, G.A. (1986) *Policy Implementation and Bureaucracy*, Chicago, IL: Dorsey Press.

Robadue, D. (ed.) (1995) *Eight Years in Ecuador: The Road to Integrated Coastal Management*, Narragansett, RI: University of Rhode Island and US Agency for International Development.

Rogers, D.L. and Whetten, D.A. (1982) *Interorganizational Coordination: Theory, Research and Implementation*, Ames, IA: Iowa State University Press.

Rolden, R. and Sievert, R. (1993) *Coastal Resources Management: A Manual for Government Officials and Community Organizers*, Manila: Department of Agriculture.

Rosier, J. (1993) 'Coastal Planning in New Zealand and the Resource Management Act', *New Zealand Journal of Geography*, 96: 2–8.

Rosier, J. (2004a) *Independent Review of the New Zealand Coastal Policy Statement. Report prepared for the Minister of Conservation*, January, Palmerston North: Massey University.

Rosier, J. (2004b) Unpublished material supplied by the author to the first edition of this book.

Rossi, P.H. and Freeman, H.E. (1993) *Evaluation: A Systematic Approach*, Newbury Park, CA: Sage.

Rubenstein, H.M. (1987) *A Guide to Site and Environmental Planning*, New York: Wiley.

Ruddle, K. and Johannes, R.E. (1983) *The Traditional Knowledge and Management of Coastal Systems in Asia and the Pacific*, Jakarta: UNESCO.

Ruitenbeek, H.J. (1991) *Mangrove Management: An Economic Analysis of Management Options with a Focus on Bintuni Bay, Irian Jaya*, Jakarta: Government of Indonesia, Dalhousie University.

Ruitenbeek, H.J. (1994) 'Modelling Economy–Ecological Linkages in Mangroves: Economic Evidence for Promoting Conservation in Bintuni Bay, Indonesia', *Ecological Economics*, 10: 233–47.

Russ, G. and Alcala, A. (1994) 'Sumilon Island Reserve: Twenty Years of Hopes and Frustrations', *NAGA*, 7(3): 8–12.

Russ, T.H. (2002) *Site Planning and Design Handbook*, New York: McGraw-Hill.

Rutledge, A.J. (1971) *Anatomy of a Park: The Essentials of Recreation Area Planning and Design*, New York: McGraw-Hill.

Saad, R. (2003) 'Sigh of relief for tourism sector', *Al Ahram*, 25 September–3 October (657).

Saeed, S., Goldstein, W. and Shrestha, R. (1998) *Planning Environmental Communication: Lessons from Asia*, Bangkok: IUCN Commission on Education and Communication.

Salam, M., Irawan, D. and Tomboelu, N. (1996) *Integrated Management Planning of Kapoposang Marine Tourism Park*, Jakarta: MREP.

Salvesen, D. and Godschalk, D.R. (1999) *Development on Coastal Barriers: Does the Coastal Barrier Resources Act Make a Difference?*, Washington, DC: Coastal Alliance.

Samaranayake, R.A.D.B. (2000) 'Sri Lanka's Agenda for Coastal Zone Management', *EEZ Journal* (through Sustainable Development International online), 5.

Satria, A. (2004) Unpublished material supplied by the author to the second edition of this book.

Satria, A.A.M. (2004) 'Decentralization of fisheries management in Indonesia', *Marine Policy*, 21(5): 465–80.

Savina, G.C. and White, A.T. (1986) 'A Tale of Two Islands: Some Lessons for Marine Resource Management', *Environmental Conservation*, 13(2): 107–13.

Schmidt, W. (ed.) (1996) *Advanced Recreation Planning and Management Course: Course Notes*, Perth: Department of Conservation and Land Management.

Schoen, R.-J. and Djohani, R. (1992) *A Communication Strategy to Support Marine Conservation Policies, Programmes and Projects in Indonesia for the Years 1992–1995*, Jakarta: World Wildlife Fund for Nature Indonesia.

Schröder, P.C. (1993) 'OCA/PAC Activities Related to Climate Change and Sea-Level Rise', in *World Coast '93 Proceedings*, Noordwijk: Coastal Zone Management Centre, The Netherlands National Institute for Coastal and Marine Management.

Scura, L.F. (1993) 'Review of Recent Experiences in Integrated Coastal Management', unpublished report to the International Center for Living Coastal Aquatic Resources Management and University of Rhode Island.

Scura, L.F. (1994) *Typological Framework and Strategy Elements for Integrated Coastal Fisheries Management*, Rome: Food and Agriculture Organisation of the United Nations.

Secretary of the Interior (1994) *Impact of Federal Programs on Wetlands. Volume II*, Washington, DC: Department of the Interior.

Seely Brown, J. and Duguid, P. (2000) *The Social Life of Information*, Cambridge, MA: Harvard Business School Press.

Senaratna, S. and Milner-Gullan, E.J. (2002) *Community Management of the Rekawa Lagoon Resources*, http://www.iucn.org/themes/sustainableuse/rekawa.html (accessed 10 April 2004).

Senge, P.M. (1990) *The Fifth Discipline: The Art and Practice of the Learning Organization*, New York: Doubleday.

Setyawan, W.B. (2002) 'The Disastrous February 2002 Flood of Jakarta, Indonesia', in *Coastal Zone Asia Pacific Conference*, Bangkok, Thailand, 11–17 May.

Shea, E.L. (2003) 'Our Shared Journey: Coastal Zone Management, Past, Present and Future', in *13th Biennial Coastal Zone Conference*, Baltimore, Maryland, 13–17 July.

Shehata, A. (1998) 'Protected Areas of the Gulf of Aqaba, Egypt: A Mechanism of Integrated Coastal Management', in *TMEMS Proceedings*, Townsville: Great Barrier Reef Marine Park Authority.

Sherwood, A. and Howarth, R. (1996) *Coasts of Pacific Islands*, Suva: South Pacific Geoscience Commission (SOPAC) Miscellaneous Report 222.

Short, A.D. (in press) 'Beach Hazards and Risk Assessment of Beaches', in *Proceedings of the World Congress on Drowning*, Dordrecht: Kluwer.

Sloan, N.A. and Sugandhy, A. (1994) 'An Overview of Indonesian Coastal Environmental Management', *Coastal Management*, 22(3): 215–34.

Small, C. and Nicholls, R.J. (2003) 'A Global Analysis of Human Settlement in Coastal Zones', *Journal of Coastal Research*, 19: 584–99.

Smith, A.H. and Homer, F. (1994) 'Collaborative Coral Reef Monitoring in the Caribbean Region', in *Proceedings of the Seventh International Coral Reef Symposium*, Guam: University of Gaum Marine Laboratory.

Smith, L.G. (1993) *Impact Assessment and Sustainable Resource Management*, Harlow: Longman.

Smith, L.G., Nell, C.Y. and Prystupa, M.V. (1997) 'The Converging Dynamics of Interest Representation in Resources Management', *Environment Management*, 21(2): 139–46.

Smith, R. (1992) 'Beach Resort Evolution', *Annals of Tourism Research*, 19: 304–22.

Smith, V.K. (1996) *Estimating Economic Values for Nature: Methods for Non Market Valuation*, Cheltenham: Edward Elgar.

Smutylo, T. (2001) 'Crouching Impact, Hidden Attribution: Overcoming Threats to Learning in Development Programs', in *Workshop on Cross-portfolio Learning*, Block Island, RI, 21–24 May, University of Rhode Island.

Smyth, D. (1991) *Aboriginal Maritime Culture in the Far Northern Section of the Great Barrier Reef Marine Park*, Townsville: Great Barrier Reef Marine Park Authority.

Smyth, D. (1993) *A Voice in All Places: Aboriginal an Torres Strait Islander Interests in Australia's Coastal Zone*, Canberra: Commonwealth of Australia.

Soby, B.A., Simpson, A.C.D. and Ives, D.P. (1993) *Integrating Public and Scientific Judgements into a Tool Kit for Managing Food-related Risks, Stage I: Literature Review and Feasibility Study*, Norwich: University of East Anglia.

Soegiarto, A. (1981) 'The Development of a Marine Park System in Indonesia', in *The Reef and Man: Proceedings of the Fourth International Coral Reef Symposium*, Quezon City: Marine Sciences Centre, University of the Philippines.

Sorensen, J. (1993) 'The International Proliferation of Integrated Coastal Zone Management Efforts', *Ocean and Coastal Management*, 21(1–3), 45–80.

Sorensen, J. (1997) 'National and International Efforts at Integrated Coastal Zone Management: Definitions, Achievements, and Lessons', *Coastal Management*, 25(1), 3–41.

Sorensen, J. (2000) 'Baseline 2000', in *Coastal Zone Canada*, St John, New Brunswick: Coastal Zone Canada Association.

Sorensen, J. (2002) *Baseline 2000 Background Report: Second Iteration*, Boston, MA: Urban Harbors Institute, University of Massachusetts, Boston.

Sorensen, J.C., McCreary, S.T. and Hershman, M.J. (1984) *Institutional Arrangements for Coastal Resources Management*, Columbia, SC: Prepared by Research Planning Institute for National Park Service, US Department of Interior.

Sorensen, J.C. and McCreary, S.T. (1990) *Institutional Arrangements for Managing Coastal Resources and Environments*, Narragansett, RI: University of Rhode Island.

Sorensen, J.C. and West, N. (1992) *A Guide to Impact Assessment in Coastal Environments*, Narragansett, RI: Coastal Resources Center, University of Rhode Island.

South Pacific Regional Environment Programme (SPREP) (1992) *Environmental Impact Assessment Training in the South Pacific Region: Meeting Report 31 August–4 September 1992*, Training Report, 5/5, disposed 1999/09/22, Samoa: SPREP.

South Pacific Regional Environment Programme (SPREP) (1993) *Project Proposal: SPREP Integrated Coastal Zone Management in the Pacific Islands Region* (Draft), Apia: SPREP.

Spalding, M. and Chape, S. (eds) (in press) *State of the World's Protected Areas*, Cambridge: IUCN, UNEP-WCMC, WCPA.

Sri Lanka Department of Census and Statistics (2005) 'Impact of Tsunami 2004 on Sri Lanka'. http://www.statistics.gov.lk/Tsunami/ (accessed 1 February 2005).

Stacey, R.D. (1992) *Managing the Unknowable: Strategic Boundaries Between Order and Chaos in Organizations*, San Francisco: Jossey-Bass.

Standards Australia and Standards New Zealand (1999) *Australian/New Zealand Standard: Risk Management*, Sydney and Wellington: Standards Australia and Standards New Zealand.

Stankey, G.H. and Wood, J. (1982) 'The Recreation Opportunity Spectrum: An Introduction', *Australian Parks and Recreation*, February, 6–14.

State of Delaware (1971) *Coastal Zone Act, Chapter 70, Title 7, Delaware Code*.

Steers, R.M., Ungson, G.R. and Mowday, R.T. (1985) *Managing Effective Organisations: An Introduction*, Boston, MA: Kent.

Stone, R. (1988) 'Conservation and Development in St Lucia', *World Wildlife Fund Letter*, 3: 1–8.

Storey, D. and Whitney, E. (2003) *JHUCCP COREMAP Final Report Highlights*, Baltimore, MD: Johns Hopkins University Center for Communication Programs.

Stratford, J. (2004) Unpublished material supplied by the author to the second edition of this book.

Susskind, L., McKearnan, S. and Thomas-Larmer, J. (2000) *Negotiating Environmental Agreements: How to Avoid Escalating Confrontations, Needless Cost and Unnecessary Litigation*, Washington, DC: Island Press.

Talukder, J. and Ahmad, M. (eds) (1992) *The April Disaster: Study on Cyclone Affected Region in Bangladesh*, Dhaka: Community Development Library.

Talukder, J., Roy, G.D. and Ahmad, M. (eds) (1992) *Living with Cyclones: Study on Storm Surge Prediction and Disaster Preparedness*, Dhaka: Community Development Library.

Tasque Consultants (1994) *Break O'Day Regional Marine and Coastal Strategy*, Hobart, Tasmania: Dorset, Break O'Day, Glamorgan and Spring Bay Councils.

Taussik, J. and Gubbay, S. (1997) 'Networking in Integrated Coastal Zone Management', in *Partnership in Coastal Zone Management: Proceedings of the Eurocoast '96 Conference* (J. Taussik and J. Mitchell, eds), Cardigan: Samara Publishing.

Taylor, S. and Roberts, G. (2004) 'Managing Samoa's Coastal Environment', *Newsletter of the New Zealand Coastal Society*, 4–5 March.

Thames Estuary Partnership (1999) *Management Guidance for the Thames Estuary Today's Estuary for Tomorrow Strategy*, London: Thames Estuary Partnership Institute for Environmental Policy University College London.

Thames Estuary Partnership (2003) *Action Plan Annual Review*, November, London: Thames Estuary Partnership.

Thomas, I. (1996) *Environmental Impact Assessment in Australia*, Sydney: Federation Press.

Thompson, G.F. and Steiner, F.R. (eds) (1997) *Ecological Design and Planning*, New York: Wiley.

Thorman, R. (1995) *Suggested Framework for Preparing Regional Environmental Strategies* (Draft Guidelines), Canberra: Australian Local Government Association.

Tickner, J.A. (ed.) (2003) *Environmental Science and Preventive Public Policy*, Washington, DC: Island Press.

Titus, J.G. (1998) 'Rising Seas, Coastal Erosion, and the Takings Clause: How to Save Wetlands and Beaches Without Hurting Property Owners', *Maryland Law Review*, 57(4): 1279–399.

Tobey, J. and Volk, R. (2003) 'Learning Frontiers in the Practice of Integrated Coastal Management', *Coastal Management*, 30: 285–98.

Tomasick, T., Suharsono and Mah, A.J. (1993) 'Global Histories: A Historical Perpective of the Natural and Anthropogenic Impacts in the Indonesian Archipelago with a Focus on Kepulauan Seibu, Java Sea', in *Colloquium on Global Aspects of Coral Reefs: Health, Hazards and History*, Miami, FL: University of Miami.

topicmap.com (2004) 'Hand-crafted Machine-generated Knowledge Interchange', http://www.topicmap.com (accessed 27 May 2004).

Torell, E. (2000) 'Adaptation and Learning in Coastal Management: The Experience of Five East African Initiatives', *Coastal Management*, 28(4): 352–64.

Torkildsen, G. (1992) *Leisure and Recreation Management*, London: E & F Spon.

Tri, N.H., Adger, N., Kelly, M., Granich, S. and Ninh, N.H. (1996) *The Role of Natural Resource Management in Mitigating Climate Impacts: Mangrove Restoration in Vietnam*, Norwich: Centre for Social and Economic Research on the Global Environment.

Turner, R.K. (1991) 'Environment, Economics and Ethics', in *Blueprint 2: Greening the World Economy* (D. Pearce, ed.), London: Earthscan.

Turner, R.K. (ed.) (1993) *Sustainable Environmental Economics and Management*, London: Belhaven.

Turner, R.K. and Adger, W.N. (1995) *Coastal Zone Resources Assessment Guidelines*, Texel: LOICZ.

Turner, R.K., Subak, S. and Adger, N. (1995) *Pressures, Trends and Impacts in the Coastal Zones: Interactions Between Socio-Economic and Natural Systems*, Norwich: Centre for Social and Economic Research on the Global Environment.

U.S. Commission on Ocean Policy (2004) *Preliminary Report of the U.S. Commission on Ocean Policy Governors' Draft*, April, Washington, DC.

Ullmann, O., Overberg, P. and Hampson, R. (2000) 'Boom on the Beach: Growth Reshapes Coasts', *USA Today*, 28 July,

UN (2002) *World Summit on Sustainable Development political declaration submitted by the President of the Summit*, A/CONF.199/L.6/Rev.2 advance unedited English version, Johannesburg, South Africa 26 August–4 September 2002: United Nations.

UN CCD (2004) *Database of Type II Initiatives*, UN Division for Sustainable Development (DSD), http://www.un.org/esa/sustdev (accessed 26 April 2004).

UN Department of International Economic and Social Affairs (1982) *Coastal Area Management and Development*, Oxford: Pergamon.

UN Population Division (2000) *United Nations World Population Prospects 2000 Revision*, New York: United Nations, Population Division.

UN Population Division (2004) *UNEP Global Environment Outlook Geo Data Portal*, http://www.geodata.grid.unep.ch (accessed 1 May 2004).

UNCED (1992) *Agenda 21 – United Nations Conference on Environment and Development: Outcomes of the Conference*, Rio de Janeiro, Brazil, 3–14 June, New York: United Nations.

Underdahl, A. (1980) 'Integrated Marine Policy: What? Why? How?' *Marine Policy*, July: 159–69.

UNEP (1982) *Environmental Problems of the East African Region*, Nairobi: UNEP.

UNEP (1995) *Guidelines for Integrated Management of Coastal and Marine Areas – With Special Reference to the Mediterranean Basin*, Nairobi, Kenya: United Nations Environment Programme.

UNEP (2002a) *UN Atlas of the Oceans*, UNEP, http://www.oceansatlas.com (accessed 5 March 2004).

UNEP (2002b) *Water Supply and Sanitation Coverage in UNEP Regional Seas – Need for Wastewater Emission Targets?*, Nairobi: World Health Organisation (WHO)/United Nations Childrens Fund (UNICEF)/Water Supply and Sanitation Collaborative Council (WSSCC).

UNEP (2002c) *Regional Seas: Strategies for Sustainable Development*, August, Nairobi: United Nations Environment Program.

UNEP (2003) *Regional Seas Strategic Directions for 2004–2007: A Global Initiative for Regional Seas Cooperation*, Fifth Global Meeting of the Regional Seas, 26–28 November, Nairobi, Kenya.

UNEP (2004) *Regional Seas: Joining Hands Around the Seas*, http://www.unep.ch/seas/main/hoverv.html (accessed March 2004).

UNEP OCA/PAC (1982) *Marine and Coastal Area Development in the East African Region*, Nairobi, Kenya: UNEP.

UNEP OCA/PAC (1986) *Action Plan for the Conservation of the Marine Environment and Coastal Areas of the Red Sea and Gulf of Aden*, Nairobi: UNEP.

UNEP/FAO/PAP (2001) *Case Study Report: Protection and Management of the Marine and Coastal Areas of Eastern Africa Project – EAF/5*, East African Regional Seas Technical Reports Series No. 10, Split, Croatia: United Nations Environment Programme, Food and Agricultural Organisation and Programme Activity Center.

UNESCO (2000) *Reducing Megacity Impacts on the Coastal Environment: Alternative Livelihoods and Waste management in Jakarta and the Seribu Islands*, Coastal Region and Small Island Papers 6, Paris: UNESCO.

UNITAR (2004) *Technical Assistance to Least Developed Countries UNFCCC Focal Point to Produce their National Adaptation Plans*, United Nations Institute for Training and Research (UNITAR), http://www.unitar.org/ccp/ldc.htm (accessed 15 March 2004).

United Nations General Assembly (2005). 'General Assembly Draft Resolution on the Tsunami to Strengthening Emergency Relief, Rehabilitation, Reconstruction and Prevention in the Aftermath of the Indian Ocean Tsunami Disaster. 14 January 2005'. Document: A/59/L.58. http://ochaonline.un.org/DocView.asp?DocID=2782.

United States Agency for International Development (1996) *Learning from Experience: Progress in Integrated Coastal Zone Management*, Washington, DC: United States Agency for International Development.

United States Agency for International Development (2005)' Indian Ocean – Earthquake and Tsunamis'. Fact sheet #28. http://www.usaid.gov/locations/asia_near_east/tsunami/ (accessed 6 February 2005).

USDA Forest Service (1973) *National Forest Landscape Management Manual*, Washington, DC: USDA Forest Service.

van Beers, C.P. and de Moor, A.P.G. (1999) *Addicted to Subsidies: How Governments Use Your Money to Destroy the Earth and Pamper the Rich*, The Hague: Institute for Research on Public Expenditure.

van Gunstern, H.R. (1975) *The Quest for Control*, London: Wiley.

van Lier, H.N., Jaarsma, C.F., Jurgens, C.F. and de Buck, A.J. (eds) (1994) *Sustainable Land Use Planning*, Amsterdam: Elsevier.

van Westen, C.-J. and Scheele, R. (1996) *Planning Estuaries*, New York: Plenum Press.

Veal, A.J. (1992) 'Definitions of Leisure and Recreation', *Australian Journal of Leisure and Recreation*, 2(4): 44–52.

Vella, L. (2002) 'CAMP Malta', in *Coastal Area Management Programmes: Improving the Implementation: MAP/PAP/METAP* Workshop Malta, 17–19 January 2002, Split: Priority Actions Programme Regional Activity Center (PAP/RAC) of the Mediterranean Action Plan.

Vestal, B., Rieser, A., Ludwig, M., Kurland, J., Collins, C. and Ortiz, J. (1995) *Methodologies and Mechanisms for Management of Cumulative Coastal Environmental Impacts*, Washington, DC: US National Oceanic and Atmospheric Administration, Coastal Ocean Program.

Vickers, G. (1995) *The Art of Judgement: A Study of Policy Making*, Thousand Oaks, CA: Sage.

Victorian Coastal Council (2000) 'Beyond the Beach', in *Coast to Coast*, Melbourne, Victoria, 6–9 March.

Wager, R.G. (1964) *The Carrying Capacity of Wild Lands for Recreation*, Washington, DC: American Society of Foresters.

Wang, H. (2004) 'Ecosystem Management and Its Application to Large Marine Ecosystems: Science, Law and Politics', *Ocean Development and International Law*, 35: 41–74.

Watson, R.A. and Pauly, D. (2001) 'Systematic Distortions in World Fisheries Catch Trends', *Nature*, 414: 534–36.

World Business Council for Sustainable Development (WBSC) (2003) *Annual Review 2003 – Reconciling the Public and Business Agendas*, Geneva: World Business Council for Sustainable Development.

Welch, D. (1991) 'Information Needs for Resource Management in Australian Marine Protected Areas, 1989–90', in *Proceedings of International Conference on Science & Management of Protected Areas*, Halifax, Nova Scotia, 14–19 May 1991.

Wells, M. and Brandon, K. (1992) *People and Parks: Linking Protected Area Management with Local Communities*, Washington, DC: World Bank.

Wenger, E. (1998) *Communities of Practice: Learning, Meaning, and Identity*, New York: Cambridge University Press.

Western Australian Planning Commission (1996) *Central Coast Regional Strategy: A Strategy to Guide Land Use in the Next Decade*, Perth: Government of West Australia.

Western Australian Planning Commission (2000) 'Western Australia tomorrow, population projection for the state, statistical divisions, planning regions and local government areas of Western Australia, population report No 4', October.

Western Australian Planning Commission (2001) *Coastal Zone Management Policy for Western Australia: For Public Comment*, Perth, Australia: Western Australian Government, Western Australian Planning Commission.

Western Australian Planning Commission (2003) *Coastal Planning and Management Manual: A Community Guide for Protecting and Conserving the Western Australian Coast*, Perth: Western Australian Planning Commission, Government of West Australia.

Western Australia Task Force (2002) *Review of the Structural Arrangements for Coastal Planning and Management in Western Australia*, May 2002, Perth: Department of Planning and Infrastructure.

White, A.T. (1986) 'Philippine Marine Park Pilot Site Benefits and Management Conflicts', *Environmental Conservation*, 13(4): 355–9.

White, A.T. (1988a) *Marine Parks and Reserves: Management for Coastal Environments in Southeast Asia. ICLARM Education Series 2*, Manila: International Center for Living Coastal Aquatic Resources Management.

White, A.T. (1988b) 'The Effect of Community-Managed Marine Reserves in the Philippines on their Associated Coral Reef Fish Populations', *Asian Fisheries Science*, 2: 27–41.

White, A. (1989a) 'Comparison of Coastal Resource Planning an Management in the ASEAN Countries', in *Coastal Zone '89* (O. Magoon, ed.), Charleston, SC: American Shore and Beach Preservation Society.

White, A.T. (1989b) 'Two Community-based Marine Reserves: Lessons for Coastal Management', in *Coastal Area Management in Southeast Asia: Policies, Management Strategies and Case Studies. ICLARM Conference Proceedings 19*, Manila: ICLARM.

White, A.T. (1995) 'Comments on Coastal Zone Management and Ecosystem Protection', in *SEAPOL Singapore Conference on Sustainable Development of Coastal and Ocean Areas in Southeast Asia: Post-Rio Perspectives* (K.L. Koh, R.C. Beckman and C.L. Sien, eds), Singapore: National University of Singapore.

White, A.T. (1996) 'Philippines: Community Management of Coral Reef Resources', in *Coastal Zone Management Handbook* (J.R. Clark, ed.), Boca Raton, FL: CRC Press.

White, A.T. (2004) Unpublished material supplied by the author to the second edition of this book.

White, A.T. and Cruz-Trinidad, A. (1998) *The Values of Philippine Coastal Resources: Why Protection and Management are Critical*, Cebu City: Coastal Resource Management Project.

White, A.T. and Samarakoon, J.I. (1994) 'Special Area Management for Coastal Resources: A First for Sri Lanka', *Coastal Management in Tropical Asia*, March: 20–4.

White, A.T. and Savina, G. (1987) 'Community Based Marine Reserves, A Philippine First', in *Proceedings of Coastal Zone '87*, Seattle, WA: American Society of Civil Engineers.

White, A.T., Barker, V. and Tantrigama, G. (1997) 'Using Integrated Coastal Management and Economics to Conserve Coastal Tourism Resources in Sri Lanka', *Ambio*, 26(6): 335–44.

White, A.T., Courtney, C.A. and Salamanca, A. (2002) 'Experience with Marine Protected Area Planning and Management in the Philippines', *Coastal Management*, 30(1): 1–27.

White, A.T., Vogt, H.P. and Arin, T. (2000) 'Philippine Coral Reefs Under Threat: the Economic Losses Caused by Reef Destruction', *Marine Pollution Bulletin*, 40(7): 598–605.

White, A., Zeitlin-Hale, L., Renard, Y. and Cortesi, L. (1994) *Collaborative and Community-Based Management of Coral Reefs: Lessons from Experience*, West Hartford, CT: Kumarian Press.

WHO (2003) *Guidelines for Safe Recreational Water Environments, Volume 1: Coastal and Fresh Waters*, Geneva: World Health Organization.

Wijetunge, J. (2005) 'Living with Tsunami: Future Directions for Coastal Land'. *Sri Lanka Daily News* 27 January 2005. http://www.dailynews.lk/2005/01/27/fea01.html (accessed 10 February 2005).

Williamson, D.N. (1978) *Landscape Perception Research: Defining a Direction*, Melbourne: Forests Commission, Victoria.

Williamson, D.N. and Calder, S.W. (1979) 'Visual Resource Management of Victoria's Forests: A New Concept for Australia', *Landscape Planning*, 6: 313–41.

Wong, P.P. (ed.) (1993) *Tourism vs. Environment: The Case for Coastal Areas*, Dordrecht: Kluwer Academic.

Wong, P.P. (2003) 'Tourism Development in Southeast Asia: Patterns, Issues and Prospects', in *Southeast Asia Transformed: A Geography of Change* (C.L. Sien, ed.), Singapore: SEAS.

Wood, C. (2003) *Environmental Impact Assessment: A Comparative Review*, Upper Saddle River, NJ: Prentice Hall.

Wood, C. and Dejeddour, M. (1992) 'Strategic Environmental Assessment of Policies, Plans and Programmes', *Impact Assessment Bulletin*, 10: 3–22.

Wood, N.J., Good, J.W. and Goodwin, R.F. (2002) 'Vulnerability Assessment of a Port and Harbor Community to Earthquake and Tsunami Hazards: Integrating Technical Expert and Stakeholder Input'. *Natural Hazards*. November: 148–57.

World Bank (1993) *Noordwijk Guidelines for Integrated Coastal Zone Management*, Washington, DC: World Bank, Environment Department, Land Water and Natural Habitats Division.

World Bank (1998) *Assessing Aid; What Works; What Doesn't, and Why*, New York: Oxford University Press.

World Commission on Environment and Development (ed.) (1987) *Our Common Future*, Oxford: Oxford University Press.

World Resources Institute (1992) *World Resources 1992–93*, New York: Oxford University Press.

World Resources Institute (2003) *Farming the Reef: Aquaculture as a Solution for Reducing Fishing Pressures on Coral Reefs*, Washington, DC: World Resources Institute.

World Summit on the Information Society (WSIS) (2003) *World Summit on the Information Society: Plan of Action*, Geneva: International Telecommunications Union.

World Tourism Organization (WTO) (2004) *Short-term Tourism Data*, January 2004, Madrid: World Tourism Organization.

World Wide Fund for Nature (1996) 'Fisheries Crisis Overshadows World Oceans Day', *World Wide Fund for Nature News Release*.

Wright, D.J. and Barlett, D.J. (eds) (1999) *Marine and Coastal Geographical Information Systems*, London: Taylor & Francis.

Yarnell, P. (1999) 'Port Administration and Integrated Coastal Management under the Canada Marine Act in Vancouver, British Columbia', *Coastal Management*, 27(4): 343–54.

Young, D. (2003) *Monitoring the Effectiveness of the New Zealand Coastal Policy Statement: Views of Local Government Staff*, March, Wellington: Conservation Policy Division, Department of Conservation.

Young, M.D. (1992) *Sustainable Investment and Resource Use: Equity, Environmental Integrity and Economic Efficiency*, Paris: UNESCO.

Zann, L.P. (1984) 'Traditional Management and Conservation of Fisheries in Kiribati and Tuvalu Atolls', in *The Traditional Knowledge and Management of Coastal Systems in Asia and the Pacific*, Jakarta: UNESCO.

Zann, L.P. (1995) *Our Sea, Our Future: Major Findings of the State of the Marine Environment Report for Australia*, Canberra: Ocean Rescue 2000 Program: Department of the Arts, Sports and Territories.

Zigterman, R. and De Campo, J. (1993) *Green Island and Reef Management Plan*, Cairns: Queensland Department of Environment and Heritage, Great Barrier Reef Marine Park Authority, Cairns City Council, Cairns Port Authority, Department of Lands.

Zube, E.H., Pitt, D.W. and Anderson, T.W. (1974) *Perception and Measurement of Scenic Resource Values in the Southern Connecticut River Valley*, Amherst, MA: University of Massachusetts.

Index